Automatic Program Development

Automatic Program Development

A Tribute to Robert Paige

Edited by

Olivier Danvy
BRICS, University of Aarhus, Denmark

Harry Mairson
Brandeis University, Waltham, MA, USA

Fritz Henglein
DIKU, University of Copenhagen, Denmark

and

Alberto Pettorossi
DISP, University of Roma 'Tor Vergata', Italy

Springer

A C.I.P. Catalogue record for this book is available from the Library of Congress.

ISBN 978-90-481-7674-8
ISBN 978-1-4020-6585-9 (e-book)

Published by Springer,
P.O. Box 17, 3300 AA Dordrecht, The Netherlands.

www.springer.com

Printed on acid-free paper

This book is dedicated to the memory
of our friend and colleague
Robert Paige (1947–1999)

Robert Paige

Preface

This book is a tribute to Robert Paige (New York, USA, 15 August 1947 – New York, USA, 5 October 1999), our accomplished and respected colleague, and moreover our good friend, whose untimely passing was a loss to our academic and research community.

In this book we have collected the revised, updated versions of the papers which appeared in two special issues of the Higher-Order and Symbolic Computation Journal [2, 3] which were published in Bob's memory. The book also includes some extra papers by colleagues and members of the IFIP Working Group 2.1 of which Bob was an active, illuminating, and stimulating member. All papers are focused on some of the research interests of Bob and, in particular, on the transformational development of programs and their algorithmic derivation from formal specifications. We hope that the reader will find this book a stimulus for continuing and deepening these research lines.

We have included in the book: (i) an obituary written by Harry Mairson, (ii) reminiscences from three colleagues, Martin Davis, Helmut Partsch, and Alan Siegel, and (iii) some personal recollections by Gary Paige, Bob's brother.

The pictures were kindly provided by Nieba Paige, Bob's wife, and their children Jane and John. They were all very kind to us.

Bob Paige's Research and Scientific Interests. Bob's scientific interests were many, and his achievements were many as well, witness his research retrospective and National Science Foundation research proposal which are included in the first part of the book. In particular, Bob strongly contributed to the development of new techniques for the derivation of algorithms and programs using the transformational methodology. According to the transformational methodology the programmer generally starts from specifications and transforms them, aided by an automatic or semiautomatic system, into runnable algorithms exhibiting high performance.

There are, basically, two kinds of transformation steps: the first kind consists in the applications of rules according to suitable strategies à la Burstall–Darlington [1], and the second kind consists in the applications of schema equivalences à la Walker–Strong [4].

Bob's many research papers provide new insights into these two approaches to transformational programming and, in particular, he contributed to the development of techniques which allow us to automatically derive algorithms and improve their efficiency, while preserving correctness. Such derivations and improvements were obtained by suitable modifications of the recursion structure and/or choices of the data structures employed.

When proposing new techniques, Bob always strove for generality, in the sense that these techniques should not be ad hoc tricks. On the contrary, they should have a large range of applicability for a large class of specifications or programs. Only general ideas could become the basis for an automatic system for program development. Bob's APTS system is indeed the incarnation of most of the techniques he proposed (see, for instance, Leonard and Heitmeyer's contribution in this book). Only general techniques may become part of a new programming methodology and this was what, ultimately, Bob was looking for and wanted to propose.

Bob's colleagues and friends are all very happy to have been working and sharing ideas with him. In particular, his colleagues at the Courant Institute in New York, the members of the IFIP Working Group 2.1, and the people at the various Universities he visited, including the University of Copenhagen and the University of Wisconsin.

All his friends and colleagues enjoyed Bob's talks and presentations. His ideas and his visions on program development and programming methodology were a source of enthusiasm, strength, and inspiration.

We know that Bob had many more scientific topics and ideas which he wanted to explore, but he did not have time to do it. He shared some of these ideas with us and he left them for us as projects to do and, most importantly, as a gift we will treasure in our lives.

Bob's dedication to research and teaching is for us all an example to follow.

Contributed Papers. In the third part of the book we have collected the following scientific papers.

– *Transformational Derivation of an Improved Alias Analysis Algorithm* by Deepak Goyal, a former PhD student of Bob Paige. The author uses various techniques developed by Bob Paige, dominated convergence and finite differencing, to derive a new algorithm for the may-alias analysis and to establish its time complexity. The new algorithm takes cubic time in the size of the input and improves asymptotically to the previous best one, which was quintic.

– *Dynamic Programming via Static Incrementalization* by Yanhong Liu and Scott Stoller. This paper describes a (semi)automatic program transformation that optimizes recursive programs amenable to dynamic programming.

– *Program Synthesis from Formal Requirements* by Elizabeth Leonard and Constance Heitmeyer. The authors describe a project to translate a requirements specification, expressed in SCR notation, into the language C. Two translation strategies are discussed. Both were implemented using the Abstract Program Transformation System (APTS), which was designed and implemented by Bob and his collaborators.

– *Derivation of Efficient Logic Programs by Specialization and Reduction of Nondeterminism* by Alberto Pettorossi, Maurizio Proietti, and Sophie Renault. The authors propose an extension of both partial evaluation and conjunctive partial deduction to specialize nondeterministic programs. To this end, they consider definite logic programs and a new set of transformation rules which extend the usual fold/unfold/define rules used for partial evaluation.

– *Universal Regular Path Queries* by Oege de Moor, David Lacey, and Eric Van Wyk. This work describes the development of an algorithm that abstracts a version of model checking for flow graphs with annotations corresponding to inferred properties. The authors show that such an algorithm can be derived systematically in the spirit of Bob Paige's work.

– *Least Reflexive Points of Relations* by Jules Desharnais and Bernhard Möller. This paper studies a generalization of the problem of finding conditions under which a function on a partially ordered set has a least fixed point, and it also studies relations on complete lattices. The authors present two main results about the existence of the least-fixed point, and a theorem of Cai and Paige for computing it.

– *Relativizations for the Logic-Automata Connection* by Nils Klarlund. This paper describes an efficient translation from logic M2L(str), a monadic logic of strings, into a variant of WS1S, i.e., weak second-order theory of one successor. The main issue is an efficient handling of the first-order and (weak) second-order quantifiers. The algorithmic treatment requires a detailed study of the corresponding relativizations of formulas.

– *Efficient Type Matching* by Somesh Jha, Jens Palsberg, Tian Zhao, and Fritz Henglein. The authors present an $O(n \log n)$-time algorithm for matching recursive types and an $O(n)$-time algorithm for matching nonrecursive types. The algorithm for recursive types works by reducing the type matching problem to the problem of finding a size-stable partition of a graph for which Paige and Tarjan provided an $O(n \log n)$ algorithm. The algorithm for nonrecursive types employs multiset discrimination techniques due to Paige, Tarjan, Cai, and Yang.

– *Aspects as Invariants* by Douglas Smith. Aspect-Oriented Programming offers tools for the modular development of systems with crosscutting features. This paper presents a way to express those features as logical invariants and then to generate the kind of code that is usually produced from manually written aspects.

– *Computational Divided Differencing and Divided-Difference Arithmetics* by Thomas Reps and Louis Rall. In this paper an approach conceptually similar to the Computational Differentiation technique is used for computing the set of finite, divided differences of a sampled function $F(x)$: $F[x_0, x_1] = (F(x_0) - F(x_1)) / (x_0 - x_1)$.

– *Program Transformations: Some Lessons from the 80s* by Dave Wile. The author presents some of the experience gained by the scientific community during that decade about designing, implementing, and using program transformation systems.

References

1. R. M. Burstall and J. Darlington: Some transformations for developing recursive programs. In *Proceedings of the International Conference on Reliable Software*, Los Angeles, USA, 465–472, 1975.
2. O. Danvy, F. Henglein, H. Mairson, and A. Pettorossi (Eds.): Special Issue in Memory of Bob Paige (Part I), Vol. 16, Nos. 1–2 of *Higher-Order and Symbolic Computation*, Springer, 2003.
3. O. Danvy, F. Henglein, H. Mairson, and A. Pettorossi (Eds.): Special Issue in Memory of Bob Paige (Part II), Vol. 18, Nos. 1–2 of *Higher-Order and Symbolic Computation*, Springer, 2005.
4. S. A. Walker and H. R. Strong: Characterization of flowchartable recursions. In *Proceedings 4th Annual ACM Symposium on Theory of Computing*, Denver, CO, USA, 1972.

University of Aarhus, Denmark — *Olivier Danvy*
University of Copenhagen, Denmark — *Fritz Henglein*
Brandeis University, Massachusetts, USA — *Harry Mairson*
Università di Roma Tor Vergata, Italy — *Alberto Pettorossi*
June 2007

Bob Paige with his colleague Cai Jiazhen (about 1986)

Contents

List of Contributors

Oege de Moor
Computing Laboratory, Oxford University,
Parks Road,
Oxford, England, OX1 3QD
oege@comlab.ox.ac.uk

Jules Desharnais
Département d'Informatique,
Université Laval,
Québec, QC, G1K 7P4 Canada
Jules.Desharnais@ift.ulaval.ca

Deepak Goyal
Calypto Design Systems, Inc.
2933 Bunker Hill Lane, Suite 202,
Santa Clara, CA 95054, USA
dgoyal@calypto.com

Constance L. Heitmeyer
Center for High Assurance Computer
Systems, Naval Research Laboratory, 5546,
Washington, DC 20375, USA
heitmeyer@itd.nrl.navy.mil

Fritz Henglein
Department of Computer Science (DIKU),
University of Copenhagen,
DK-2100 Copenhagen, Denmark
henglein@diku.dk

Somesh Jha
Computer Sciences Department,
University of Wisconsin,
Madison, WI 53706, USA
jha@cs.wisc.edu

Nils Klarlund
Google, Inc.
76, 9th Avenue
New York, NY 10011, USA
klarlund@google.com

David Lacey
ClearSpeed Technology plc
3110 Great Western Court,
Hunts Ground Rd, Bristol, BS34 8HP, U.K.
david.lacey@clearspeed.com

Elizabeth I. Leonard
Center for High Assurance Computer
Systems, Naval Research Laboratory, 5546,
Washington, DC 20375, USA
leonard@itd.nrl.navy.mil

Yanhong A. Liu
State University of New York
at Stony Brook,
Stony Brook, NY 11794, USA
liu@cs.sunysb.edu

Harry Mairson
Computer Science Department,
Brandeis University,
Waltham, MA 02254, USA
mairson@brandeis.edu

Bernhard Möller
Institut für Informatik,
Universität Augsburg,
D-86135 Augsburg, Germany
moeller@informatik.uni-augsburg.de

Gary D. Paige
Dept. of Neurobiology and Anatomy,
University of Rochester, 601 Elmwood Ave.,
Box 603,
Rochester, NY 14642, USA
gary_paige@urmc.rochester.edu

Robert Paige (1947–1999)
Department of Computer Science,
New York University, 251 Mercer Street,
New York, NY 10012, USA
paige@cs.nyu.edu

Jens Palsberg
Computer Science Department, UCLA
4531K Boelter Hall,
Los Angeles, CA 90095-1596, USA
palsberg@ucla.edu

Helmuth Partsch
Faculty of Computer Science,
University of Ulm,
D-89069 Ulm, Germany
Helmuth.Partsch@uni-ulm.de

Alberto Pettorossi
University of Roma Tor Vergata,
Via del Politecnico, 1
I-00133 Roma, Italy
pettorossi@info.uniroma2.it

Maurizio Proietti
IASI-CNR,
Viale Manzoni, 30
I-00185 Roma, Italy
proietti@iasi.rm.cnr.it

Louis B. Rall
Dept. of Mathematics,
University of Wisconsin,
480 Lincoln Dr.,
Madison, WI 53706, USA
rall@math.wisc.edu

Sophie Renault
European Patent Office, Patentlaan 2
P.O. Box 5818
NL-2280 HV Rijswijk, The Netherlands
srenault@epo.org

Thomas W. Reps
Comp. Sci. Dept., University of Wisconsin,
1210 W. Dayton St.,
Madison, WI 53706, USA
reps@cs.wisc.edu

Douglas R. Smith
Kestrel Institute,
3260 Hillview Avenue,
Palo Alto, California 94304, USA
smith@kestrel.edu

Scott D. Stoller
State University of New York
at Stony Brook,
Stony Brook, NY 11794, USA
stoller@cs.sunysb.edu

Eric Van Wyk
Department of Computer Science and
Engineering, University of Minnesota,
Minneapolis, Minnesota 55455, USA
evw@cs.umn.edu

David S. Wile
Teknowledge Corp.
4640 Admiralty Way #1010,
Marina del Rey, CA 90292, USA
dwile@teknowledge.com

Tian Zhao
Department of Electrical Engineering and
Computer Science, University of Wisconsin,
Milwaukee, WI 53211, USA
tzhao@cs.uwm.edu

Part I

Robert Paige's Research: A Retrospective and A Proposal

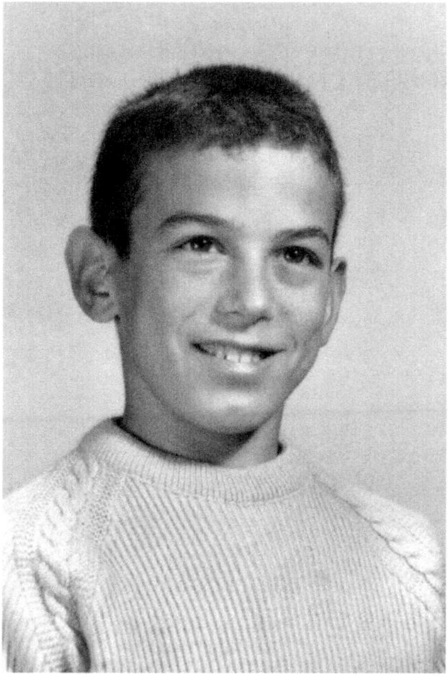

The young Bob Paige (about 1958)

Research Retrospective on Transformational Development of Programs

Robert Paige (1947–1999)

Department of Computer Science, Courant Institute, New York University,
251 Mercer Street, New York, NY 10012, USA. paige@cs.nyu.edu

Robert Paige wrote this retrospective of his research work for the IFIP Working Group 2.1 *meeting n.* 53 *which unfortunately he could not attend due to illness. This paper, delivered at the meeting in his absence, is focused on the transformational development of programs. The reader will find it useful to refer to the related paper "A National Science Foundation Proposal", also by Robert Paige, which is included in this book.*

The group [the IFIP Working Group 2.1] was exciting in the 1970s, when we were groping for direction and divided by different orientations. I guess it was in this atmosphere that combined purpose with uncertainty where I found my own voice. The common goal was a transformational program development methodology that would improve productivity of designing and maintaining correct software. The emphasis was on algorithmic software.

We differed as to how to achieve this goal, and my approach was out on a limb. Based on a few transformations, the most exciting of which was Jay Earley's iterator inversion combined with high level strength reduction, and also on an overly optimistic faith in the power of science to shed light on this subject, I believed that algorithms and algorithmic software could be designed scientifically from abstract problem specifications by application of a small number of rules, whose selection could be simplified (even automated in some cases) if it could be guided by complexity. Most all others (including the SETL crowd at Courant) disagreed, and accepted the notion that algorithm design was *inspired*, and that the most significant steps in a derivation were unexplainable *eureka* steps.

I knew that my goals were ambitious and with little supporting evidence. In fact the case was too flimsy even to begin work on the components of a program development methodology. It seemed better to gather facts first, to uncover compelling examples that might lead to a theory, and to put together this theory only after the pieces were sufficiently understood and developed.

In order to test the viability of a new transformational idea, I tried to demonstrate how it could be used to improve some aspect of a "conventional" area of Computer Science; e.g., databases, algorithms, programming languages, etc. (but not transformational programming itself). A conservative, neutral test would be determined by an evaluation of the improvement (independent of the transformatonal idea) by members of the conventional area. A more subjective, but still useful clinical test could be made by me checking whether the new idea could improve the quality of education or facilitate my research. (I used the class room not only as a laboratory for clinical tests, but also as a way for me to learn and retool in new subject areas.) After developing the transformational idea further, it could be evaluated directly by the program transformation community, e.g., at IFIP WG 2.1 meetings, other conferences, or through the publication process.

Perhaps the most important component of a transformational methodology is a wide spectrum language for expressing all levels of abstraction from problem specifications down

O. Danvy et al. (eds.), Automatic Program Development – A Tribute to Robert Paige, 3–6.

to hardware-level implementations. Since I was a student at New York University with Jack Schwartz as my adviser, naturally I went with SETL, and found it convenient to use in my thesis work in finite differencing (from 1977–1979). As it turned out I was teaching full time (4 courses per year) at Rutgers starting Fall 1977, and tried lots of research ideas out in the class room early on.

Despite serious semantic flaws in SETL (some of which were fixed much later in SETL2), a small subset of the language, augmented with a few standard abstract notations found in the mathematical preliminaries chapters of standard Computer Science texts, provided students with a simple, powerful, and sometimes executable mathematical notation. SETL's small but powerful repertoire of abstract operations proved convenient in being widely reusable in modeling a variety of computer science concepts without the need for language extension.

In algorithms courses the built-in abstractions of SETL allowed data structures and rudimentary strategies of many algorithms to be specified perspicuously and without extraneous implementation detail. Default implementations of SETL's abstract operations allowed such specifications to be studied as executable prototypes, and tested empirically. Such specifications could also be used as the starting points for derivations of more efficient implementations. This made algorithm understanding easier, even (especially) with weak students who were overwhelmed by the detail of Mix-like or Algol-like implementation-level specifications. It seemed that this approach allowed me to cover much more material and convey much greater understanding.

In database courses SETL notations seemed better than the standard query languages. It facilitated teaching different data models, and different levels of description including implementations at the level of file systems and indexes. A Philips Research Technical Report by Lacroix and Pirotte [15], which compared over 100 database queries in 9 different query languages plus English was instructive. SETL variants sketched out by Koenig and me seemed to have greater clarity. A uniform wide-spectrum notation facilitated description and analysis of file and database organization, and made it easier to describe mappings from query to physical level and from one data model to another.

The wide-spectrum nature of the language made it easy to illustrate transformations at and between various levels of abstraction. The first transformation that I developed was finite differencing (a generalization of conventional strength reduction plus Earley's Iterator Inversion), whose goal was to speedup programs by replacing costly repeated computations by less expensive differential counterparts.

Differencing is a mathematical idea with an old history. The chain rule mechanism and other features of finite differencing can be described independently of any particular language, but the wide-spectrum nature and simplicity of SETL seemed ideal for illustrating the power and broad applicability of the transformation in a separate implementation design. Using SETL finite differencing, one can easily explain how to put together and analyze (using a natural notational form of worst case and amortized complexity) so very many efficient algorithms in a systematic way. This made teaching algorithms much easier.

One of the best examples of how surprising connections can be made, due to finite differencing, took place at the IFIP WG 2.1 meeting in Chamrouse (France). We were all trying to derive one of those efficient max sequence algorithms, and the use of finite differencing at a major step suggested a similar step (to compute nested collective minimum values) in the derivation of Dijkstra's Single Source Shortest Path algorithm.

During these early years there was also one interesting objective test of finite differencing in the area of databases. In his Turing Award address Codd said that integrity control was an important open problem in relational databases. Successful use of finite differencing to solve view maintenance and integrity control was reported in [14] and by Paige at the IFIP WG 2.1 meeting in Toulouse (France), in 1983.

In Compiler classes at Rutgers I also developed a crude form of dominated convergence in order to derive workset algorithms found in Hecht's book [11] for solving global program analysis problems specified by fixed points. Work on this transformation progressed as it was seen to be progressively more useful in deriving an increasingly wider range of algorithms. And I continued to use it in combination with finite differencing in my compiler lectures in order to derive the more difficult algorithms. However, it was not until my collaboration with Cai that a comprehensive investigation and development of this transformation was first completed in [4].

It became apparent from the beginning that I was looking for a transformational program development methodology whose final step would be data structure selection beyond which lay conventional compiler optimization. I also favored an approach limited to composable simple data structures as are found in SETL's method of basings and Wiederhold's file system design methodology instead of an expert system style as proposed by Barstow and Kant. Despite some ingenious ideas, SETL data structure selection was largely ad hoc, not amenable to formal complexity analysis, overrelying on hashing, missing a linked list, and never proven (either analytically or empirically) to be an improvement over naive unoptimized data structures.

Unfortunately SETL data structure selection was not good enough to produce high performance code by a complexity driven approach. What was needed were data structures guaranteed to support associative access (e.g., set membership testing) in unit worst-case time for search arguments and set elements of arbitrary type. Also needed were principles for selecting such data structures. It has taken many years to develop this transformation, and a first comprehensive treatment will appear in Deepak Goyal's thesis (expected in Fall 1999) [9]. Only recently was the transformation equipped with a suitable read method to create the desired data structures. And I still do not understand how to eliminate enough overhead explaining even very simple forms of data structure selection (described in several papers cited in the references) to use this transformation conveniently in a standard algorithms class.

Nevertheless, I believe that it is an essential final stage of and underpinning to a transformational methodology that (1) begins with dominated convergence to compute fixed-point specifications, then (2) applies finite differencing to implement repeated expensive high-level expressions more efficiently by exploiting the differential maintenance of program invariants, and, finally, (3) uses data structure selection by real-time simulation to implement associative access operations in unit worst case time on a RAM.

Over the years we have uncovered extensive evidence that our methodology will scale up and be capable of improving the productivity of large-scale system implementation. Three kinds of evidence have been considered. Our SETL-based specification languages and the transformations that implement them have been shown to be effective in terms of: (1) explaining complex algorithms, (2) discovering new algorithms, and, most importantly, (3) improving the productivity of efficient implementations of algorithms.

[The illustration of these three points has been done by Robert Paige himself in the Sections 2.1.2 and 2.1.3 of the paper entitled: *A National Science Foundation Proposal*, which is included in this book.]

Over the past few years, I have felt that, finally, there was enough evidence and know-how to put together much of our work into a formal program development methodology that could achieve my goals of 20 years ago. Although I was too sick to carry this work out, I am happy that my student Deepak Goyal has done it in his Ph.D. thesis, which should be finished some time over the summer 1999. Deepak ought to present it at a future IFIP Working Group 2.1 meeting. Annie Liu, who is on Deepak's Committee, can tell you more, or you may look at Deepak's home page, which is in the Ph.D. Student area of our New York University Department home page. I think you will like it.

For anyone interested, Deepak will be working in John Field's group at IBM Thomas J. Watson Research Center, Yorktown Heights, NY, starting November 1st. Until then, he will be at New York University.

References

(The references which were not cited in the text have been included for making it easier to access the relevant literature.)

1. B. Bloom: Ready simulation, bisimulation, and the semantics of CCS-like languages. Ph.D. Thesis, Massachusetts Institute of Technology, 1989.
2. B. Bloom and R. Paige: Transformational design and implementation of a new efficient solution to the ready simulation problem. *Science of Computer Programming*, 24(3), 189–220, 1995.
3. J. Cai and R. Paige: Binding performance at language design time. In *Proceedings of the 14th Annual ACM Symposium on Principles of Programming Languages*, M.J. O'Donnell (Ed.). ACM Press, München, West Germany, 85–97, 1987.
4. J. Cai and R. Paige: Program derivation by fixed point computation. *Science of Computer Programming*, 11(3), 197–261, 1989.
5. J. Cai and R. Paige: Towards increased productivity of algorithm implementation. In *Proc. ACM SIGSOFT FSE*, 71–78, 1993.
6. J. Cai and R. Paige: Using multiset discrimination to solve language processing problems without hashing. *Theoretical Computer Science*, 145(1–2), 189–228, 1995.
7. C.-H. Chang and R. Paige: From regular expressions to DFA's using compressed NFA's. *Theoretical Computer Science*, 178(1/2), 1–36, 1997.
8. W. F. Dowling and J. Gallier: Linear-time algorithms for testing the satisfiability of propositional Horn formulae. *J. of Logic Programming*, 1(3), 267–284, 1984.
9. D. Goyal: A Language Theoretic Approach To Algorithms. *Ph.D. Thesis*, Department of Computer Science, New York University, NY, 169 pages. 2000.
10. D. Goyal and R. Paige: The formal reconstruction and improvement of the linear time fragment of Willard's relational calculus subset. In *Algorithmic Languages and Calculi*, R. Bird and L. Meertens (Eds.), Chapman & Hall, 382–414, 1997.
11. M. S. Hecht: *Flow Analysis of Computer Programs*, Elsevier North-Holland, 1977.
12. J. P. Keller and R. Paige: Program derivation with verified transformations — A case study. *Comm. on Pure and Applied Mathematics*, 48(9/10), 1053–1113, 1996.
13. D. E. Knuth: *The Art of Computer Programming, Volume II: Seminumerical Algorithms.* Addison-Wesley, 1969.
14. S. Koenig and R. Paige: A transformational framework for the automatic control of derived data. In *Proceedings of the 7th International Conference on Very Large Data Bases.* Cannes, France, 306–318, 1981.
15. M. Lacroix and A. Pirotte. Example queries in relational languages. Technical Note 107, Philips Research Laboratory, Brussels (Belgium), 1977.
16. R. Paige *Formal Differentiation.* UMI Research Press, 1981.
17. R. Paige and F. Henglein: Mechanical translation of set theoretic problem specifications into efficient RAM code — A case study. *Journal of Symbolic Computation*, 4(2), 207–232, 1987.
18. R. Paige and S. Koenig: Finite differencing of computable expressions. *ACM Transactions on Programming Languages and Systems*, 4(3), 402–454, 1982.
19. R. Paige, R. E. Tarjan, and R. Bonic: A linear time solution to the single function coarsest partition problem. *Theoretical Computer Science*, 40(1), 67–84, 1985.
20. R. Paige and Z. Yang: High Level reading and data structure compilation. In *Proceedings of the Twenty-Fourth Annual ACM Symposium on Principles of Programming Languages*, N. D. Jones (Ed.). ACM Press, Paris, France, 456–469, 1997.
21. J. H. Reif and H. R. Lewis: Symbolic evaluation and the global value graph. In *Proceedings of the Fourth Annual ACM Symposium on Principles of Programming Languages*, R. Sethi (Ed.), ACM Press, 104–118, 1977.

A National Science Foundation Proposal

Robert Paige (1947–1999)

Department of Computer Science, Courant Institute, New York University, 251 Mercer Street, New York, NY 10012, USA. paige@cs.nyu.edu

Summary. The objectives of this research are to improve software productivity, reliability, and performance of complex systems. The approach combines program transformations, sometimes in reflective ways, to turn very high-level perspicuous specifications into efficient implementations. These transformations will be implemented in a metatransformational system, which itself will be transformed from an executable specification into efficient code. Experiments will be conducted to assess the research objectives in scaled-up applications targetted to systems that perform complex program analysis and translation.

The transformations to be used include (i) dominated convergence (for implementing fixed-points efficiently), (ii) finite differencing (for replacing costly repeated calculations by less expensive incremental counterparts), (iii) data structure selection (for simulating associative access on a RAM in real time), and (iv) partial evaluation (for eliminating interpretive overhead and simplification). Correctness of these transformations, of user-defined transformations, and of the transformational system itself will be addressed in part. Both the partial evaluator and components of the transformational system that perform inference and conditional rewriting will be derived by transformation from high-level specifications. Other transformations will be specified in terms of Datalog-like inference and conditional rewriting rules that should be amenable to various forms of rule induction.

Previously, Cai and Paige in [12] used an ideal model of productivity free from all human factors in order to demonstrate experimentally how a transformation from a low-level specification language into C could be used to obtain a fivefold increase in the productivity of efficient algorithm implementation in C in comparison to hand-coded C. However, only small-scale examples were considered. The proposed research includes a plan to expand this model of productivity to involve other specification languages (and their transformation to C), and to conduct experiments to demonstrate how to obtain a similar fivefold improvement in productivity for large-scale examples of C programs that might exceed 100,000 lines.

The proposal lays out extensive evidence to support the approach, which will be evaluated together with its theoretical underpinnings through substantial experiments. If successful, the results are expected to have important scientific and economic impact. They are also expected to make interesting, new pedagogical connections between the areas of programming languages, software engineering, databases, artificial intelligence, and algorithms.

Keywords: partial evaluation, program transformation, software productivity, software performance, data structure selection, language translators.

1 Introduction

Program Transformations is about semantics-based analysis and manipulation of programs. Over the last 20 years we have made contributions to the area by developing two distinct tracks: (1) general-purpose problem specification and its transformation to efficient programs, and (2) special-purpose specification of systems that implement the program

O. Danvy et al. (eds.), Automatic Program Development – A Tribute to Robert Paige, 7–27.

analysis and transformations used in Track (1). The long term goal is to combine both tracks.

Previously, using the approach of Track (1) we were able to demonstrate effective translations of high-level specifications of algorithms into high performance codes limited to small-scale examples. In [12] Cai and Paige developed an ideal model of productivity free from human factors. Within this model they gave experimental evidence that their transformational approach to program development leads to at least a fivefold improvement in productivity of efficient algorithm implementation in C as compared to hand-coded C. However, those experiments only considered small-scale examples of procedureless programs no more than 10 pages long. In the proposed grant period, we plan to extend our specification languages and the transformations that implement them in order to specify and develop efficient large-scale systems with a similar improvement in productivity.

Previously, using the approach of Track (2) we were able to demonstrate effective development of large-scale systems limited to inefficient prototypes. The complex translation of a statically typed variant of SETL into C used in the productivity experiments mentioned above was specified in RSL (Rule Specification Language), a high-level language implemented by the APTS program transformation system [37]. The translation suffered from two major sources of inefficiency. First, APTS is implemented in SETL2 [49], which runs 30 times slower than C at best. Second, APTS only provides an interpretive implementation for RSL. Hence, the translation of SETL to C was bogged down to 24 lines per minute on a SPARC 2. In the proposed grant period we plan experiments to test the feasibility of a radical new transformational methodology that combines reflective forms of partial evaluation (to eliminate interpretive overhead) and data structure selection (to replace the naive SETL2 runtime) used in Track (1) in order to gain a 300-fold speedup in RSL execution.

By combining improvements to both tracks, we plan to demonstrate a five- to tenfold improvement in productivity for implementing large-scale systems with more than 100,000 lines of C. Our automated program development methodology is most effective in development of systems with high algorithmic content, which are among the most difficult to construct and maintain by hand. Included in this class of systems are those that implement complex program analysis and transformation, which is the application domain for the research proposed here. An example is the SETL-to-C translator, an RSL specification which our methods are expected to speedup by a factor of 300.

Section 2 of this proposal describes a cohesive research project that should take 3–5 years to complete for three Ph.D. students and the Principal Investigator. Section 3 details a number of experiments, including a novel, reflective combination of partial evaluation and data structure selection, that are expected to be completed within the 3-year funding period. Although this proposal only seeks funding for one Ph.D. student, two other student participants will be funded by other sources—one by New York University (NYU) fellowship and another by an Office of Naval Research (ONR) grant that partly overlaps with this proposal.

2 Background and Objectives

2.1 Track (1): Background

Specification Languages and Transformations

Within Track (1) we consider an implementation language (e.g., C) that serves as a conventional RAM model of computation. We also consider three successively more abstract specification languages, each implemented in terms of the next successively lower-level language by a distinct transformation. By associating syntactic constructs with asymptotic complexity, we are able to predict precisely how each transformation can improve program running time and space. Consequently, the selection of transformations can be guided by complexity considerations, and the three specification languages are made *computationally transparent* (i.e., amenable to formal algorithmic analysis).

Low SETL, our lowest level specification language, is a statically typed, executable variant of SETL2 [49], a pointerless, block structured, imperative language augmented with a repertoire of primitive set operations such as membership testing, element addition and deletion, arbitrary choice, for-loops through a set, map application, indexed map assignment, and so forth. Its perspicuousness rests on copy/value semantics, and its ability to navigate through data directly rather than by indirect location formulas using pointers or cursors. Associative access and nondeterministic selection and search contribute to its readability and succinctness. The data structure selection transformation [6, 36, 38] is based on a low-level theory of data structures in which set and map operations (such as associative access) are simulated on a RAM in real time; i.e., a conventional physical structure written in C is selected for implementing each primitive set operation (for search arguments and type-compatible set or map domain elements of any data type) in unit worst-case time.

An important aspect of this work is that the type system (fragments of which are found in [7, 24, 28, 40]) is a strongly typed variant of the Curry/Hindley type discipline for the λ-calculus [17,26]. It is parametric since it contains type variables, but currently is not polymorphic. The type system together with the data structure selection transformations make it possible to analyze Low SETL programs for their worst-case time and space complexity. This allows programmers to be guided by complexity considerations, which is essential to the production of high performance systems. Using Low SETL as a solid foundation, the other two specification languages obtain computational transparency by the way they are mapped into Low SETL.

High SETL is a statically typed, imperative, executable superset of Low SETL augmented with such abstract operations as set comprehension, quantification, and a variety of operations on binary relations. Computational transparency is obtained by implementing High SETL expressions in Low SETL in two ways. Firstly, the cost of fresh evaluations of high-level expressions can be determined by implementing them directly into Low SETL. More interestingly, we can use our finite differencing transformation [31, 35] to evaluate repeated costly High SETL expressions by more inexpensive incremental counterparts in Low SETL. We determine the cost of these differential calculations by associating precise amortized complexities with an eager strategy for maintaining equality invariants $E = f(x_1, \ldots, x_k)$ within worst-case sequences of modifications to variables $x_1, \ldots, x_k$; i.e., each time a modification to $x_1, \ldots, x_k$ occurs, variable E is updated to reestablish the invariant. High SETL allows us to avoid having to write error-prone bookkeeping operations that maintain invariants differentially. By associating amortized complexity with maintenance of basic invariants, and by closure rules that allow us to determine the cost of maintaining collections of interdependent invariants, we can generate languages of High SETL invariants that can be maintained differentially with precise complexities [8].

The highest level specification language is SQ2+ [9], a nonexecutable, statically typed, functional subset of High SETL augmented with least and greatest fixed-points. These fixed-point expressions abstract error-prone, iterative looping constructs. Computational transparency is obtained for SQ2+ by associating fixed point operations with precise implementations in High SETL using our dominated convergence transformation [9]. Dominated convergence computes fixed points in terms of High SETL loops generating "chaotic", finitely converging sequences [16] that are less expensive than Tarski sequences [50]. Both fresh and differential calculations of fixed-point expressions have been considered. We also developed specification languages that are subsets of SQ2+ with guaranteed worst-case execution complexities in time and space for each polynomial degree [8, 10].

Over the years we have uncovered extensive evidence that our methodology will scaleup and be capable of improving the productivity of large-scale system implementation. Three kinds of evidence have been considered. The specification languages and the transformations that implement them have been shown to be effective in terms of (1) explaining complex algorithms, (2) discovering new algorithms, and, most importantly, (3) improving the productivity of efficient implementations of algorithms. In order to enhance the credibility of this proposal, we will present some of this evidence in the following two subsections.

Algorithm Explanation and Discovery

Most mainstream research in Program Transformations emphasizes principles for constructing derivations with illustrations drawn from known algorithms. The purpose has been to uncover common patterns of abstraction in specification and transformation that could form the basis of a useful methodology for making algorithm design and program development easier. Short, elegant transformational proofs (of correctness integrated with analysis) document implementations of complex algorithms, lend greater confidence in the correctness of complex codes, and provide greater assurance in the reliable modification of such codes.

Perhaps the first examples of nontrivial algorithms being derived by finite differencing were presented in [31, 34]. Included among these examples is a SETL specification of Dijkstra's naive Bankers Algorithm and its transformation into Habermann's efficient solution. This derivation was done without knowledge of Habermann's solution, and there were no dead ends. Finite differencing homed right in on a solution matching Habermann's time/space bounds.

It is well known that the construction of optimizing compilers is a costly labor-intensive task. Can this labor be reduced by our methods? To answer this question in part, we showed how easy it was to specify dozens of programming language and compiler analysis problems in SQ2+, to simplify these specifications, and to transform them by dominated convergence into High SETL prototypes [9]. In [8] we showed how the constant propagation algorithm of Reif and Lewis [42] could be expressed as set-theoretic equations in a subset of SQ2+ that could be mapped into RAM code guaranteed to run in linear time in the size of the program dataflow relation.

The use of notation has been regarded as a burden to algorithm designers ever since Knuth came out with Mix [30]. But can notation also help satisfy the needs of the algorithm community—precise algorithmic analysis and succinct exposition? An SQ2+ specification of the Single Function Coarsest Partition Problem and its transformation by dominated convergence, finite differencing, and real-time simulation was used to derive a new linear time solution [39]. That algorithm paper was selected for publication in a special issue of TCS as a best paper from ICALP. In [40] a much improved explanation of the algorithmic tool called Multiset Discrimination in [11] was obtained by specifying the algorithm in Low SETL and using its type system to formally explain and analyze the low level implementation that would be obtained by real-time simulation. The earlier presentation of this algorithmic device involved so much indirection from pointer-based primitives, that most readers were confused. Some of these earlier readers agreed that the Low SETL presentation clarified their understanding.

The viability of a transformational methodology can be demonstrated by using it to "explain" or "prove" well-known algorithms. However, if in the course of such formal explication no new deep structure is uncovered that leads to improved solutions, then its impact on algorithmics and programming productivity is likely to be limited. More powerful evidence favoring a transformational methodology is provided if it can be demonstrated to help facilitate the discovery of new algorithms. Our success with algorithm discovery may be attributed to our reliance on complexity in both specification and transformation.

Our first instance of algorithm discovery by transformation was reported in [38], where we used all three transformations to turn an SQ2+ specification of Horn Clause Propositional Satisfiability into a new linear time pointer machine algorithm. Previously, Dowling and Gallier found a linear time algorithm [19] that relied heavily on array access. In [4] we used dominated convergence and finite differencing to derive a Low SETL executable prototype from an SQ2+ specification of ready simulation. We then showed informally how the Low SETL prototype could be turned into an algorithm that runs five orders of magnitude faster than the previous solution in [3]. All three transformations, but especially real-time simulation (where types were shown to be useful in modeling complex data structures), were involved in the discovery of a new improved solution to the classical problem of DFA minimization [28]. Finite differencing was used extensively in [15] to derive a new improved solution to the classical problem of turning regular expressions into DFA's. Perhaps our most convincing paper-and-pencil result was in [24], where Goyal and Paige used Low SETL specifications and real-time simulation to improve Willard's time bound for query processing from linear expected to linear worst-case time without degrading space. Willard's original algorithm was extremely difficult, and involved over 80 pages of proofs. Our transformational approach yielded much shorter but also more precise constructive proofs that led to an implementation design. Here is a first successful example of scalingup.

Two summers ago, Ph.D. student Deepak Goyal designed and implemented "practical" algorithms in Java at Microsoft. He believes that his use of Low SETL and the data structure selection transformation as part of a programming methodology increased his productivity and improved the quality of the code that was produced.

Experimental Foundations for Productivity Improvement

Perhaps the most compelling evidence that our transformational methodology will scaleup and provide a dramatic improvement in the productivity of large high-performance complex systems may be found in the experiments by Cai and Paige [12]. In that paper we developed a simple but conservative model of productivity. Within that model we demonstrated a fivefold improvement in productivity of high performance algorithm implementation in C in comparison to hand-coded C.

Those experiments tested an approach to producing C programs by writing programs in a simple variant of Low SETL, and compiling them into high performance C. The high performance of the C code produced by the SETL-to-C translator is based on the translator's ability to simulate associative access (e.g., finite set membership or finite map application) on a RAM in real time.

Measuring productivity improvement depended on two assumptions. The first is that one line of Low SETL takes no more time to compose than one line of C. This assumption is not controversial. Our experience is that one line of Low SETL can be produced faster than a line of C. The generally lower-level of discourse in C creates an intellectual gap between the program and the mathematical function it computes. For example, in order to implement SETL's element deletion operation (`s less:= x`) efficiently in C would require at least 10 carefully chosen C operations. The need to access data through pointers and cursors in C creates a greater level of indirection (which complicates understanding) than in SETL, where data is accessed directly through values.

The second assumption is that the C code generated automatically from Low SETL has roughly the same number of lines as equivalent hand-coded C. We found that the generated C was between 10% and 30% larger than hand-coded C, and concluded that the errors in the two assumptions would cancel each other out.

Suppose we measure programming productivity in a given programming language as the number of pretty-printed source lines produced per unit time. Suppose also that productivity decreases as the conceptual difficulty in understanding a program grows. Then in our investigation, which is restricted to highly algorithmic programming (as is found in complex language systems and environments), conceptual difficulty is roughly reflected in the size of the dataflow relation, which can be expected to grow nonlinearly with the number of source lines. Thus, programming productivity $P(L)$, as a function of the number of program source lines L, increases as L decreases.

Let L_2 be the size of a C Program C_2 compiled from a Low SETL specification S_1 of size L_1, and let C_3 be a handcrafted C program equivalent to C_2. By assumption one, we can use the same productivity function $P(L)$ for programs written in both Low SETL and C. By assumption two, we know that the size of C_3 is roughly the same as the size L_2 of C_2. Then the improvement in productivity by using our automated program development methodology versus handcrafted programming is given by the time $L_2/P(L_2)$ to manually produce C_2 divided by the time $L_1/P(L_1)$ to manually produce S_1, which is

$$(L_2/L_1)(P(L_1)/P(L_2)) > L_2/L_1$$

since $P(L_1)/P(L_2) > 1$ whenever $L_1 < L_2$.

It should be emphasized that our experiments did not measure productivity directly, which would have required difficult and costly analysis of human factors such as programming expertise and intelligence. Instead we measured productivity improvement by exploiting the two assumptions mentioned above to obtain an objective, inexpensive, and credible framework for conducting comparative productivity experiments that could avoid all human factors. The ratio of lines of C code generated automatically from Low SETL divided by the number of lines of Low SETL being compiled gives a lower bound on productivity improvement. Every algorithm that we tested in [12] yielded ratios that exceeded 5, and that grew as the input size grew.

Although we found that the C code generated automatically from Low SETL had runtime performance comparable to good hand code (whose running time, excluding I/O, was at least 30 times faster than SETL2 running time executed by the standard SETL2 interpreter), only small-scale examples were used. The largest C program was about 10 pages. Low SETL lacked procedures, and the type system was highly restricted. The SETL-to-C production rate was too slow—about 24 lines of C per minute on a SPARC 2. Finally, high-level SETL input had to be translated into low-level C input at compilation time.

The hypothesis that productivity improvement will scaleup by our methods is based on the methodology and the application domain. Real-time simulation is applied uniformly to each instruction of a program regardless of program size. Program analysis and transformation algorithms have a complex combinatorial nature, which fits our model of productivity.

2.2 Track (1): Scaleup Objectives

In order for Track (1) to be useful in developing systems, the three specification languages need to be reconstructed. The foundation for this reconstruction is a full-scale design of Low SETL. To this end we plan to enrich the functionality of Low SETL and extend its type system. Formal semantics for Low SETL need to be worked out along with the type system as it was in the earlier typed SETL variant found in [7]. And in order to support computational transparency, we need to incorporate complexity assertions within a formal rule system along the lines of Goyal and Paige [24].

In [40] tagged alternation, user defined types, and recursive subtypes were proposed for Low SETL. Polymorphism and higher order functions are also needed. Thus far, we have successfully modelled listlike data structures in the type system. The next step is to show how the type system can be used to model more realistic hybrid data structures built up from arrays and lists. We believe that our solutions might benefit the implementation of vectors in Java.

We also need to develop a type inference model that augments the model found in [7] and the inference algorithm found in the SETL-to-C translator [12] to handle the new type system. The powerful batch read feature found in [40] allows external input in string form to be validated and converted to complex data structures in linear time in the length of the string for any list of variables with any signature in the type system. This needs to be augmented with interactive input/output. Modules, procedures, and type conversions across procedure and module boundaries need to be added too.

Perhaps the main unresolved problem in implementing Low SETL has to do with its copy/value semantics. This is a hard problem that has been the major source of inefficiency in two generations of SETL compilers. The strategy of SETL1 [47] and SETL2 is to implement lazy copies. That is, assignment of large objects (e.g., sets and tuples with arbitrary depth of nesting) or incorporation of a large object into another large object is implemented by only copying pointers. Since a large object may be shared under this strategy, it is first copied whenever it is updated in order to avoid any side effect. With this approach hidden copies can potentially degrade program performance from $O(f(n))$ expected time to $O(f(n)^2)$ actual time. In [12] we observed 30,000-fold slowdowns in SETL2 performance due to unnecessary hidden copies.

Schwartz [45,46] developed an interesting but complicated value flow analysis for SETL1 [47] in order to detect when hidden copies could be avoided, and update operations could be performed in place. The difficulty of the analysis seems to stem, in large part, from the fact that SETL1 did not implement reference counts, so that the analysis had to prove that an r-value was unshared. It was never implemented, and a completely different naive approach that was eventually used proved to be unsatisfactory.

In SETL2 dynamic reference counts of all references (these are used for default boxed implementations, and are not part of the SETL2 language) to each tuple or set value are maintained. Highly restricted circularity of references ensures that when a value has a reference count greater than 1, then that value is *shared*, and cannot be updated unless it is first copied. In order to perform element addition S *with*:= a (which adds element a to set S) in place, the location that stores S must have a reference count of 1 and be different from the location that stores a. Otherwise, a hidden copy of S is made, and the update is performed on the copy. Unfortunately, the backend of Snyder's SETL2 compiler introduces so many compiler-generated temporary variables (which do not get garbage collected until the end of scope) that practically all data is shared at runtime.

In [12, 36], we solved this problem by assuming that all updates were performed on unshared values, and so, could be updated in place. Any program that violated this assumption was considered "erroneous" (in the sense of Ada). This approach obtained some credibility in [12, 24], where SQ2+ and High SETL specifications were transformed into Low SETL programs guaranteed not be erroneous a priori. However, this approach may not be satisfactory for manual programming directly in Low SETL in scaledup applications. Recently, Goyal and Paige solved this problem using dynamic reference counts, dead code analysis, and analysis of when large data *must* share the same location [25]. A very local implementation of this approach for SETL2 was shown to speedup APTS runtime by a factor of 10. However, dynamic reference counts do incur a constant factor overhead in running time, so it would be interesting to see if more powerful and complicated analysis of when data *may* share the same location could be used in order to detect a wide range of contexts where dynamic reference counts can be avoided.

Development of a robust Low SETL specification language would serve as the foundation for the other two higher level specification languages. In [7] variants of High SETL and SQ2+ were formulated within a simple type system, which was only crudely associated with complexity in an ad hoc way. We propose to reconstruct both specification languages with an enriched type system based on Low SETL.

Scaledup applications planned during the grant period include a reimplementation of APTS modules for pattern matching and inference in High SETL to be compiled into C. The batch read algorithm will be written in High SETL, translated into C, and benchmarked relative to the hash-based SETL2 read method. Finally, we plan to implement Willard's RCS queries in High SETL to be compiled into C.

Developing better scientific means of measuring productivity and productivity improvement are important but extremely difficult. The model of productivity improvement developed in [12] was necessarily ideal, and only used for the Low SETL specification language.

We would like to explore how to extend this model in order to test productivity improvement in C for High SETL and other specification languages.

2.3 Track (2): Background

Specification Language and Implementation

Track (2) has to do with the methodology needed to implement Track (1). RSL (Rule Specification Language) is a high-level language for specifying language translators, analyzers, and program transformations used in Track (1). RSL specifications are compiled and executed using the APTS metatransformational system, which was built by Cai and Paige to implement this methodology [37]. Appendix A explains the methodologies of Tracks (1) and (2) by illustrating an actual APTS transcript of automatic program development using the SETL-to-C translator.

APTS was designed to implement complex program transformations and program analysis for *any* deterministic context free language. It is a collection of modules, each performing an independent task implemented by an interpreter for a different language paradigm, including logic-based inference, conditional rewriting, finite differencing, commands, and syntax. RSL is a single integrated transformational language with programming-in-the-large features such as an Ada-like library, separate compilation, and fine-grained incremental compilation. APTS is entirely written in only 15,000 lines of SETL2 source code (including comments), and it contains no foreign tools. It can be extended by call-in and call-out capabilities relative to compatible SETL2 modules.

Compilers written in RSL make use of the following APTS components. The Rule Database (RDB) contains inference rules that define relations storing program properties, e.g., type or dataflow. These relations may be defined over a variety of domains including conventional domains (booleans, integers, strings), the domain of abstract program points (i.e., simple contexts), program terms (the syntactic values that occur at program points), and Lisplike S-expressions (a general domain used to embed arbitrarily deeply nested types amenable to first order unification). Inference rules are specified in a language similar to Datalog [51] augmented with function symbols and primitive pattern matching predicates. The APTS inference engine analyzes the program (being compiled) for properties specified in the RDB, and it outputs the Program Database (PDB) of finite relations (sets of tuples that represent ground terms) storing the program's properties.

The Transformation Database (TDB) contains conditional meaning-preserving program transformations of two kinds—rewriting or finite differencing. The transformation engine selects a transformation T by matching T with a portion of the program, and by ensuring that the applicability condition for T, when instantiated with the current PDB, is satisfied. It then applies transformation T to the program to obtain a new transformed program. Consequently, the PDB must be updated to be consistent with the new program and the RDB.

The inference and transformation engines make use of the efficient bottom-up pattern matching algorithm of Cai, Paige, and Tarjan [13]. The inference engine used to calculate RDB relations [6] combines this bottom-up pattern matcher with RETE style pattern matching [22] and seminaive evaluation of Datalog [1,2,51]. Part of the signature of a relation allows seminaive evaluation to calculate relations as in a simple addition system or with built-in unification.

The use of APTS as a crucial vehicle in the proposed research may be justified simply by convenience—we have access to its source code, and the source language is SETL2, which is not hard to rewrite into Low SETL. However, its functionality also compares favorably with other systems. The APTS logic-based inference method for program analysis implementation was influenced by the earlier Mentor and Centaur systems [5,18], which were among the first systems to break away from attribute grammars and use a more general logic-based approach to program analysis. Reliance on foreign tools such as Lex, YACC, and Prolog makes Centaur powerful but inefficient. Unlike Centaur's reliance on Prolog's general-purpose inference engine, our implementation has been designed specifically for high performance program analysis.

The goal of using incremental computation as part of a transformational environment for APTS was influenced by the Synthesizer Generator [43]. It uses a first-order functional language called SSL to specify syntax, attribute equations, as well as program transformations. In the SG, program analysis is performed by attribute evaluation and incremental attribute evaluation relative to program editing modifications [44]. Although efficient attribute reevaluation is an important benefit of attribute grammars, this approach limits navigation to the syntax tree, which makes global analysis inefficient. Thus, it is often the case that an escape from the attribute grammar formalism is warranted, and C procedures are introduced.

Although RSL includes a command language, which, like SSL, can operate directly on syntax trees, the RSL subcomponents for logical inference, conditional rewriting, and built-in finite differencing are more abstract and declarative. The RSL style discourages direct reference to the syntax tree or its structure, and encourages reference to the tree indirectly by pattern matching. The algorithms that implement those subcomponents are highly sophisticated; e.g., the on-line preprocessing algorithm for multipattern tree matching [13].

Refine [41] is a robust, commercially available transformational system, that offers a language like SSL for manipulating syntax trees. KIDS [48] (which is built on top of Refine) implements rewriting and inference on top of Refine, and is highly regarded as a well-engineered system in the transformational community. The directed inference mechanism used by KIDS uses the full power of first order theorem proving, which is needed for program synthesis tasks. However, this approach trades efficiency and automation for generality, and is probably not well suited to scaledup applications, where automatic program analysis is essential.

2.4 Track (2): Scaleup Objectives

The most pressing and perhaps most difficult open problem in APTS system research is in devising an automatic scheme to maintain the consistency of PDB relations incrementally after a program is changed by transformation. The idea is to automatically generate an Incremental Rule Database (IRDB) from the Batch RDB and a program transformation. This would allow the inference engine to recalculate the PDB efficiently by executing the IRDB each time a program transformation is applied. Some combination of differential techniques and partial evaluation is needed to solve this crucial but difficult research problem.

The current version of APTS handles this problem automatically for extensional relations such as *monotone* and *type* (which are invariant with respect to equivalence-preserving expression replacement). However for intensional relations such as *free* and *bound* variables, *control flow*, and most subtype relations, APTS requires the user to annotate each conditional rewriting rule R with instructions on how to update the PDB each time R is applied.

Solving this problem fully automatically would eliminate one of the most difficult and error-prone aspects of RSL programming. Of course, a good algorithmic solution must depend on a clean formal semantics, leading to the practical integration of analysis and transformation. More generally, it would also enhance a top-down stepwise refinement framework in which global analysis is reserved for the highest level, perspicuous specifications, and local analysis is sufficient to select and justify transformations whose application propagates semantic facts to lower-level implementations.

The only related work we are aware of is that of Emma van der Meulen [52], who gave initial solutions to incremental conditional rewriting for the ASF+SDF system [29], a descendant of Centaur. Her methods were based on reducing primitive recursive schemes to strongly noncircular attribute grammars, and either applying the Reps, Teitelbaum, Demers algorithm [44], or using a batch-oriented approach with lazy incremental updates. We would be seeking a sharper, more practical solution within our RETE-based inference strategy that exploits the fact that a transformation preserves semantics.

Another major problem has to do with ensuring reliability of the program development process. Currently, all transformations used within APTS preserve program semantics, but this is only proved on paper outside the system. Since APTS is capable of unbridled production of dangerously large quantities of code without any human intervention or oversight, the absence in APTS of machine-assisted metalevel support to prove transformations correct is a major deficiency that needs to be overcome. We expect that unfunded participants in the area of computational logic from the University of Catania and the University of L'Aquila will investigate how to verify our transformations by mechanical proof checking and theorem proving. Recently, formal verification of our implementation design of Willard's query processor [24] has been achieved with paper and pencil in Cantali's Thesis [14] at the University of Catania. Cantali's immediate future plans are to pass his proofs through the ETNA set theoretic mechanical proof checker.

Within the proposed research we plan to consider two other aspects of formal correctness. First, we will derive APTS system components that implement pattern matching and inference. Second, we plan to investigate various forms of rule induction to prove properties of RSL inference and conditional rewriting rules that implement particular kinds of program analysis and transformation.

The main objective of Track (2) research is to improve the speed of RSL execution in scaled-up applications. Unless otherwise stated, our analysis will not include running time for input and output methods. Slow speed results from levels of interpretive overhead, the fact that RSL is fully specified in its own formalism, interfacing between modules, and the fact that APTS is implemented in SETL2. Slow speed of SETL2 results from dynamic memory allocation, dynamic typing, the high cost of redundant expressions, a model of computation based on associative access, hidden copies, etc.

In order to speedup RSL execution, we will reconstruct the existing SETL-to-C translator (written in RSL) into a robust compiler, called the Low SETL-to-C Accelerator, for the full Low SETL language. Next, we will rewrite APTS in Low SETL, and compile it into C using the SETL Accelerator. Based on previous experiments [12], this should speedup APTS by a factor of 30.

We will also build a self-applicable partial evaluator for Low SETL, written in Low SETL itself. RSL will be used to generate the Low SETL abstract syntax tree for the partial evaluator. This would allow us to partially evaluate the Low SETL version of APTS together with any RSL specification to produce a seamless Low SETL program equivalent to the original RSL specification, but running about 10 times faster. We could then compile this Low SETL specification into C using the Low SETL Accelerator to gain another factor of 30 in speed. In this way we could speedup the RSL specification of the SETL Accelerator by a factor of 300.

Since RSL syntax is fully specified in the RSL syntax formalism, the power of RSL can sometimes be used to analyze and implement itself more succinctly than in SETL. In this regard, we plan to replace the current APTS SETL modules for finite differencing and for RSL compilation by more perspicuous RSL modules. We also plan to make RSL statically typed, and to equip the RSL compiler with an RSL specification for type analysis. Partial evaluation would then be used to undo the extra level of interpretive overhead that would otherwise make this kind of bootstrapping unreasonably inefficient.

Partial evaluation [27] may be the most widely applicable and potentially practical transformation around. It is based on a general software engineering principle in which highly parameterized programs are concretized by simplification when a subset of the parameters is fixed. It offers an attractive generic implementation strategy that traces a portion of the computation for which expressions can be evaluated (for any datatype and operation in the language). Consequently, partial evaluation must be implemented in full accordance with language semantics. A partial evaluator can turn an interpreter with fixed program input into a compiled program, or it can turn a partial evaluator with fixed interpreter input into a compiler. Amazingly, it can turn a partial evaluator with fixed partial evaluator input into a compiler generator. Partial evaluation has been worked out for various programming languages within the major language paradigms.

An off-line partial evaluator for a low-level subset of dynamically typed SETL2 has been built with a finite, uniform, congruent division and polyvariant specialization based on the method for flowchart languages found in [27]. It includes interprocedural analysis for control flow, dataflow, and live variables, which is used for various optimizations including procedure unfolding during specialization. So far the partial evaluator has been used successfully to implement the First Futamura Projection on several interesting examples, including an interpreter for a simple imperative language. Minor modification is underway to implement the Second Futamura Projection, which is critical for automatic transformation of the APTS interpreter into a compiler.

However, the partial evaluator has been implemented using a mixture of Standard ML and SETL2 programs. We need to rewrite it more perspicuously as a mixture of RSL and Low SETL specifications. The partial evaluator needs to be extended to handle modules (including module variables of static extent). It also needs to be rewritten to partially evaluate Low SETL. Finally, it should be said that designing a good self-applicating partial evaluator for Low SETL may be extremely difficult, and alternative approaches known to work and achieve the same ends will be taken if need be.

One possible alternative to get a good generating extension is to directly implement a compiler generator (cogen, also known as generating extension generator) [32], whose functionality is the same as the result of double self-application of a partial evaluator as formulated by the Third Futamura Projection. This new approach avoids several technical problems of self-applicating approach, which arise in statically typed languages such as Low SETL. The relationship of a compiler generator and a partial evaluator is like that of a compiler and an interpreter, so it is not too difficult to write a compiler generator by hand.

In addition, it is also difficult to design a good binding-time analysis, especially for separately compiled modules (which is an open problem in itself). For the specific application here, however, manually annotating and refining the binding-time of variables and operators would be a practical alternative, giving the developers a finer control over the generated code.

3 Research Plan for Automated Software Manufacturing

Within the grant period we intend to turn the rudimentary ideas found in [12] and outlined above into a practical technology offering a five- to tenfold improvement in productivity for large-scale examples of over 100,000 lines of C. In order to accomplish the preceding goals, we will combine *data structure selection* by real time simulation of a set machine on a RAM [7, 36] and *partial evaluation* [27]. We plan to use the software components described in the preceding section to automatically build other components that would be the essential tools for a new practical program development technology. These new components are described below.

Let P_L be a program P implemented in language L. Let R denote RSL, and let S stand for Low SETL implemented in the standard runtime environment of dynamically typed SETL2. Using the transformational products in the project, including the SETL Accelerator written in RSL (denoted by A_R), the APTS system written in Low SETL (denoted by $APTS_S$) and the partial evaluator for Low SETL written in Low SETL (denoted by PE_S), we propose a series of transformational experiments that, in scaledup applications, test the feasibility of combining partial evaluation with our own transformations (sometimes in complex reflective ways), test speedups predicted for transformed code, and test productivity improvement for C implementations.

From experience we have found that the runtime performance of any program P_R is roughly 10 times slower than the runtime performance of an equivalent S program P_S. We would like to test whether partial evaluation of RSL yields similar speedups. Based on the experiments reported in [12] and unpublished independent experiments by Snyder (the designer and implementer of SETL2), SETL2 programs P_S should run 30 times slower than equivalent C programs P_C. (We assume here, that P_S would have no hidden copies, or else it might run many more times slower.) Our experiments [12] showed that the SETL-to-C translator produced C codes that matched the 30-fold speedup observed in hand-coded C. Since partial evaluation and data structure selection are completely independent, we plan to test the hypothesis that P_C will run 300 times faster than P_R regardless of whether P_C is produced mechanically or by hand.

We will first apply A_R to PE_S to give us a partial evaluator of Low SETL in C, i.e., $PE_C = A_R(PE_S)$, which should be 30 times faster than PE_S. Next, we will apply PE_C to APT_S and specialize it w.r.t. A_R to give us a Low SETL Accelerator in Low SETL, i.e., $A_S = PE_C(APTS_S, A_R)$, which should be 10 times faster than A_R. Self-application of A_S produces an equivalent translator $A_C = A_S(A_S)$ from S to C that is written in C and should run 300 times faster than A_R.

The next set of experiments relate to generating a fast generic C read method. First, we apply A_C to the read method written in Low SETL, denoted by $read_S$, to give us a generic C read method, i.e., $read_C = A_C(read_S)$. To get a specialized version of the read method for a fixed type signature sig (which includes a type assignment for the input variables and subtype constraints), we first apply PE_C to $read_S$ and sig to obtain a Low SETL version of the specialized read routine for signature sig, i.e., $read_S^{sig} = PE_C(read_S, sig)$. We further apply A_C to $read_S^{sig}$ to get its C version equivalent, i.e., $read_C^{sig} = A_C(read_S^{sig})$.

In a similar way, we can improve the speed for APTS. Applying A_C to $APTS_S$ will yield a C version of APTS, i.e., $APTS_C = A_C(APTS_S)$. We get a compiler version of APTS by specializing PE_S w.r.t. $APTS_S$ using PE_C, i.e., $CAPTS_S = PE_C(PE_S, APTS_S)$. This can further be converted to C, i.e., $CAPTS_C = A_C(CAPTS_S)$. Now, our transformations can be done quickly using $CAPTS_C$ and A_C, i.e., $P_S = CAPTS_C(P_R), P_C = A_C(P_S)$ for any program P_R.

Among the tools generated for free from the three implemented tools PE_S, A_R, and $APTS_S$, we believe that $APTS_C$, $CAPTS_C$, and even A_C are likely to exceed 100,000 lines of C. Furthermore, programs P_C that result from compiling substantial RSL programs P_R into Low SETL programs P_S by $CAPTS_C$, which are further translated into C by A_C, can easily form codes of 100,000 lines or more.

A successful production of $APTS_C$ would, for the first time, yield C modules implementing efficient forms of extremely difficult algorithms and subsystems that would be useful to the programming language community. These include, (1) the fastest known preprocessing algorithm for bottom-up multipattern tree matching [13], (2) an inference engine combining our fast pattern matching algorithm with RETE-syle forward chaining [22] to implement logic-based program analysis and computation, and (3) a bottom-up conditional rewriting engine that makes use of our fast pattern matching algorithm to implement source program transformation.

As a final application Willard's Predicate Retrieval theory [53–55] deals with a large database query optimization, and serves as an attractive scaled-up application for implementing and verifying a difficult query compiler. Conceptually and technically difficult, an RCS query compiler has resisted all previous attempts at an implementation. Nevertheless, we believe that our implementation plans will succeed. RCS queries expressed in High SETL will be transformed into semantically equivalent programs in Low SETL using an RSL specification W_R. We expect the speed for this transformer to be considerably improved by the transformation $W_C = A_C(CAPTS_C(W_R))$. The Accelerator A_C will turn Low SETL versions of these queries into C for execution.

4 Results from Prior NSF Support

The present grant proposal stems from an ongoing feasibility study sponsored by the National Science Foundation under the SGER program within the Software Engineering and Languages area of CISE/CCR. The ongoing NSF grant has award number CCR-9616993, funding amount $100,000., and support period September 1, 1996 to August 31, 1999. The title is "Improving Productivity of Algorithm and System Implementation in Scaled Up Applications".

Results from the currently funded grant have been reported earlier in this proposal. They include considerable progress in the design of Low SETL, in the implementation of a SETL partial evaluator, and in the investigation of the hidden copy problem. Publications include a POPL 97 paper [40] on the formal semantics and algorithmic development of a batch reading method for Low SETL. This paper was coauthered by the Principal Investigator and Ph.D. student Zhe Yang, and presented at POPL by Yang. As a result of this work Yang won a prestigious BRICS fellowship to study last year with Olivier Danvy at the University of Aarhus in Denmark. This year he has returned to NYU to work on his Thesis.

While at BRICS, Yang worked on the type encoding problem in languages with Hindley–Milner type system. This work is partly motivated by the effort to type the generic read routine for Low SETL [40]. This routine has its input arguments dependently typed; i.e., one of the arguments provides the type signature for the remaining inputs. Results of this work were reported in an ICFP 98 paper [56] that was presented by Yang. The ICFP 98 paper formulated this kind of problem in terms of type-indexed values, and developed several general approaches to program with them within a Hindley–Milner type system, the basis for many popular functional languages such as ML. These approaches are based on encoding types as higher-order polymorphic functions, whose types reflect the encoded types themselves. For the general functional programming community, this paper provides programming techniques for writing dependently-typed programs using commonly available languages. We are also further convinced that a Hindley–Milner type system with some variations can provide the type basis for Low SETL.

As an application of the above type encoding paper, we solved the problem of implementing type-directed partial evaluation *natively* in ML [57]. Although native implementations of type-directed partial evaluators have been reported to be several magnitudes faster than meta-language implementations, previously, they were only implemented using untyped languages such as Scheme, which cannot guarantee runtime type safety.

Another publication [24], coauthored with Ph.D. student Deepak Goyal, demonstrated how the Low SETL type system could actually be used to improve the runtime complexity of Willard's database query processing method. This work is unusual in the sense that a difficult algorithmic improvement was not obtained by the usual combinatorial arguments but by algebraic and logical reasoning using theoretical programming language concepts. Goyal has given invited talks on this work at SUNY Albany, the University of Copenhagen, and the University of Aarhus. As noted earlier, these also inspired a Bachelor's Thesis in the area of Mechanical Verification at the University of Catania in Italy. This year Goyal and I coauthored another paper [25] on how to avoid hidden copy operations in an imperative language with large datatypes and copy/value semantics with a lazy copy strategy, such as Low SETL. Goyal presented the work at the Static Analysis Symposium 98 (which took place in Pisa, Italy), and gave invited talks at several universities in Italy.

Goyal is the unique holder of the prestigious Dean's Dissertation Fellowship in the Computer Science Department at NYU, so he will not need academic support for next year. Zhe will be funded by an ONR grant that partly overlaps with the current proposal. This proposal seeks funding for one Ph.D. student in addition to Goyal and Yang for a total of three students working full-time on the project. The principal investigator will also be on sabbatical next year, and will remain at NYU in order to make good progress with the ambitious work proposed here.

5 Conclusion

There is compelling experimental evidence that SETL Accelerator A_C will compile programs P_S into C codes P_C whose performance is comparable to hand-coded C for small examples. We also believe that the performance of C codes P_C produced by our methods will actually grow in competitiveness with hand-coded C as the number of source C lines grow. If this is confirmed, then our automated software manufacturing technology may not only work for scaled-up applications, but may allow us to build high performance systems that cannot even be built under current technology.

An important feature of this technology is the natural way in which it can evolve as each of its separate basic components (PE_S, A_R, or $APTS_S$) is improved. Improvements are of two kinds, both of which lead to a merger of Tracks (1) and (2) as part of a bootstrapping process. One kind of improvement has to do with progressive elevation of the partial evaluation language, the Accelerator source language, and the implementation languages of PE_S and $APTS_S$ from Low SETL to High SETL to either SQ2+ or RSL. Another kind has to do with combining the logic-based relational style of RSL with the functional set-based style of SQ2+ into a single very high-level specification language.

Appendix A: Transcript of an APTS Derivation

This section illustrates our specification languages and transformational methodology by tracing through an actual transcript of mechanical program development in APTS. The transcript demonstrates how a formal specification of live code analysis (for an imperative well structured programming language), written in only a few lines of SQ2+, can be turned automatically into a C program of several hundred lines with worst-case running time and space linear in the input space. We also show how live code analysis can be specified in RSL, and used as part of the SQ2+-to-C compiler.

The SQ2+ specification of live code analysis appears pretty-printed by APTS just below:

```
program useless;
  assume oneone (instof);
  assume onemany (iuses);
  assume manyone (compound);
  assume disjoint (range instof, range compound);
  read (instof, usetodef, iuses, compound, crit);
  print (clfp (crit, live + instof [usetodef [iuses [live]]] +
                          compound [live], live));
end program;
```

where `crit` is a set of initial live statements (`read` and `print`), `iuses` is a one-to-many map from statements to uses of variables, `usetodef` is a many-to-many map from uses to definitions (left-hand-side occurrences of variables) of variables that can reach these uses along definition-clear program paths, `instof` is a one-to-one map from definitions to their enclosing statements, and `compound` is a many to one map from statements to immediately enclosing compound statements (i.e., if-statements and while loops). Assumptions are given to improve the quality of the compiled code. Image set expression $f[s]$ yields the image of set s under binary relation f; i.e., $\{y \mid \exists x \in s : [x, y] \in f\}$. The conditional least fixed-point expression `clfp(...)` computes the smallest set `live` (with respect to set containment) that includes set `crit` and satisfies equation

`live` = `live` ∪ `instof` [`usetodef` [`iuses` [`live`]]] ∪ `compound` [`live`].

For example, in the following program,

```
program test;
 read(b);
 if c2 then c := y;
 elseif c3 then a := y;
 end if;
 print(a);
end program;
```

let statements `print(a)` and `read(b)` belong to set `crit`. Then all but assignment `c := y` would be live.

Based on the SQ2+ type inference system described in [7], the APTS inference engine computes types for each program expression. It also computes other program properties, such as monotone expressions, and bound and free variables of expressions. These properties are all stored as relations in the the APTS PDB (Program Database).

The binary `type` relation is defined over the domain of program terms (the first component) and s-expressions representing types (the second component). The binary relation `mono` (see below) is defined over program terms. RSL code for two of the inference rules used to compute types are shown below,

```
match(%expr, .x%) | null(z, type(%expr, .x%, z)) ->
  bind(.t, newatom(t)) and type(%expr, .x%, .t);
match(%expr, .x + .y%) | type(%expr, .x%, .t) ->
  type(%expr, .x + .y%, .t) and type(%expr, .y%, .t);
```

The first rule states that for every program point `p` in the SQ2+ specification with syntactic category `expr` (i.e., for every occurrence of an expression) such that `type` is not defined for the term `r` stored at `p`, create a new type variable `t`, and make `t` the type for `r` (i.e., store pair `[r,t]` in the `type` relation). The second rule is more complicated. It states that for every program context `p1+p2`, where `p1` and `p2` are program points containing terms `t1` and `t2` respectively, if the `type` relation contains pair `[t1,t]`, then perform the following actions. Obtain a most general unifier `t'` of `t` and the types of `t2` and `t1+t2` if these types already exist, and let `t'`=`t` otherwise. Then store pairs `[t1,t']`, `[t2,t']`, and `[t1+t2,t']` in relation `type` after deleting each of the types for `t1`, `t2`, and `t1+t2` that may exist.

Inference rules in APTS may be performed in any order. There is a notion of "safe" rules (under a closed world assumption) to allow for limited forms of negation and built-in predicates, and a semantics similar to the logic databases found in [51]. More details about the inference rules and how they are implemented in APTS can be found in [6].

The conditional rewriting transformation `nminfp` can turn the functional live code analysis specification above into a simple imperative program that computes the conditional least fixed-point by dominated convergence. The transformation is displayed in APTS using the "help" feature.

```
>:help nminfp;
lhs code to be matched
    print (clfp (.w, .s + .k, .s));
rhs replacement code
      .x := .w;
      while exists .z in (.k - .x) loop
        .x with := .z;
      end loop;
      print (.x);
pattern expansion code is
    readvar (.x) and
    genvar (.z) and
    subst (.k, .s, .x)
database action code is
    type(.x,[set, .t]) and
    type(.z, .t) and
    type(%expr,.k-.x%, [set, .t])
enabling predicate
    mono (% expr, .s + .k %, .s) and
    type (.s, [set, .t])
```

This rule applies to any program point that is matched by the left-hand-side pattern and that satisfies the enabling predicate. Nonlinear pattern matching is carried out between the left-hand-side pattern (essentially a sentential form in the SQ2+ grammar) and the program. Variables preceded by a period are pattern variables that match program points. For matching to succeed, all occurrences of the same pattern variable in a pattern must match the same terms.

An environment *env* that maps pattern variables to program points is created as a side effect of matching. Matching the left-hand-side of `nminfp` with the SQ2+ specification results in the following environment:

env(`.w`) = `crit`
env(`.s`) = `live`
env(`.k`) = `instof [usetodef [iuses [live]]]` + `compound [live]`

Next, the enabling predicate pattern is instantiated relative to environment *env*. Predicate `mono(live + instof[usetodef[iuses[live]]] + compound[live], live)` and `type(live, [set, .t])` which results from substitution, must then be matched against the mono and type relations stored in the PDB. That is, the pair:

`[live + instof [usetodef [iuses [live]]] + compound [live], live]`

must be stored in relation `mono`, and relation `type` must store pair `[live, [set,` τ`]]` for some s-expression τ that represents a type. Both conditions are satisfied. Since variable live has type `[set t5]`, the environment is extended so that *env*(`.t`) = `t5`.

At this point the environment is extended further by executing the pattern expansion commands. Command `readvar(.x)` requires the user to supply an identifier from the terminal. In response we will supply the string `livest`, after which *env*(`.x`) = `livest`. Command `genvar(.z)` creates a new identifier automatically. In this case the system supplies `x1`, after which *env*(`.z`) = `x1`. Command `subst(.k, .s, .x)` is a general substitution mechanism that replaces all occurrences of *env*(`.s`) in *env*(`.k`) by *env*(`.x`). In this case, substitution will result in *env*(`.k`) = `instof[usetodef[iuses[livest]]]` + `compound[livest]`.

The right-hand-side replacement code in the `nminfp` transformation can now be instantiated relative to environment *env*, and used to replace the SQ2+ subtree matched by the left-hand-side. Finally, the database action code indicates how to update the type relation in the PDB for the new terms that are introduced by tree replacement. The High SETL program just below results from dominated convergence,

```
program useless; -- Assumptions elided
5  read (instof, usetodef, iuses, compound, crit);
6  livest := crit;
7  while exists x1
         in instof [usetodef [iuses [livest]]] +
     compound [livest] - livest loop
8    livest with := x1;
   end loop;
9  print (livest);
end program;
```

The SQ2+-to-C translator proceeds to the next phase of compilation, which applies finite differencing (see [20, 21, 33, 48] for related work). In order to expose opportunities for finite differencing and also to regularize the program into a simplified form for which a limited number of finite differencing rules can have wide utility, the translator turns the program into a normal form. This is done by applying a group of conditional rewriting rules exhaustively bottom-up until there is no further change. Consequently, line 7 is replaced by the following equivalent code:

```
7  while exists x1
         in {x2 in instof [usetodef [iuses [livest]]]
              + compound [livest] | x2 notin livest} loop
```

The algorithm used to implement group transformations is based on the incremental linear pattern matching preprocessor found in [13]. Regardless of the number of rewriting transformations belonging to a group, matching can proceed bottom-up so that the exact subset of individual rules whose left-hand-sides match a given program subtree can be computed in unit time, and presented in linear time in the subset size. Efficient incremental preprocessing of groups with respect to rule addition and deletion is also supported.

Next, the finite differencing transformation automatically detects invariants of the form `x = e` (where `e` is a program expression and `x` is a new variable uniquely associated with `e`) that should be maintained and exploited in order to avoid the costly repeated calculation of the truth set at the top of the while-loop at line 7. The following invariants are detected automatically by bottom-up analysis of the truth set expression; their left-hand-side variables are all supplied by the user:

```
uses = iuses [livest]
defs = usetodef [uses]
livedf = instof [defs]
livecf = compound [livest]
liveall = livedf + livecf
work = {x in liveall | x notin livest}
```

After the six invariants are detected, a stream processing transformation based on Goldberg and Paige [23] inserts code that aims to establish the invariants on entry to the while loop using a minimal number of loops. In this case stream processing generates the code appearing on lines 7 through just before line 29 below. A chain rule is used to perform the necessary bookkeeping operations needed to reestablish these invariants just before they are falsified by the modification to livest at line 45. Consequently, the costly truth set expression at the top of the while loop is made redundant, and is replaced by variable `work`. The Low SETL program that results from finite differencing appears just below.

```
program useless; -- Assumptions elided
5    read (instof, usetodef, iuses, compound, crit);
6    livest := crit;
7    livecf := { };
8    livedf := { };
9    defs := { };
10   uses := { };
11   for x3 in livest loop
12     if compound (x3) notin livecf then
13       livecf with := compound (x3);
       end if;
14     for x5 in iuses { x3 } loop
15       for x9 in usetodef {x5} | x9 notin defs loop
16         livedf with := instof (x9);
17         defs with := x9;
         end loop;
18       uses with := x5;
       end loop;
     end loop;
19   work := { };
20   liveall := { };
21   for x11 in livedf loop
22     if x11 notin livest then
23       work with := x11;
       end if;
24     liveall with := x11;
     end loop;
25   for x12 in livecf loop
26     if x12 notin livest then
27       work with := x12;
       end if;
28     liveall with := x12;
     end loop;
29   while exists x1 in work loop
```

```
30      if compound (x1) notin livecf then
31        if compound (x1) notin livest then
32          work with := compound (x1);
          end if;
33        liveall with := compound (x1);
34        livecf with := compound (x1);
        end if;
35      for x6 in iuses {x1} loop
36        for x9 in usetodef {x6} | x9 notin defs loop
37          if instof (x9) notin livest then
38            work with := instof (x9);
            end if;
39          liveall with := instof (x9);
40          livedf with := instof (x9);
41          defs with := x9;
          end loop;
42        uses with := x6;
        end loop;
43      if x1 in liveall then
44        work less := x1;
        end if;
45      livest with := x1;
      end loop;
46    print (livest);
end program;
```

Finite differencing often causes intermediate invariants and code in the untransformed program to become useless. The next step of the SQ2+-to-C translator performs a dead code analysis and elimination transformation. First inference rules are used to compute control flow, data flow, types, and an abstraction of the program that allows live code to be calculated. These consist of around three pages of RSL specifications, including the following two axioms to determine the initial live statements:

```
match(%statement, print(.x);%) -> live(%statement, print(.x);%);
match(%statement, read(.x);%) -> live(%statement, read(.x);%);
```

Next, the following two inference rules

```
-- a use of variable z is live if it occurs free in expression y
-- that is reached from a definition of z along a path clear of
-- other definitions to z
   reach(.z, .y) and freevar(.y, .z) -> live(.z);
-- a program component that encloses a live subcomponent is also live
   live(.x) and neq(pred(.x), nil) -> live(pred(.x));
```

are used to determine those program statements that contribute either directly or indirectly (via dataflow and control flow) to the print statement at line 46. This step is carried out in logic programming fashion by calculations involving only relations in the PDB and without reference to the program syntax tree. Analysis determines that statements at lines 10, 19, 42, and 44 are dead. These statements are removed from the program, and control structures are subsequently simplified using RSL conditional rewriting rules.

The final transformation, which makes use of concepts found in [45], but relies mainly on the type/subtype inference mechanism of [7], implements all set and map datatypes using array and list data structures. This transformation rests on the discovery of finite universal sets, called *bases*, to be used for data sharing and for creating aggregate data structures that serve to simulate associative access (e.g. `x in s`) in real time; i.e., each such access is implemented by a unit time array or pointer access. The final C code with 342 labeled statements runs in time linear in the size of the `usetodef` relation. Similar to the benchmarks reported in [12], it runs 30 times faster than the SETL2 program running under Snyder's "stlx" interpreter, which indicates performance comparable to good hand-coded C.

References

1. F. Bancilhon: Naive evaluation of recursively defined relations. In *On Knowledge-Base Management Systems,* M. Brodie and J. Mylopoulos (Eds.). McGraw-Hill, 165–178, 1986.
2. R. Bayer: Query evaluation and recursion in deductive database system, Unpublished Manuscript, 1985.
3. B. Bloom: Ready simulation, bisimulation, and the semantics of CCS-like languages. Ph.D. thesis, Massachusetts Institute of Technology, Sept. 1989.
4. B. Bloom and R. Paige: Transformational design and implementation of a new efficient solution to the ready simulation problem. *Science of Computer Programming,* 24(3), 189–220, 1995. http://cs.nyu.edu/cs/faculty/paige/papers/readysc.ps.
5. P. Borras, D. Clement, T. Despeyroux, J. Incerpi, G. Kahn, B. Lang, and V. Pascual: Centaur: the system. Rapports de Recherche 777, INRIA, 1987.
6. J. Cai: A language for semantic analysis. Technical Report 635, Courant Institute, New York University, 1993.
7. J. Cai, P. Facon, F. Henglein, R. Paige, and E. Schonberg: Type transformation and data structure choice. In *Constructing Programs From Specifications,* B. Möller, (Ed.). North-Holland, Amsterdam, 126–164, 1991. http://cs.nyu.edu/cs/faculty/paige/papers/subtype.ps.
8. J. Cai and R. Paige: Binding performance at language design time. In *Proc. Fourteenth ACM Symp. on Principles of Programming Languages*, 85–97 1987.
9. J. Cai and R. Paige: Program derivation by fixed-point computation. *Science of Computer Programming*, 11:3, 197–261, 1989. http://cs.nyu.edu/cs/faculty/paige/papers/fixpoint.ps.
10. J. Cai and R. Paige: Languages polynomial in the input plus output. In *Algebraic Methodology and Software Technology,* M. Nivat, C. Rattray, T. Rus, and G. Scollo, (Eds.,) Workshops in Computing, Springer-Verlag, Conference Record of the Second AMAST, 287–302, 1992.
11. J. Cai and R. Paige: Using multiset discrimination to solve language processing problems without hashing. *Theoretical Computer Science*, 145(1–2), 189–228, 1995. http://cs.nyu.edu/cs/faculty/paige/papers/hash.ps.
12. Cai J. and R. Paige: Towards increased productivity of algorithm implementation. In *Proc. ACM SIGSOFT*, 71–78, 1993. http://cs.nyu.edu/cs/faculty/paige/papers/prod.ps.
13. J. Cai, R. Paige, and R. Tarjan: More efficient bottom-up multi-pattern matching in trees. *Theoretical Computer Science*, 106(1), 21–60, 1992. http://cs.nyu.edu/pub/tech-reports/tr604.ps.Z.
14. A. Cantali: Using ETNA to prove correctenss and complexity of a linear time implementation of a subset of Willard's RCS. Bachelor's thesis, University of Catania, Catania, Italy, 1997.
15. C.-H. Chang and R. Paige: From regular expressions to DFA's using compressed NFA's. *Theoretical Computer Science*, 178(1–2), 1–36, 1997. http://cs.nyu.edu/cs/faculty/paige/papers/cnnfa.ps.
16. P. Cousot and R. Cousot: Constructive versions of Tarski's fixed-point theorems. *Pacific J. Math.*, 82(1), 43–57, 1979.
17. H. Curry: Modified basic functionality in combinatory logic. *Dialectica*, 23, 83–92, 1969.
18. V. Donzeau-Gouge, G. Huet, G. Kahn, and B. Lang: Programming environments based on structured editors: The mentor experience. In *Interactive Programming Environments.* McGraw-Hill, 1984.

19. W. Dowling and J. Gallier: Linear-time algorithms for testing the satisfiability of propositional Horn formulae. *J. Logic Programming.* 1(3), 267–284, 1984.
20. J. Earley: high-level iterators and a method for automatically designing data structure representation. *J. of Computer Languages*, 1(4), 1976, 321–342.
21. A. Fong and J. Ullman: Induction variables in very high-level languages. In *Proc. 3rd ACM Symp. on Principles of Programming Languages*, 104–112, Jan. 1976.
22. C. Forgy: RETE, a fast algorithm for the many patterns many objects match problem. *Artificial Intelligence,* 19(3), 17–37, 1982.
23. A. Goldberg and R. Paige: Stream processing. In *Proceedings of the ACM Symposium on LISP and Functional Programming,* ACM, 53–62, 1984.
24. D. Goyal and R. Paige: The formal reconstruction and improvement of the linear time fragment of Willard's relational calculus subset. In *Algorithmic Languages and Calculi.* R. Bird and L. Meertens (Eds.) Chapman & Hall, 382–414, 1997. http://cs.nyu.edu/phd_students/deepak/lrcs.ps.
25. D. Goyal and R. Paige: A new solution to the hidden copy problem. In *Proc. 5th International Static Analysis Symposium.* G. Levi (Ed.) LNCS 1503, Springer, 327–348, Sept. 1998. http://cs.nyu.edu/phd_students/deepak/copy.ps.
26. R. Hindley: The principal type-scheme of an object in combinatory logic. *Trans. Amer. Math. Soc.*, 146, 29–60, 1969.
27. N. Jones, C. Gomard, and P. Sestoft: *Partial Evaluation and Automatic Program Generation.* Prentice-Hall, 1993.
28. J. Keller and R. Paige: Program derivation with verified transformations—a case study. *Comm. on Pure and Applied Mathematics*, 48 (9–10), 1053–1113, 1996. http://cs.nyu.edu/cs/faculty/paige/papers/ltmjform.ps.
29. P. Klint: The ASF+SDF meta-environment user's guide, Version 26. Technical report, Centrum voor Wiskunde en Informatica, 1993.
30. D. Knuth: *The Art of Computer Programming.* Vol. 3, Addison-Wesley, 1968–1972.
31. S. Koenig and R. Paige: A transformational framework for the automatic control of derived data. In *Proc. 7th Intl. Conf. on VLDB*, 306–318, Sept. 1981.
32. J. Launchbury and C. K. Holst: Handwriting cogen to avoid problems with static typing. In *Draft Proceedings, Fourth Annual Glasgow Workshop on Functional Programming.* C.K.H.R. Heldal and P. Wadler (Eds.) Workshops in Computing, Skye, Scotland, Springer-Verlag, 210–218, 1991.
33. Y. Liu: Principled strength reduction. In *Algorithmic Languages and Calculi.* R. Bird and L. Meertens (Eds.) Chapman & Hall, 357–381, 1997.
34. R. Paige: *Formal Differentiation.* UMI Research Press, 1981.
35. R. Paige: Programming with invariants. *J IEEE Software*, 3(1), 56–69, 1986.
36. R. Paige: Real-time simulation of a set machine on a RAM. In *Computing and Information.* N. Janicki and W. Koczkodaj (Eds.). Vol. II of *ICCI 89*, Canadian Scholars' Press, Toronto, 69–73, May 1989. http://cs.nyu.edu/cs/faculty/paige/papers/realtime.ps.
37. R. Paige: Viewing a program transformation system at work. In *Programming Language Implementation and Logic*, M. Hermenegildo and J. Penjam, (Eds.), LNCS 844, Springer-Verlag, Berlin, 5–24, Sept. 1994. http://cs.nyu.edu/cs/faculty/paige/papers/viewing.ps.
38. R. Paige and F. Henglein: Mechanical translation of set theoretic problem specifications into efficient RAM code-a case study. *Journal of Symbolic Computation*, 4(2), 207–232, 1987.
39. R. Paige, R. Tarjan, and R. Bonic: A linear time solution to the single function coarsest partition problem. *Theoretical Computer Science*, 40(1), 67–84, 1985.
40. R. Paige and Z. Yang: high-level reading and data structure compilation. In *Proc. 24th ACM Symp. on Principles of Programming Languages*, 456–469, 1997. http://cs.nyu.edu/phd_students/zheyang/papers/read.ps.

41. Refine user's guide version 3.0, 1990.
42. J. Reif and H. Lewis: Symbolic evaluation and the global value graph. In *Proc. 4th Annual ACM Symp. on Principles of Programming Languages*, 104–118, 1997.
43. T. Reps and T. Teitelbaum: *The Synthesizer Generator: A System for Constructing Language-Based Editors.* Springer-Verlag, New York, 1989.
44. T. Reps, T. Teitelbaum, and A. Demers: Incremental context-dependent analysis for language-based editors. *ACM TOPLAS*, 5(3), 449–477, 1983.
45. J. Schwartz: Automatic data structure choice in a language of very high-level. *CACM*, 18(12), 722–728, 1975.
46. J. Schwartz: Optimization of very high-level languages, Parts I, II. *J. of Computer Languages*, 1(2–3), 161–218, 1975.
47. J. Schwartz, Dewar, R., Dubinsky, E., and Schonberg, E.: *Programming with Sets: An Introduction to SETL.* Springer-Verlag, New York, 1986.
48. D. Smith: Kids—a semi-automatic program development system. *IEEE Transactions on Software Engineering*, 129–136, 1990.
49. K. Snyder: The SETL2 programming language. Technical Report 490, Courant Insititute, New York University, 1990.
50. A. Tarski: A lattice-theoretical fixpoint theorem and its application. *Pacific J. of Mathematics*, 5, 285–309, 1955.
51. J. Ullman: *Principles of Database and Knowledge-Base Systems.* Computer Science Press, 1988.
52. E. van der Meulen: Incremental Rewriting. Ph.D. thesis, CWI, Amsterdam, The Netherlands, 1994.
53. D. E. Willard: Predicate Retrieval Theory. Tech. Report 83-3, SUNY Albany, USA, 1983.
54. D. E. Willard: Quasi-linear algorithms for processing relational data base expressions. In *Proceedings of the 9th ACM Sigact-Sigmod-Sigart Symposium on Principles of Database Systems*, 243–257, 1990.
55. D. E. Willard: Applications of range query theory to relational data base join and selection operations. *J. Computer and System Sci.*, 52, 157–169, 1996.
56. Z. Yang: Encoding types in ML-like languages. In P. Hudak and C. Queinnec, eds., *Proceedings of the 1998 ACM SIGPLAN International Conference on Functional Programming* (Eds.), Baltimore, Maryland, USA, ACM Press, 289–300, 1998.
57. Z. Yang: A native ML implementation of type-directed partial evaluation. In *Proceedings of the 1998 APPSEM Workshop on Normalization by Evaluation, NBE '98.* O. Danvy and P. Dybjer, (Eds.). Göteborg, Sweden, May 8–9, 1998, NS-98-1 in BRICS Notes Series, BRICS, Department of Computer Science, University of Aarhus, May 1998.

Part II

Robert Paige: Brother, Friend, Colleague

Bob Paige with his brother Gray (fall 1996)

A Song for My Brother

Gary D. Paige

Department of Neurobiology and Anatomy, University of Rochester,
601 Elmwood Ave., Box 603, Rochester, NY 14642, USA
gary_paige@urmc.rochester.edu

I am grateful to all of you who have contributed to this book in honor of my brother, Robert Paige. I remain deeply touched by the impact that his career and his life has had among his colleagues. There is no doubt that he would have been flattered, humbly surprised, and enormously grateful for this effort. I have put together some personal recollections and thoughts relevant to our familial lives, and which might provide some additional insight into who Rob was and what he meant to each of us in our own ways. It is in the form of four sections, much like a sonata or concerto. I'll explain that shortly. Oh, yes, and I will refer to my brother as Rob, not Bob, just as I have since shortly after birth; such is the dynamic of family relations.

Early Foundations

Sibling relationships are special at many levels. The most obvious and unique attribute is that my brother and I shared half our genes. The only way to beat that ratio is to have been identical twins. Unlike twins, however, there is no understood relationship between how shared and unshared genes play out, interact, and produce the people we become. Further, we were born into the same family, environment, and culture, though I followed by five years (enough to make a difference). Such are the "nature" and "nurture" attributes of brotherhood. Without speculating further, I can tell you that we shared many elements of our being, including mutual abilities, tastes, talents, and foibles, as well as such intangibles as political, religious, and sociological outlook. We also differed in recognizable ways. Indeed, I have friends and colleagues who look more like Rob than I do.

Being far apart in age, I was more an early mascot than a competitor, and that lack of sibling rivalry held for the duration of our lives. We generally reveled in our respective paths, accomplishments, and nuances—playing with them and discussing their meaning and implications. The age gap also meant that we were often affected differently by the same events.

Growing Up

Early events shape and reflect our outlook and character, not always in predictable ways. We spent our initial years in a post-World War II suburb of New York, ensconced into a New Deal political and social environment heavily imbued with the predominant liberal Jewish movement characteristic of much of the community. In 1960, however, the family moved to Phoenix when Rob was at the vulnerable age of 13. The above cultural attributes turned upside down, and the transition was difficult, to say the least. After four years we moved to California with no regrets. Rob spent a semester or two at Arizona State University before transferring to Occidental College in Los Angeles, majoring in chemistry of all things, though the seeds of computer science were clearly planted there. By 1967, Rob had experienced enough of the Southwest, and headed back—and in fact, home—to New York where he had

O. Danvy et al. (eds.), Automatic Program Development – A Tribute to Robert Paige, 31–34.

always remained in heart and mind despite a few forays away. That first move hurt, and in my view deeply affected his outlook thereafter.

Let me back up for a moment. Before we left New York in 1960, Rob had discovered music, taking up the trumpet and rapidly advancing in ability, reflecting an inherent talent. He proved to be a fine performer and even quite the "ham". But music was much more than that. Music provided an expressive outlet, sure, but also a very personal creative outlet, a source of peace and introspection, and indeed, something of a sanctuary. Music is a fundamental language that in those so sensitized, bypasses cortex and heads directly to more emotional and primordial attributes of the mind and brain. I say all this with some conviction, because it would not be long before I, too, would discover my own version of music in Phoenix, having taken up a complementary instrument, the trombone. Here marked an interesting difference between us. While Rob loved to perform, whether for family, friends, schoolmates, or a microphone, I, in contrast, approached seizure threshold at the very thought, stifled by nearly pathological shyness. Oddly, as we later entered the mainstream of our professional careers, this difference dissipated as we converged closer to some middle ground. That synergy remains mysterious.

Late in our Phoenix period, Rob entered a regional music contest and won a scholarship to attend the National Music Camp at Interlochen, Michigan, for the summer of his senior year of high school. That was a major positive turning point for him. By the following year, I had done sufficiently well that, with his encouragement, we attended the camp together. The experience was monumental for me as well. As a humorous aside, that summer witnessed a major strike in this country's air carriers, and so imagine Rob and I (all of 17 and 12) traveling by Greyhound bus for four days to Chicago and then by train to reach Interlochen. It is a good thing we were not kidnapped or otherwise eaten by trolls. But what an adventure, as was the trip home by train. I still recall sharing the Rockies by sunrise together. After that came a key shift in our relationship. The following year I attended camp alone. Rob was off to college and we were, for the first time and thereafter, separated. However, Rob remained closely in the loop for me. He had discovered earlier that Interlochen had established an arts-oriented high school in the early 1960s, too late for him, but perhaps not for me. With his insight and encouragement, I applied, but only after some understandably agonizing chats with our parents. I would be effectively leaving home at 14, and would require a substantial scholarship in order for the family to afford it all. The gambit ultimately worked out, and two weeks after returning from camp I once again headed east. I will forever attribute this most extraordinary and influential portion of my life to Rob for his efforts on my behalf. I told him so, in person, just two weeks before his death.

Adult Life

I must begin this movement where the last one left us—with music. Keep in mind that we were separated by over 2000 miles. Nevertheless, our intense passion for music secured a continued bond. In particular, we both relished jazz. Miles Davis was a mutual hero, and on my visits home, we wasted no time enjoying the clubs of Los Angeles to witness the best and most exhilarating live club-dates for all manner of jazz, including Miles. That habit would continue for over a dozen years, through his move back to New York, and through mine to Irvine and then to Chicago. No matter where we were, we traded visits and headed to the clubs, eventually to include Chicago blues in the mix. By the way, the title of this essay/sonata is taken from a similar one by the famous jazz artist and composer Horace Silver.

What happened to our own musical outlet—our instruments? Well, we both ultimately quit. By late college or a bit beyond, neither of us could play as well as we could when we were 13. That proved fatally painful. Our mind's ears could not be replicated in kind. The same, I have learned, applies to champion ice skaters, and many others who move on to alternative careers and cannot maintain the ability despite the passion and history behind it. Rob later tried Cello, but ultimately that, too, went by the wayside. We were both condemned to being most enthusiastic of fans, and that change affected our artistic outlook as well. No longer focused by our instruments, we took a shine to all the arts, and frequently visited art museums and galleries, attended theater, and the ballet—just about anything we could sink our teeth into that reflected the fine arts.

Oh, did I mention food? We shared a rather adventurous yen for dining. Those visits noted above included all manner of ethnic excursions, the more exotic and eclectic the better. There were only a few misses. We did not order seconds on the "half sheep head" at a cutesy place in Little Italy, nor the sea cucumbers in nearby Chinatown—the things resemble that which jumps out the end of a caterpillar, having stepped on the opposite side; just imagine huge ones. Well, most efforts were rewarded by terrific culinary experiences and the events and chats associated with them. All provided fond memories. Yes, we seemed bound by music, food, and conversation—not a bad combination, eh? Did genes have influence? I rather think so.

After all the above, I finally arrive at our careers. Both of us landed firmly within the academic cocoon—better to preserve our sensitive and idealistic natures, and to provide options as to how we spend our time and with whom and when. The creative element of research and writing proved directly related to our beloved music. What might interest you is how often conversation focused on style as well as substance across our different fields. He was forever curious about the National Institutes of Health (NIH) system of checks and balances, peer review, and boundaries of ethical behavior encoded in how biomedical research is conducted and funded in the United States. I must admit, the NIH system, with all its peculiarities, has proven to be a remarkable success for over half a century. The grant application mill is indeed stressful, but in fact it does provide a rational critical structure that ensures some semblance of fairness and openness (transparency). Nevertheless, none of this trumps human nature and its ramifications. The system just filters it all a bit. Another element that Rob always admired was the process and implementation of experiments. The elegance and arguments entwined with theory, computation, and modeling can be compelling, and yet something important is missed without testable implications backed by tangible experiments. The outcome can bring a house of cards tumbling down, a school of thought trashed, albeit not without anguish and squawking by true believers. In the end, the scientific method has sharp teeth, and the truth tends to win in time. It is humbling. I think Rob saw this, and the concept resonated with him. He lamented that there was insufficient experimentation in computer science. I wonder what you think of all this.

Somewhere in the mid-1990s we had the idea of witnessing each other present our work to our respective audiences. We managed to find a time for Rob to visit the Computer Science Department at Rochester the same week I gave a Grand Rounds presentation in Neurology. What a treat for both of us! At last, as adults, we shared our respective professional lives in real time, on stage, as a member of the audience, or, perhaps more realistically, as a fly on the wall. We so enjoyed that visit, but also learned from the experience.

As to personal life, we both married within a few years, and both had two children. We lived far apart, and regrettably could not share the kind of childhood family interplay that we ourselves experienced early in life. There were occasional holiday visits back and forth, but clearly not enough of them. And ultimately, it became too late.

Illness

I learned of Rob's illness by phone while in Seattle visiting the University of Washington. We spoke about the early pathology report and the options that would ultimately lead to a definitive diagnosis of mesothelioma. This was not really the beginning of illness, but just the diagnosis; the condition no doubt lurked for years, stimulated by a shockingly unlikely and prolonged exposure to asbestos. Over the next roughly three years, we would speak frequently, and visit during some of his hospitalizations related to procedures. We also visited on holidays more so than before. Time was our enemy.

I was the "doctor in the family", a losing role that I hated. Like the man who knew too much of Hitchcock fame, I knew too much. Each signpost of progression or treatment outcome was a harbinger for the next for me. Worse, I was incapable of offering the superhuman task that we all craved—a means of saving Rob. I could at best offer occasional guidance and opinion, and perhaps catalyze some useful contacts. Arguably the most personal contribution once again reflected our most enduring and significant bond—music. Early on, during a visit in Sloan Kettering, I gave him a portable CD player and a half dozen selected discs, something that could be an ever-present comfort, offering some semblance of peace, comfort, and yes, distraction from all the rest. Without a word exchanged, he arose from bed, complete with tubes and chest drains, and sat on my lap to share a hug and some tears. But every time I gave a squeeze he would yelp in pain, as somehow I always managed to aggravate a delicate spot over a drain, and then we laughed out loud anyway.

I had dreamed of a future when Rob and I might alternate family visits with occasional jaunts as a twosome to share our musical passions and our yen for conversation over spooky food. Over the last few years, Rochester has developed a world-class jazz festival that I have voraciously consumed regularly. I can sense Rob's presence in my mind. I cannot help imagining—we would have laughed, cried, conversed, and ate, while reveling in the varied performances for hours, and over days.

Gary D. Paige
Rochester, May 2007

Robert Paige: Researcher and Teacher

Harry Mairson

Computer Science Department, Brandeis University, Waltham, MA 02254, USA
mairson@brandeis.edu

Bob Paige, a professor of computer science and a leading researcher in the area of programming languages and transformational programming, died October 5, 1999, at his home in Manhattan. He was 52 years old.

Bob had mesothelioma, a type of cancer, which he fought courageously and successfully for several years.

For the last fourteen years of his life, he was a professor in the Department of Computer Science at New York University. He also served on the faculty at Rutgers, Purdue, Wisconsin, Yale, the University of Copenhagen and the University of Aarhus in Denmark.

Robert Paige was born in Brooklyn in 1947. He was an accomplished musician in his youth, showing great promise as a trumpet player, but turned down a professional orchestral career to attend Occidental College, where he earned his B.S. degree in 1968. His subsequent work in the emerging computer industry was followed by joining NYU in 1969 as a staff member, where he built one of the first time-sharing systems for multiuser mainframe computers. This project led him to a research career in computer science; he became a graduate student at NYU, and earned his Ph.D. in 1979.

The fundamental theme of Paige's research was automatic programming of complex computer systems. These systems typically include many clever "hand crafted" algorithms whose efficiency scientists try to optimize; in contrast, Paige worked on fundamental methodology for automating the creation of these algorithms, largely without human intervention. His thesis introduced the novel use of "finite differencing," usually used in numerical calculations, as a means of automatic program synthesis. He intended his technology to support the high productivity needed in software design, while minimizing human error. Paige's technology could synthesize programs involving hundreds of thousands of lines of code. This seminal research also produced the best known algorithms for many ubiquitous programming problems.

Dr. Paige was the author of many research papers, covering related topics in programming languages, compilers, algorithms, and database design. An invited speaker at conferences and university seminars around the world, he also served regularly as a reviewer of research projects for major government agencies. He was a devoted mentor of Ph.D. students who today hold research positions at leading universities and research centers.

Bob was more than a good scientist. More important, he was a lovely person. His students and many of his colleagues regarded him with an affection and respect that was far more than professional courtesy. His generosity and his encouragement were an inspiration to those around him. When I was a junior faculty member, and I knew him some but not well, he came up after a conference talk I gave and said, "I like your stuff and think it's really good—and I know a program officer who funds me, and would be very interested. Come on—let me introduce you to him!" The result was an invitation to a meeting of principal investigators, several of whom were friends and colleagues who had never uttered a word to

O. Danvy et al. (eds.), Automatic Program Development – A Tribute to Robert Paige, 35–36.

me about this funding source—their expressions said, "how did you find out about this?" Years later, I am still stunned by the great generosity that he showed me. I am sure that many other colleagues can tell similar stories.

The articles appearing in this long-awaited book are testament to the professional impact that Bob had on the research community around him, and also to the affection that members of that community had for him. As scientists, we all hope to have a lasting impact on our respective fields. Most of us must be satisfied with the knowledge that we put a couple of really good bricks in the foundational walls of our disciplines. More important is our inspiration to the colleagues and students around us, and the force of that inspiration which sustains the ongoing work of science. This Festschrift is also a testament to Bob's having done that so well. It is the hope of the Editors that this volume stands as witness of many gifts—gifts of the spirit, gifts of personal and professional inspiration, gifts of shared experience–that Bob Paige bestowed on his family, friends and colleagues. Bob is survived by his wife Nieba, and his children Jane and John, who are now 19 and 15 years old, respectively. As much as Bob is missed, those who we truly love never really leave us.

Harry Mairson
Waltham, May 2007

An Appreciation of Bob Paige

Martin Davis

Department of Computer Science, Courant Institute, New York University,
3360 Dwight Way, Berkeley, CA 94704-2523, USA
martin@eipye.com

As a senior faculty member of the Courant Institute at NYU since 1965 and a charter member of its Computer Science Department, founded in 1969, I knew Bob Paige as a graduate student and later as a colleague. My own area of expertise is mathematical logic, and so was only very peripherally connected with Bob's work. But being directly involved with decisions about Bob's faculty appointment and his promotions, it was important that I understand what he was doing.

It became clear that Bob was not looking for easy problems. Keenly aware of the problems associated with the development and maintenance of reliable and efficient software systems, and following up on his dissertation work with Jack Schwartz, he devoted himself to transformational methods. His work required not only that he build and maintain elaborate systems, but also that he be able to think in an abstract and theoretical manner. His important insight that loop invariants are related to the classical notion of differencing is really striking. I loved his demonstration in which his system transformed a simple n^2 sorting program into an efficient $n \log n$ program.

When I retired to emeritus status in 1996, Bob was already struggling with his fatal disease. Although I moved to California, I kept getting reports of his courageous stance and of his continuing to work while his strength ebbed. His tragic very premature death is a great loss to the Courant Institute and to computer science. He is very much missed.

Martin Davis
Berkeley, May 2007

O. Danvy et al. (eds.), Automatic Program Development – A Tribute to Robert Paige, 37.

Bob Paige and the IFIP Working Group 2.1

Helmuth Partsch

Chairman of IFIP WG 2.1 (March 1988 – May 1993)
Faculty of Computer Science, University of Ulm, D-89069 Ulm, Germany
Helmuth.Partsch@uni-ulm.de

Bob Paige's first contact with the IFIP Working Group 2.1 was the meeting #27 in Wheeling, West-Virginia, in 1980 which he attended as an invited observer. For me, this was also the first occasion in which I met him — however, without really getting into contact with him. His favorite interest and central Ph.D. topic on "finite differencing" was already known to WG 2.1 through Micha Sharir, a colleague of Bob's from the Courant Institute, New York University. Unlike Bob, Micha had already participated as an observer in the two previous meetings #24 in Summit, New Jersey in 1978 and #25 in Brussels, Belgium in 1979. Therefore, from the groups (restricted) perspective, Micha seemed to be the one who genuinely represented the topic of finite differencing, certainly not knowing that — at this time — Bob had already done much more work in this area than Micha. Also, Micha was a more convincing "salesman", and therefore he was the one to become a member of WG 2.1 at the Wheeling meeting — the same meeting, by the way, when I became a member of the group.

Then Bob participated in the next meeting #28 near Nijmegen, The Netherlands, in May 1981, again as an observer. During that meeting, on one of the evenings, we had the first somewhat longer private conversation, because Bob wanted to know many details about our work within the CIP project at the Technical University of Munich. On this occasion we also had the opportunity to talk about personal subjects, and for me it was the first occasion to get to know Bob and learn something about him as a person. Afterwards, he missed the meetings #29 (Newbury, Great Britain, January 1982), #30 (New York City, fall 1982), and #31 (Munich, Germany, summer 1983), probably because of limited traveling funds. Maybe he was also a little disappointed because he was not considered for membership at the Nijmegen meeting. In fact, I do not even know whether he got an invitation for the next meeting at all.

The next occasion to meet him was during a workshop on "Program Transformation and Programming Environments" which was organized by Peter Pepper in Munich in fall 1983. Although this workshop was (formally) outside the IFIP context, a large number of people somehow related to WG 2.1 (i.e., members and observers) had been invited to this workshop. A very small party with only a few guests (Bob himself, Martin Feather, and Dave Wile) at my house after the workshop was the next opportunity to learn more about Bob — about his childhood in New York, his life at that time, his scientific interests, but also about his deep love for Nieba who later became his wife.

On this occasion I also learned that Bob was a rather pragmatic person who was always able to find a very simple solution to any kind of problem: long after midnight, when Dave, Martin and I were still talking, he got tired; and since Dave and Martin did not want to go back to their hotel yet, without further ado Bob simply decided to lay down on the carpet, in one corner of the room and immediately fell deeply asleep. In the morning, when I managed to wake him up for driving him, Dave, and Martin back to the hotel, his reaction gave us the impression that for him it was the most normal thing to sleep on the floor.

It took six years then, after the Nijmegen meeting, until he came back to the group as an observer. He attended the meeting #37 in Montreal, Canada, in 1987 and the meeting #38 in Rome, Italy, in March 1988. Because I could not attend the Montreal meeting, for me the Rome meeting was the first occasion to see Bob again. The Rome meeting, organized by Alberto Pettorossi, was in some sense important for both of us: it was my first meeting

O. Danvy et al. (eds.), Automatic Program Development – A Tribute to Robert Paige, 38–39.

as chairman of WG 2.1 (succeeding Peter King) and it also was the meeting at which Bob finally became a member of the group.

From then on — according to my records — he participated in quite a number of meetings under my chairmanship (#39: Chamrousse, France, January 1989; #40: Lost Valley Ranch, Colorado, September 1989; #44: Augsburg, Germany, September 1992; #45:Winnipeg, Canada, May 1993 — the meeting at which I resigned from chairmanship). In between these meetings we met twice more: in May 1991, on the occasion of the IFIP Working Conference "Constructing Programs from Specifications", in Pacific Grove, California, and in September 1991, at the Workshop on "Parallel Algorithm Derivation and Program Transformation" which Bob organized at the New York University.

Also afterwards, under the chairmanship of my successor, Doug Smith, he participated in Working group meetings (#46: Renkum, The Netherlands, January 1994; #50: Le Bischenberg, France, February 1997). This last meeting, Le Bischenberg, which was held in conjunction with the IFIP Working Conference on "Algorithmic Languages and Calculi", was when I first learned about Bob's illness. At this time, he still had hope to be able to defeat the cancer. In fact, however, it was his last Working Group Meeting and also the last time I saw him.

Bob loved his wife and his children. His family was at least as important to him as his scientific work. I remember very well when I first met him together with his wife and children, on the grounds of the Asiloar Conference Center (at the Working Conference in Pacific Grove). I was deeply impressed by the obvious great pleasure and pride with which he introduced his family to me.

Bob was a typical scientist — in a very positive sense. On the one hand, he only wanted to make a scientific presentation when he really had something to say; otherwise, he stayed in the background and attended presentations by others very carefully, always eager to learn something new. On the other hand, he was able to talk about a scientific topic which attracted his interest at any time and at any occasion — even late at night, or immediately after waking up. And when he gave a talk about an interesting subject, he just seemed to live for his presentation and the message he wanted to convey. He nearly forgot everything around him — sometimes even "hard reality" in the form of time constraints imposed by the chairman. His scientific contributions covered the full range — from purely theoretical issues to very practical ones. He never was satisfied with the theoretical result alone, but always was interested in its practical implementation.

Also typical of a scientist he had a "vision" and deep scientific respect for topics which were not his own. For instance, when organizing the workshop at Courant Institute in New York in 1991, Bob had the vision to bring together two scientific "communities" (with an obvious overlap of interests). Unfortunately, this experiment failed, mainly because a large number of participants did not realize the scientific potential in Bob's vision. Bob was rather depressed by this. Particularly disappoiting for him was the fact that one half of the participants systematically ignored the other by simply not being present when someone from the other half gave a talk.

Not necessarily typical of every scientist, Bob was also a humble and, in particular, very persistent person: At that workshop in New York in 1991 he had no support at all from NYU. But that was no reason for Bob to give up. As if it were the most normal thing in the world, he simply took care of absolutely everything by himself — even of lunch catering. He, his wife, and his friends had prepared small lunch packages and offered them in a very nice picnic kind of atmosphere, partly outside the building. This was, by the way, also the second time I had the opportunity to meet his family and confirmed the impression about their harmonic relationship I had got at the first time.

In his contribution "In Memoriam, Bob Paige" on the WG 2.1 homepage, Allen Goldberg writes: "I'm sure many of the group's members feel a deep sense of personal as well as professional loss". I wholeheartedly agree.

Helmuth Partsch
Ulm, May 2007

Remembrances of Bob Paige

Alan Siegel

Department of Computer Science, Courant Institute, New York University,
251 Mercer Street, New York, NY 10012, USA
siegel@cs.nyu.edu

I first met Bob in 1969. He had just come to New York to join the systems staff at the Courant Institute. We had a CDC6600 back then, but our computing services were in the dark ages. Bob was hired to debug a homegrown time-sharing system that didn't work. He turned out to be a perfect hire; Bob thoughtfully discarded the code and built the system afresh.

Our acquaintance was casual for years. I was just aware that he was an outstanding systems implementor who was knowledgeable about music and film. His name also appeared at the top of the Institute chess ladder. That was no mean feat; the Institute had a number of Russian students back then, and most were strong players.

Time passed, and our relationship grew. But we really got to know each other after we both became faculty members at NYU. Somehow, we started having discussions of all sorts: computer science, math, advising students, programming languages, research, chess, parenthood, children and family, growing up, music, film, restaurants, life, and doing the right thing. Whatever the topic, Bob always had something insightful to say.

I have to confess that Bob was a hard study. His interests were broad, and he often used the power of metaphor and literary allusion to make a point. I recall several discussions where I struggled to keep up with what he was saying.

As I came to understand over time, Bob had extremely high standards, and it is fair to suggest that there were times when he himself did not meet them. This was never a reason for him to give up, but rather to work harder. For example, Bob had felt that his mathematics background was inadequate, so he took some math classes. I never completely understood just why he thought that mathematics was so important to his research, but it is fair to say that mathematical ideas and paradigms influenced his work. Bob was especially proud of his programming version of finite differencing, which he invented to maintain program invariants and to generate efficient code.

Very few systems researchers have as mathematical and high a level perspective about programming languages as Bob had. His was visionary. His research pursued very long term problems that almost no one thought could be solved. He started as an army of one, and built a system that semiautomatically transforms programs with great power and efficiency. This work covered many areas of expertise, and no matter what was needed, Bob stood ready to accept the challenges. It is no accident that a number of basic algorithms and algorithmic ideas came out of his research. And Bob carried this intensity and passion into his teaching and advisement. I think it fair to say that Bob caringly enriched his students with his uniquely big-picture approach to language research.

You can tell a great deal about Bob from his students. I do not think anyone in our department ever managed to attract a brighter group than Bob's. Jeff Ullman once commented that although he (Jeff) had had a very large number of students, there were very few for whom he can claim to have made a major impact. Jeff was actually commenting about how

O. Danvy et al. (eds.), Automatic Program Development – A Tribute to Robert Paige, 40–42.

bright his students were, and how little guidance they needed. With Bob, the circumstances were very different. It is probably a reasonable analogy to suggest that Bob's systems views were to conventional programming perspectives as the ideas of Nimzovitch (whom he greatly admired) were to conventional chess. So it was inevitable that his students would receive something very special. All benefited from his highly abstract perspective and desire to turn idealized possibilities into efficient pragmatic practice. For example, Bob's perspective about the methodology of high-level algorithm design is evident in Jiazhen Cai's work with Tarjan on extensions to problems in planarity testing. Similarly, Fritz Henglein's work on the Y2K problem bears eloquent testament to the training Bob provided his students.

I should add that Bob's students also became members of his extended family, and this special relationship has persisted to this very day.

Our conversations covered many topics, and would often drift from research to students to life and matters of family. I remember Bob coming into my office bursting with pride. "My boy", he said. "Johnny has just drawn his first picture. He is so practical; he drew a potato". Bob was like that; a mix of love, good humor, and pride. He was also practical. One day he told me how Janie got into trouble in middle school. It seems that she had organized an unauthorized field trip for some friends and classmates. I can guarantee that New York City public school administrators would not let parents plan such an activity without lots of consent forms being signed, and it is a certainty that Janie's actions did not go over very well with the school. But Bob was glowing. He was pleased that Janie had the nerve to plan the adventure, and ever so proud of her strong sense of independence. On the other hand, he knew that she needed no encouragement to become a rebel, so he pretended to be a little upset with her. As for Nieba, whenever I would comment on Bob's aesthetic observations and his insightful literary interpretations, he would reply, "Oh that's really Nieba's department. She has a much better sense about these things than I have".

Then there were the people, places, and the like. You never know who would be visiting the Paige apartment. Many of their art works came with a story. And there were the deals. My refrigerator came from a Bob Paige contact. The price was unbeatable. My wife and I were once let in at the head of a two-hour line when Paul Prudhomme located his traveling summer restaurant in New York City. The shortcut was a payback for Bob's helping the State of Louisiana with some decisions about fostering computer science research. And it was official; Governor Edwards's office had phoned Paul Prudhomme to explain that Bob was a friend of Louisiana. I seem to recall a personally fished (pried?) abalone dinner that Bob never found the time to collect out in southern California. I think it was in return for his suggestions about how to teach some of the more challenging material in Aho, Hopcroft, and Ullman.

Bob also knew his wines. I remember being treated to remarkable tastings from places he had visited, and being offered some imports of especially successful vintages. Now Nieba is an excellent cook, but they both enjoyed fine restaurants. Bob, of course, knew where to go and what to order, but there was always a personal element to these things. For example, Bob not only knew where the best nouvelle French cuisine was to be found, but also knew the chef. Similarly, he knew about the best chamber groups in the city, and knew the musicians as well.

Bob was a master musician; he had turned down a professional career in trumpet to work in computer science. He took up new instruments with enthusiasm. I remember when he started playing the cello, which was motivated by his appreciation of Yo Yo Ma. He began studying the piano to learn it with Janie. It was fun to watch him struggling to realize his musicianship and interpretive skills with instruments he could not quite master. You could tell that he enjoyed the challenge, and did not mind that he did not always meet his standards. Curiously, I never heard him play the trumpet. In retrospect, I regret not asking him why he no longer played it. Were the challenges of new instruments more exciting? Was there a sense of loss over skills dulled by the passage of time? I have no idea what Bob's answer would have been, but do know that in his words, I would have learned something new about Bob Paige, about myself, and about the human condition in general.

Our discussions were like that. Somehow, Bob could communicate insights that I cannot recreate despite the unlimited opportunity to revise every syllable of this description of him and his thoughts. The book "Tuesdays with Morrie" was a pale imitation of what Bob had on offer. I regret not transcribing our conversations to revisit from time to time, and to share with others. On the other hand, I believe that a tape recorder would have been needed to get the job done; his thinking was just too rich to absorb in full detail. Yet the gist of these conversations remain with me. They are alive, and his spirit continues to challenge me even though I cannot give his thoughts the force that he could through extemporaneous discourse.

That kid I had met back in 1969 grew up. Over time, he acquired a wisdom for the ages. Even as his disease progressed and took over his body, Bob's mind stayed sharp, and his words remained uplifting. Bob was like that, and he still is.

Alan Siegel
New York, May 2007

Bob Paige (mid-1980s)

Bob Paige with his wife Nieba (1985)

Bob Paige with his daughter Jane (1990)

Bob Paige with his son John (1993)

Part III

Contributed Papers

Transformational Derivation of an Improved Alias Analysis Algorithm

Deepak Goyal

Calypto Design Systems, Inc.*
dgoyal@calypto.com

Summary. In this paper we use a program transformational approach to obtain an asymptotically improved *may–alias* analysis algorithm. We derive an $O(N^3)$ time algorithm for computing an intraprocedural flow sensitive may–alias analysis, where N denotes the number of edges in the program control flow graph (CFG). Our algorithm improves the previous $O(N^5)$ time algorithm by Hind et al. [20]. Our time complexity improvement comes without any deterioration in space complexity. We also show that for a large subclass of programs in which the in-degree and out-degree of all CFG nodes is bounded by a constant, our algorithm is linear in the sum of the number of edges in the CFG of the program and the size of the output, i.e., the size of the computed alias information, and is therefore asymptotically optimal. Our transformational algorithm derivation technique also leads to a simplified yet precise analysis of time complexity.

Keywords: intraprocedural flow analysis, may–alias analysis, program transformation.

1 Introduction

Alias analysis (also called Pointer Analysis) of programs has been the subject of considerable research for over a decade [19] as it is often an important prerequisite to many compiler optimizations in languages such as C/C++/Java. In this paper we present a transformational derivation of an $O(N^3)$ time intraprocedural flow sensitive alias analysis algorithm [5, 9], where N denotes the number of edges in the *program control flow graph* (CFG). A program control flow graph is a graph-based representation of a program in which each node represents an assignment statement in the program and edges represent the transfer of control from one statement to another. Our algorithm vastly improves the existing $O(N^5)$ time algorithm by Hind et al. [20]. Our time complexity improvement comes without any deterioration in space complexity. The space complexity of both our algorithm and the previously known algorithm is $O(N^3)$ in the worst case. For a large subclass of programs in which the in-degree and out-degree of nodes in the CFG is bounded by a constant, we show that the time and space complexity of our algorithm is linear in the sum of the number of edges in the CFG of the program and the size of the output, i.e., the size of the computed may–alias information. Thus, our algorithm is asymptotically optimal for this class of input programs. Our time complexity analysis makes the assumptions that the number of variables in a statement is bounded by a constant and the number of dereferences (the $*$ operator in the language C) being applied in any expression is also bounded by a constant. These assumptions are commonly made in the time complexity analysis of alias analysis algorithms. Hind et al.'s $O(N^5)$ time result [20] relies on the same assumptions.

* The work in this paper was done when the author was a graduate student at New York University. This paper was originally submitted when the author was a Research Staff Member at the IBM T.J. Watson Research Center.

O. Danvy et al. (eds.), Automatic Program Development – A Tribute to Robert Paige, 49–70.

We derive our algorithm by using two high-level program transformations, dominated convergence [6,8,13] and finite differencing [27]. The dominated convergence transformation provides a generalized iteration scheme for computing fixed points, and the finite differencing transformation replaces expensive repeated computations by cheaper incremental counterparts. These transformations may be viewed as schema transformations which are applied when the program matches a given form, i.e., matches a schema.

Hind et al. specified their alias analysis as a least fixed point computation on the CFG of a program. We first show how their specification can be transformed into a naive $O(N^6)$ time algorithm using the dominated convergence transformation. Next, we show how a primitive form of finite differencing can be used to transform the naive algorithm into a workset-based $O(N^5)$ time algorithm. This algorithm turns out to be identical to the one obtained using Kildall's strategy [23]. Next, we use a second dominated convergence transformation and a more advanced finite differencing transformation to derive an asymptotically improved $O(N^3)$ time algorithm.

We also show that our transformational algorithm derivation technique leads to a simplified yet precise analysis of time complexity. The time complexity analysis of such algorithms is nontrivial. In the past, the time complexity analysis of alias analysis algorithms has often been ignored, and in some cases even been incorrect or imprecise. For example, a simplified version of the may–alias analysis problem was studied in [11] and the time complexity of a $\Theta(N^5)$ time algorithm was incorrectly stated to be $O(\omega \times N^3)$ where ω is the loop-connectedness parameter [22] of the CFG. Actually, the time complexity of the may–alias analysis is independent of the loop-connectedness parameter. A refined and extended version of this work was published recently [20], in which the time complexity of the algorithm was stated to be $O(N^6)$ time. Although this time complexity analysis is *correct*, it is not precise, and a more careful analysis reveals that the time complexity of this algorithm is $\Theta(N^5)$.

Even though we focus primarily on the alias analysis problem in this paper, the techniques used are quite general and can be used to obtain improved algorithms for other problems, e.g., Escape Analysis [12].

1.1 Background

This work has been inspired by the work of Bob Paige, who, in the late 1970s, set upon the goal of establishing a transformational program development methodology that could help in the formal design of algorithms. The main transformations used by Paige were finite differencing, dominated convergence, and data structure selection.

Finite Differencing

The first transformation developed by Paige was finite differencing, which was a generalization of strength reduction [2], and Earley's Iterator Inversion [16]. The goal was to speed up programs by replacing costly repeated computations by their more efficient incremental counterparts.

Paige showed in his Ph.D. thesis [27] that the finite differencing rules could be used to transform many abstract program specifications into asymptotically more efficient implementations. The first examples of nontrivial algorithms being derived by finite differencing were presented in [27,28].

Dominated Convergence

Dominated convergence (also known as Cousot's *chaotic iteration* [13]) is a generalized iteration schema for computing fixed points. The study of program analysis problems such as live variable analysis, constant propagation, etc., which could be specified as least or greatest fixed point computations on monotonic functions led Cai and Paige to the use of dominated convergence as a second transformation technique for program derivation in [6, 8]. Cai and Paige applied the dominated convergence and finite differencing transformations to the constant propagation problem. They showed how four increasingly accurate specifications of the constant propagation problem (by more accurate specifications, we mean specifications that could find more constants) could be systematically transformed into worst-case linear time implementations on a uniform cost sequential RAM [1]. Later, in [17], it was shown that these algorithms could also be implemented in worst-case linear time on a pointer machine.

In [7], Cai and Paige used these transformations to define a linear-time language L_1 such that any problem expressible in L_1 could be automatically compiled into programs with time and space complexities linear in the sum of the input space and the output space. They showed that problems like cycle testing, constant propagation, and unreachable code elimination could all be expressed in L_1, and thus, automatically transformed into linear time programs.

Data Structure Selection

Paige demonstrated the finite differencing and dominated convergence transformations on programs written in a SETL-like [32, 33] set-theoretic language. Consequently, an important final transformation in his methodology was the *data structure selection* transformation which was required to efficiently implement set-theoretic associative access operations such as membership testing. The problem of data structure selection had been extensively studied in the SETL project and had led to the development of many ingenious ideas such as the idea of *basings* developed by Schwartz et al. [14, 31]. The idea was intended to reduce (and possibly eliminate) hashing for implementing associative access operations. Paige refined the rudimentary ideas of basings into a *data structure selection* transformation which could ensure that set-theoretic operations such as membership testing could be implemented in constant time.

Paige used his three stage methodology based on finite differencing, dominated convergence, and data structure selection, to derive improved algorithms for problems such as the Single Function Coarsest Partition Problem [29], and the Ready Simulation problem [4]. The alias analysis algorithm presented in this paper can also be transformed using the data structure selection transformation into a worst-case $O(N^3)$ time algorithm in which all associative access operations can be performed in $O(1)$ time without the use of hashing [17]. However, the details of the data structure selection transformation are beyond the scope of this paper and will not be discussed here. For our purposes, it will suffice to assume that associative access operations such as set membership test and set element addition and deletion of $O(1)$ sized data can be performed in $O(1)$ time using hashing. The algorithm of Hind et al. [20] also uses hashing and their time complexity analysis relies on the same assumption.

1.2 Outline of the Rest of the Paper

In Section 2 we present an overview of the set-theoretic programming language notation used in the rest of the paper. In Section 3 we present an introduction to alias analysis and describe Hind et al.'s [20] compact representation of alias information. In Section 4 we describe Hind et al.'s alias analysis and its formulation as a least fixed point computation. In Section 5 we present a naive algorithm for computing the above least fixed point, and prove that the time complexity of this algorithm is $O(N^6)$. In Section 6 we use a simple finite differencing transformation to transform the naive algorithm into Kildall's workset algorithm [23] with a time complexity of $O(N^5)$. Finally, in Section 7 we show how a dominated convergence transformation coupled with an advanced finite differencing transformation can be used to transform the workset algorithm into a new $O(N^3)$ time algorithm.

This chapter is an abbreviated version of [18]. Some proofs and details that have been skipped in the chapter may be found in [18].

2 Notation

This section describes the SETL-like notation that will be used throughout this paper. SETL [32, 33] is a programming language based on finite set theory [35]. Besides the usual elementary datatypes, SETL includes builtin datatypes for finite sets, tuples, and maps. A set is an unordered collection of distinct values, while a tuple is an ordered collection of values. A map is a set of ordered pairs (i.e., 2-tuples). The elements of sets and tuples may belong to any datatype (including sets and tuples). We use SETL set comprehension expression, which in the most general form, is written as

$$\{E(x_1,...,x_k) : x_1 \in S_1, x_2 \in S_2(x_1), ..., x_k \in S_k(x_1,...,x_{k-1}) \mid K(x_1,...,x_k)\},$$

and denotes the set of values $E(x_1,...,x_k)$ evaluated over all k-tuples $[x_1,...,x_k]$ belonging to the values satisfying the condition,

$x_1 \in S_1$ & $x_2 \in S_2(x_1)$ & ... & $x_k \in S_k(x_1,...,x_{k-1})$,

and also satisfying the boolean expression $K(x_1,...,x_k)$. The expression

$\{[x_1,\ldots,x_k] \in \times_{i=1}^{k} S_i \mid K(x_1,\ldots,x_k)\}$

is a short form for

$\{[x_1,\ldots,x_k] : x_1 \in S_1,\ldots,x_k \in S_k \mid K(x_1,\ldots,x_k)\}$

When the expression $K(x_1,...,x_k)$ is the constant *true*, it may be elided.

Let F be a map and S be a set. Expression $\#S$ is used to denote the cardinality of S. An arbitrary element can be nondeterministically selected from set S by the expression $\ni S$. If S is empty, then $\ni S$ returns the value *undefined*. The expression $\{\}$ stands for the empty set and the empty map and the expression $[]$ stands for the empty tuple. The conditional expression $x \in S$ is used for set membership test. A map can be seen as a binary relation. Note that the first components of the ordered pairs in a map need not be distinct. Expression $F\{x\}$ is used for the image of element x in map F, i.e., $\{y : [x,y] \in F\}$. The expression $domain(F)$ denotes the set containing the first components of the pairs in F, i.e., $domain(F) = \{x : [x,y] \in F\}$. The assignments S *with* $:= x$ and S *less* $:= x$ stand for the modification of set S by the insertion and deletion of element x respectively. The extended image set $F[S]$ stands for the set $\cup_{x\in S} F\{x\}$. We use $\circ$ as the map composition operator, i.e., for maps F and G, $F \circ G = \{[x,z] : [x,y] \in G, [y,z] \in F\}$. Furthermore, for any map F, we define F^0 to be the identity map over the domain of F and map F^i to be the composition $F \circ F^{i-1}$ of maps F and F^{i-1} for $i > 0$. The for-loop control structure

```
for x ∈ S loop
        block(x)                                        (1)
endloop
```

is used to execute a sequence of statements, denoted by *block(x)*, for each element x of set S. While executing the loop, the elements of set S are selected nondeterministically and without repetition. The iteration of the loop proceeds through the initial value of S on entry to the loop (as if the iteration were on a copy of S) and is not affected by modifications to S within *block(x)*.

2.1 Time Complexity of Set Operations

In order to define the time complexity of associative access operations, we need to define the *Size* of a value.

$$Size(x) = 1 \text{ if } x \text{ is an elementary type (such as int, char etc.)}$$
$$Size(S) = 1 + \Sigma_{x \in S} Size(x) \text{ if } S \text{ is a set} \quad (2)$$
$$Size(F) = 1 + \Sigma_{x \in domain(F)}(Size(x) + \Sigma_{y \in F\{x\}} Size(y)) \text{ if } F \text{ is a map}$$

For the rest of the paper, we will assume that sets are implemented by linking the elements together in a doubly linked list. In addition, the elements of the set are also inserted into a hash table. The domains of maps are implemented similarly. In addition, each element in the domain of a map contains a pointer to the set of elements that constitute its image under the map. Tuples are implemented as arrays. For an $O(1)$-sized element x, set S and map F, the hash table-based implementation allows operations such as $x \in S$, S *with* $:= x$, S *less* $:= x$, $F\{x\}$ to be performed in $O(1)$ time under the assumption that hashing of $O(1)$-*sized* data can be done in $O(1)$ time. The linked list implementation of the elements of the set allows iteration over a set in time proportional to the cardinality of the set.

3 An Introduction to Alias Analysis

Consider the following assignment statement in the language C.

$$a \;=\; \&b \quad (3)$$

The effect of the assignment is to assign to a the address of b. After execution of statement (3), the lvalues of expressions $*a$ and b (i.e., the locations referred to by expressions $*a$ and b) are the same. In such a case, $*a$ and b are said to be *aliased.* We call expressions such as $*a$ and b, *access-paths* [25]. More precisely, an access-path is the lvalue of an expression constructed from variables and the pointer dereference operator $*$. The aliasing of $*a$ and b is represented by the alias pair $(*a, b)$.

Let p_1 be a program point, i.e., a point between successive statements in a program P, and x and y be access paths constructed from variables in P. Access paths x and y are said to be *may–aliases* at program point p_1 if there exists an execution of program P in which x and y are aliased when control reaches program point p_1. An *alias relation* R is a set of alias pairs. A relation R is said to be a *may–alias* relation at program point p_1 iff R contains all (but not necessarily only) alias pairs (x, y) such that x and y are may–aliases at p_1. Thus, if R is a may–alias relation at p_1 and $(u, v) \notin R$, then access paths u and v can never be aliased whenever control reaches p_1 in any possible execution of program P. By definition, if R is a may–alias relation at p_1, then any superset of R is also a may–alias relation at p_1.

Let R_1 and R_2 be two may–alias relations at a program point p_1. We say that R_1 is more precise than R_2 if $R_1 \subset R_2$. A may–alias relation R at p_1 is said to be *precise* if it contains exactly the set of may–aliases at p_1. It has been shown that even under the assumption that all paths of a program are feasible execution paths, i.e., that all branches of conditionals can be taken, the computation of a precise may–alias relation at a program point is undecidable [24,30]. Consequently, all may–alias analysis algorithms compute some overapproximation of the precise may–alias relation at every program point. It does not make sense to compare the time complexity of different may–alias analysis algorithms unless they compute may–alias information of comparable precision. The algorithm presented in this paper computes precisely the same may–alias information that is computed by Hind et al. [20], but is more efficient than their algorithm.

Since the alias analysis involves the computation of a may–alias relation at every program point, Hind et al. use a graph-based representation for may–alias relations, which is briefly described below. In the following sections, the terms "alias" and "may–alias" will be used interchangeably.

3.1 Graph-based Representation of Alias Information

Hind et al. [20] use a graph-based representation of alias information in which storage locations are associated with names, and are referred to as *named objects* [9,21]. The names are either program variable names or new names created on demand for storage locations created through storage allocation statements such as `malloc` in the language C. This graph-based representation of may–alias relations is called an *alias graph*. In order to get an alias graph representation of a may–alias relation R, the relation R has to be transformed into a set R' of alias pairs such that all alias pairs in R' are of the form (a, b), $(*a, b)$ or $(b, *a)$ where a and b are named objects. The transformation from R to R' is done by creating new named objects as necessary. For example, an alias pair $(**c, *d)$ in R is replaced by the alias pairs $(*c, t)$, $(*t, u)$, $(*d, v)$, and (u, v), where t, u, and v are newly created named objects. Next, all pairs of the form (a, b) in R', where a and b are named objects, are removed, and either all occurrences of a in the remaining alias pairs are replaced by b, or b by a. After this, all remaining alias pairs in R' are of the form $(*a, b)$ or $(b, *a)$.

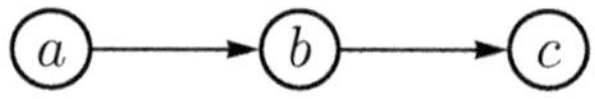

Fig. 1. An example of an alias graph.

An alias graph G is constructed from R' as follows. The set of vertices in G contains a distinct vertex for each distinct named object in some alias pair in R'. The set of edges in G contains a directed edge from the vertex corresponding to a to the vertex corresponding to b for every alias pair of the form $(*a, b)$ or $(b, *a)$ in R'. We use the notation $[a, b]$ to represent an edge from a to b in the alias graph. An alias graph can also be thought of as a points-to graph [34] where an edge from vertex a to vertex b represents that the named object corresponding to vertex a may point to the named object corresponding to vertex b. An example of an alias graph is shown in Figure 1. Let E denote the set of edges and V denote the set of vertices in an alias graph G representing a may–alias relation R. The set of edges E can equivalently be considered as a map from V to V. The set of alias pairs in R can be computed using Definition 1. For a named object a, we will use a_i to denote the access-path obtained by i applications of the pointer dereference operator $*$ to a, e.g., $a_2 \stackrel{def}{=} **a$.

Definition 1 Let G be an alias graph, and let E denote the set of edges and V denote the set of vertices in G. The set of aliases of a_i (denoting i applications of the dereference operartor $*$ to a) in G is defined to be the set of vertices in G reachable by a path of length i from vertex a, and is given by $E^i[\{a\}]$, where E^0 denotes the identity map over the set of vertices V, and E^i denotes the composition of maps E and E^{i-1} for $i > 0$.

Using Definition 1, we can verify that the alias graph in Figure 1 represents a may–alias relation R which contains the alias pairs

$\{(*a, b), (*b, c), (**a, c), (a, a), (b, b), (c, c)\}$,

and all other alias pairs that can be inferred by symmetry (if $(u, v) \in R$, then $(v, u) \in R$) and congruence (if $(u, v) \in R$, then $(*u, *v) \in R$). The alias graph representation of may–alias relations assumes transitivity. Therefore, the conversion of a may–alias relation to an alias graph may involve some loss in precision.

For the remainder of the paper, we will use alias graphs to represent may–alias relations, and use Definition 1 to compute the may–alias relations explicitly.

4 Hind et al.'s May–Alias Analysis Specification

Hind et al. [20] compute the may–alias analysis over the *sparse evaluation graph* (SEG) [10], which is a sparse representation of the CFG of a program. In this paper we deal with two kinds of graphs, namely, alias graphs which represent may–alias relations, and SEGs which compactly represent the CFG of a program. In order to avoid confusion, we shall henceforth use the term *vertex* for the vertices of an alias graph and the term *node* for the vertices of the SEG.

Informally, an SEG contains only the subset of the nodes in the CFG which are considered "interesting", and the edges required to connect these nodes. The "interesting" nodes of the CFG are the nodes at which alias information is potentially modified, i.e., nodes representing assignments to variables declared as pointers and storage allocation and deallocation statements such as `malloc` and `free` in the language C, and the nodes where alias information arriving along multiple paths in the CFG is combined. An SEG may be constructed from the CFG as follows.

1. Mark all nodes at which alias information is potentially modified.
2. Pick an unmarked node u which has exactly one incoming edge (say, from node v). Merge node u with its predecessor node v. Repeat Step 2 until no unmarked nodes having exactly one incoming edge remain.

Assuming that the CFG has exactly one *entry* node, i.e., a node with no incoming edges, the only nodes remaining in the SEG when Step 2 terminates, are the unique *entry* node, the marked nodes, and the nodes which have more than one incoming edge. The nodes in the SEG which have more than one incoming edge are called *join* nodes. The edges coming into a join node may come either from marked nodes or from other join nodes. Except possibly for join nodes, each node in the SEG is associated with some pointer assignment statement. For join nodes that do not represent any other pointer assignment statement, we associate the trivial pointer assignment statement $p = p$, where p is an arbitrary pointer variable in the program.

The may–alias information at each program point, i.e., on entry to, and on exit from each node n of the SEG, can be computed as a traditional dataflow analysis [2]. Equations (4) and (5) below define the relationship between the alias information flowing into and out of a node n.

Two alias graphs are computed for each node n of the SEG, one immediately before node n and the other immediately after node n. Let I_n be the set of edges in the alias graph computed immediately before node n, and let O_n be the set of edges in the alias graph computed immediately after node n. Let $Pred(n)$ denote the set of predecessor nodes of node n in the SEG, i.e., all nodes from which there is an SEG edge to node n. Similarly, let $Succ(n)$ denote all the successors of node n, i.e., the set of all nodes to which there is an SEG edge from node n. The set of edges I_n is the union of the sets O_l computed at the predecessor nodes l of node n [20]:

$$I_n = \bigcup_{l \in Pred(n)} O_l \,. \tag{4}$$

The set of edges O_n is defined as a function of the pointer assignment statement associated with node n and the set of edges I_n. Let node n of the SEG correspond to the assignment statement $p_i = q_j$, where p_i is an access-path with i levels of pointer dereferencing from named object p, q_j is an access-path with j levels of pointer dereferencing from named object q, and as a special case, q_{-1} denotes $\&q$. We denote the set of edges O_n by $F_n(I_n)$ which is defined as

$$F_n(I_n) \stackrel{def}{=} (I_n - Must(I_n, I_n^i[\{p\}])) \,\cup\, (I_n^i[\{p\}] \times I_n^{j+1}[\{q\}]), \tag{5}$$

where expression *Must* is defined by Equation (6) below.

$$Must(I_n, S) = \begin{cases} I_n & \text{if } S = \{\} \\ \{[p,q] \in I_n \mid p \in S\} & \text{if } \#S = 1 \text{ and } S = \{p\} \\ & \text{for some vertex } p \text{ in the alias} \\ & \text{graph computed before node } n \\ \{\} & \text{if } \#S > 1 \end{cases} \tag{6}$$

The intuition behind Equation (5) is explained in more detail in [18].

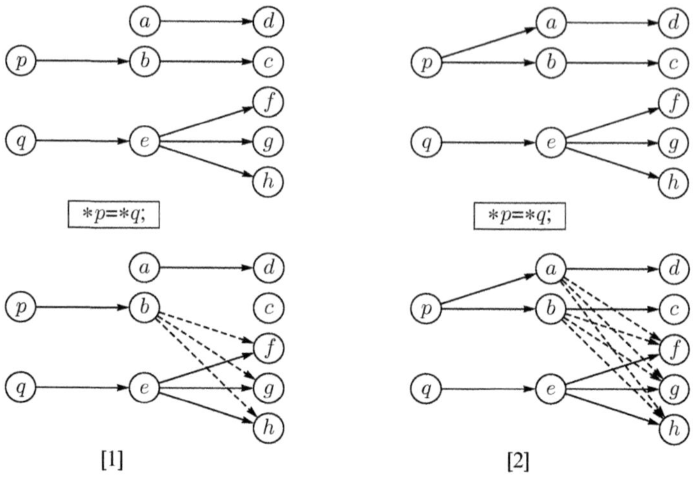

Fig. 2. Examples illustrating the computation of an alias graph immediately after a node n of the SEG corresponding to the pointer assignment $*p = *q$.

Example 1 Figure 2 illustrates the computation of the alias graph according to Equation (5). In Figure 2 [1], $I_n^1[\{p\}] = \{b\}$ and $I_n^2[\{q\}] = \{f, g, h\}$. The assignment $*p = *q$ causes the removal of edge $[b, c]$ and the addition of edges $[b, f]$, $[b, g]$, and $[b, h]$. However, in Figure 2 [2], $I_n^1[\{p\}] = \{a, b\}$. In this case, p may be pointing to either a or b. If p points to b, then a will continue to point to d after the assignment and if p points to a, then b will continue to point to c after the assignment. In this case, we cannot remove either $[b, c]$ or $[a, d]$ from the resulting alias graph since we are trying to compute a conservative approximation to the may–alias relation. Therefore, in this case, no edges are removed and edges $[a, f]$, $[a, g]$, $[a, h]$ $[b, f]$, $[b, g]$, and $[b, h]$ are added.

Equations (4) and (5) may be combined into the single equation

$$I_n = \bigcup_{l \in Pred(n)} F_l(I_l), \tag{7}$$

for a node n, where $F_l(I_l)$, as defined by Equation (5), is the set of edges in the alias graph computed immediately after node l. In other words, the computed may–alias information must be a fixed point of the following system of equations:

$$\begin{aligned} I_1 &= \textstyle\bigcup_{l \in Pred(1)} F_l(I_l) \\ I_2 &= \textstyle\bigcup_{l \in Pred(2)} F_l(I_l) \\ &\dots \\ I_N &= \textstyle\bigcup_{l \in Pred(N)} F_l(I_l), \end{aligned} \tag{8}$$

where the numbers $1 \dots N$ are used to denote the N nodes of the SEG. The may–alias information that we seek to compute, is in fact, the least fixed point of the system of equations (8). All other fixed points of the system of equations (8) are also safe approximations of the may–alias relations at every node n, although they are not as accurate (precise) as the least fixed point. Let

$$G_k(I_1, \dots, I_N) \stackrel{def}{=} \bigcup_{l \in Pred(k)} F_l(I_l), \tag{9}$$

and

$$\mathcal{F}([I_1, \ldots, I_N]) \stackrel{def}{=} [G_1([I_1, \ldots, I_N]), \ldots, G_N([I_1, \ldots, I_N])]. \quad (10)$$

Then, the may–alias information that we seek is the least fixed point of function $\mathcal{F}$ with respect to $[I_1, \ldots, I_N]$, and is denoted by $\text{LFP}_{\leq}(\mathcal{F})$ where the relation $\leq$ is the component-wise containment relation over the subset lattice, i.e.,

$$[I_1, \ldots, I_N] \leq [I'_1, \ldots, I'_N] \;\; \textit{iff} \;\; \forall k = 1, \ldots, N, I_k \subseteq I'_k. \quad (11)$$

5 A Naive Algorithm

We first review a few basic definitions and concepts from lattice theory that underlie the computation of least fixed points (for more details see [8]). This background material may be found in any introductory text on lattice theory; for example, Birkhoff [3].

5.1 Definitions

A *poset* $(L, \leq)$ is a reflexive, transitive, antisymmetric binary relation $\leq$ on a nonempty set L. A poset $(L, \leq)$ has a minimum element $\mathbf{0}$ iff $\forall x \in L \; \mathbf{0} \leq x$.

Let $\mathcal{G} : T \longrightarrow T$ be a function from poset $(T, \leq)$ to $(T, \leq)$. Function $\mathcal{G}$ is said to be monotone (respectively antimonotone) if for every two elements $x, y \in T$ such that $x \leq y$, it is the case that $\mathcal{G}(x) \leq \mathcal{G}(y)$ (respectively $\mathcal{G}(y) \leq \mathcal{G}(x)$).

Theorem 1, originally due to Tarski and Kleene [26, 36], shows how to compute fixed points of monotone functions over a finite poset with a minimum element.

Theorem 1 *Let $(T, \leq)$ be a finite poset with a minimum element $\mathbf{0}$, and $\mathcal{G} : T \longrightarrow T$ be a monotone function. The set $\{\mathcal{G}^i(\mathbf{0}) : i = 0, 1, \ldots\}$ is finite and the least fixed point of $\mathcal{G}$ (denoted by $LFP_{\leq}(\mathcal{G})$) exists, and is equal to $\mathcal{G}^k(\mathbf{0})$ where k is any nonnegative integer for which $\mathcal{G}^k(\mathbf{0}) = \mathcal{G}^{k+1}(\mathbf{0})$.*

Theorem 2 (Dominated Convergence) [8] *Let $(T, \leq)$ be a finite poset with minimum element $\mathbf{0}$, and $\mathcal{G} : T \longrightarrow T$ be a monotone function. Let $s_0, \ldots, s_i, \ldots$ be any sequence such that:*
(1) $s_0 = \mathbf{0}$;
(2) $s_{i+1} \in T$ *and* $s_i \leq s_{i+1} \leq \mathcal{G}(s_i)\}$, *for* $i = 0, 1, \ldots$
Then we conclude the following:
(1) *If there exists an integer $k \geq 0$ such that $s_k = \mathcal{G}(s_k)$, then $s_k = LFP_{\leq}(\mathcal{G})$.*
(2) *If for all i, $s_i < s_{i+1}$ whenever $s_i \neq \mathcal{G}(s_i)$, then there exists $k \geq 0$ such that $s_k = \mathcal{G}(s_k)$.*

It can be seen that the sequence $(\mathbf{0}, \mathcal{G}(\mathbf{0}), \mathcal{G}^2(\mathbf{0}), \ldots)$ computed in Theorem 1 is a special case of the sequence $(s_0, s_1, s_2, \ldots)$, for $i = 0, 1, \ldots$ where $s_{i+1} = \mathcal{G}(s_i)$.

5.2 The Naive Algorithm

Let S be the the set of all possible points-to edges, i.e., all pairs of named objects in the program. Thus, $I_n \in 2^S$. Let us use $(2^S)^N$ to denote the N-way cross-product $2^S \times 2^S \times \ldots \times 2^S$. Let $\leq$ be the relation defined by (11). Then, $((2^S)^N, \leq)$ is a finite poset with a minimum element $[\{\}, \{\}, \ldots, \{\}]$. In order to show that expression $\mathcal{F}$, as defined by Equation (10), is monotonic, we must show that each expression G_i is monotonic. Since the union operator $\cup$ is monotonic, it suffices to show that each expression F_n is monotonic. Using the facts that 1) the expression $E(F, S) = F[S]$ is monotonic in both parameters F and S, 2) the composition of monotonic expressions is monotonic, and 3) if expression E_1 is monotonic in some parameter and E_2 is antimonotonic in the same parameter, then expression $E_1 - E_2$ is monotonic in that parameter, it is easy to show that expression $Must(I_n, I_n^i[\{p\}])$ is antimonotonic in its parameter I_n and that expression $F_n(I_n)$ defined by Equation (5) is monotonic.

From now on, we use I to denote the N-tuple $[I_1, \ldots, I_N]$. Let $I_\perp$ denote the minimum element $[\{\}, \{\}, \ldots, \{\}]$ of the poset $((2^S)^N, \leq)$. It follows from Theorem 1 that $\mathrm{LFP}_{\leq}(\mathcal{F})$ can be computed by an iterative scheme which computes the values $I_\perp, \mathcal{F}(I_\perp), \mathcal{F}^2(I_\perp), \ldots$. Such an iteration is guaranteed to terminate at the fixed point in a finite number of steps.

Theorem 2 suggests an alternative scheme for computing the fixed point. Since each expression G_i is monotonic in each of the parameters $I_1, \ldots, I_N$, it follows that

$$\begin{aligned} & [I_1, \ldots, I_k, \ldots, I_N] \\ \leq\ & [I_1, \ldots, G_k(I_1, \ldots, I_N), \ldots, I_N] \\ \leq\ & [G_1(I_1, \ldots, I_N), \ldots, G_k(I_1, \ldots, I_N), \ldots, G_N(I_1, \ldots, I_N)], \end{aligned} \tag{12}$$

or, in other words

$$I \leq [I_1, \ldots, G_k(I), \ldots, I_N] \leq \mathcal{F}(I) \tag{13}$$

This suggests an iterative scheme in which an arbitrary component I_k of I satisfying the condition $I_k \neq G_k(I)$ can be nondeterministically selected, and its value updated to $G_k(I)$. Algorithm (14), given below, is based on this idea.

$$\begin{aligned} & \forall i = 1 \ldots N\ I_i := \{\} \\ & \texttt{while}\ \mathit{exists}\ n \in \{k \in 1 \ldots N \mid I_k \neq G_k(I)\}\ \texttt{loop} \\ & \quad I_n := G_n(I) \\ & \texttt{endloop} \end{aligned} \tag{14}$$

In Algorithm (14), we assume that for a set S, predicate *exists* $x \in S$ has the side effect of assigning an arbitrary value in S to x if S is nonempty. From Equation (13) it follows that the sequence of I's obtained at the end of each iteration satisfy the conditions of Theorem 2 and are also strictly increasing. Thus, Algorithm (14) is guaranteed to converge to the least fixed point.

5.3 Time Complexity

The time complexity of Algorithm (14) can be computed as follows. Assume that the SEG is transformed into an equivalent SEG in which each node has a maximum in-degree and out-degree of 2. This can be done easily by introducing dummy nodes. Let N denote the number of nodes in the transformed SEG. Note that the number of nodes in the transformed SEG is proportional to the number of edges in the original SEG. Let O denote the set of all distinct named objects in the program and let V denote the cardinality of O. The distinction between N and V is important. Informally, N is a measure of the size of the program, i.e., the number of statements in the program, whereas V is a measure of the number of distinct variables in the program. Since a variable must appear in at least one statement, and we assume that the number of variables in any statement is bounded by a constant, $V = O(N)$. Since each vertex in an alias graph corresponds to a distinct named object, the number of vertices in an alias graph is bounded by V and the number of edges in an alias graph is bounded by V^2. For a pointer assignment statement $p_i = q_j$, the subscripts i and j refer to the number of applications of the pointer dereference operator $*$. For the purpose of time complexity analysis, we assume (like Hind et al.) that i and j are bounded by a constant.

Lemma 1 *For any* SEG *node k, expression $G_k(I)$ can be computed in $O(V^2)$ time.*

Lemma 2 *Expression $\{k \in 1 \ldots N \mid I_k \neq G_k(I)\}$ can be computed in $O(NV^2)$ time.*

Lemma 3 *The number of iterations in the while loop of Algorithm* (14) *is at most* (NV^2).

For the proofs, look at [18]. It follows from the lemmas that the time complexity of Algorithm (14) is $O(N^2V^4)$ time. The space complexity of this algorithm can easily be seen to be $O(NV^2)$. Since $V = O(N)$, the time complexity of the naive algorithm is $O(N^6)$ and the space complexity $O(N^3)$.

6 Derivation of Kildall's Workset Algorithm

In this section we show how a finite differencing transformation can be applied to Algorithm (14) to derive Kildall's Workset Algorithm with a time complexity of $O(N^5)$.

Finite differencing has to do with replacing costly repeated computations with their more efficient incremental counterparts. In Algorithm (14) we see that the expression $\{k \in 1 \ldots N \mid I_k \neq G_k(I)\}$ needs to be recomputed in each iteration. This costly recomputation can be avoided by maintaining the invariant

$$Workset = \{k \in 1 \ldots N \mid I_k \neq G_k(I)\} \tag{15}$$

on entry to the while loop, and incrementally maintaining this invariant with respect to the assignment $I_n := G_n(I)$. In order to maintain Invariant (15), we note that expression $G_k(I) = \bigcup_{l \in Pred(k)} F_l(I_l)$ depends only on the I_l values for the predecessors l of node k. Consequently, any change to I_n can only change the value of expressions $G_s(I)$ where s is a successor of node n, i.e., $s \in Succ(n)$. Thus, the code to reestablish Invariant (15) with respect to the assignment $I_n := G_n(I)$ is given by

```
Workset less := n
for s ∈ Succ(n) loop
   if I_s ≠ G_s(I) and s ∉ Workset then
      Workset with := s
   endif
endloop
```

Putting all this together, we get the following algorithm.

```
 1 ∀i = 1...N I_i := {}
 2 Workset := {k ∈ 1...N | I_k ≠ G_k(I)}
 3 while exists n ∈ Workset loop
 4    I_n := G_n(I)
 5    Workset less := n
 6    for s ∈ Succ(n) loop                          (16)
 7       if I_s ≠ G_s(I) and s ∉ Workset then
 8          Workset with := s
 9       endif
10    endloop
11 endloop
```

Algorithm (16) is similar to the general algorithm proposed by Kildall in his seminal paper on data flow analysis [23], and is the same as the algorithm presented in [20].

6.1 Time Complexity of the Algorithm

Lemma 4 *A* SEG *node n may be inserted into Workset at most V^2 times.*

The proof can be found in [18]. It follows from the lemma that the number of iterations of the while loop in Algorithm (16) is bounded by NV^2.

Lemma 5 *The time complexity of executing the while loop (lines* 3–11) *in Algorithm* (16) *once is $O(V^2)$.*

The proof can be found in [18]. Thus, the time complexity of Algorithm (16) is $O(NV^4)$ time. Since $V = O(N)$, the time complexity of the workset algorithm is $O(N^5)$. As before, the worst-case space complexity is $O(N^3)$.

7 Going Beyond Kildall's Workset Algorithm

In this section we show how we can use dominated convergence and finite differencing to improve the algorithm further. We exploit the fact that there is some redundancy in the recomputation of expression $G_k(I)$ with respect to a small modification to one of the components of I on which $G_k(I)$ depends.

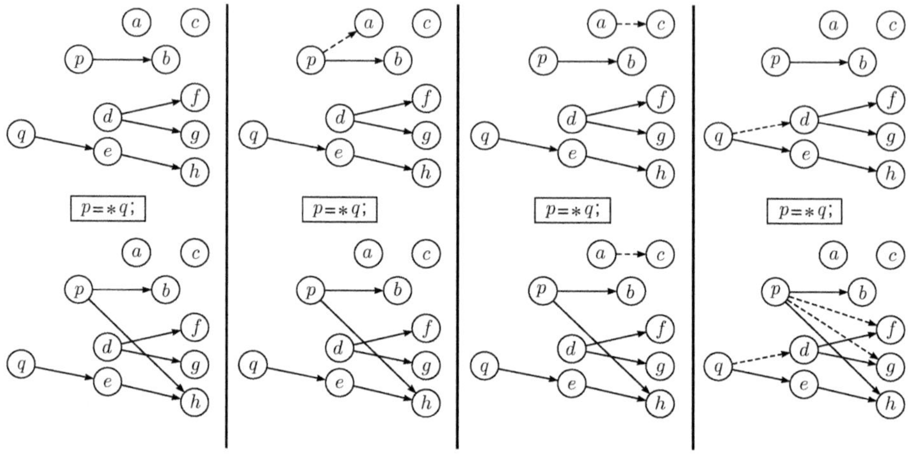

Fig. 3. Examples illustrating the potential effects of adding a single edge to the incoming alias graph.

For example, consider the case when SEG node k has a single predecessor node n. Then, $G_k(I) = F_n(I_n)$. Let node n be associated with the assignment statement $p = *q$ where p and q are named objects in the input program. Let a, b, c, d, e, f, g, and h be other named objects in the program. Figure 3 illustrates how expression $G_k(I) = F_n(I_n)$ changes with respect to the addition of one edge to I_n. Figure 3[1] shows the set of edges I_n and the corresponding set of edges $F_n(I_n)$. In Figure 3[2] we see that the addition of edge $[p, a]$ causes no change in the value of expression $F_n(I_n)$. In Figure 3[3] we see that the addition of edge $[a, c]$ causes the addition of edge $[a, c]$ to the old value of expression $F_n(I_n)$. Finally, in Figure 3[4], we see that the addition of edge $[q, d]$ causes three edges $[q, d]$, $[p, f]$, $[p, g]$ to be added to the old value of expression $F_n(I_n)$. Although, in the worst case, the addition of one edge to I_n can result in the addition of up to $O(V^2)$ edges to the old value of expression $F_n(I_n)$, we note that in many cases the change in the value of expression $F_n(I_n)$ may be small.

We shall attempt to improve Algorithm (16) by eliminating redundancy in the recomputation of expression $F_n(I_n)$ with respect to addition of edges to I_n. In Section 5.2 we used a dominated convergence transformation that allowed one component of tuple $[I_1, I_2, \ldots, I_N]$ to be selected and updated at a time. For any k such that $G_k(I) \neq I_k$, I_k was updated by changing its value to $G_k(I)$. The new transformation involves selecting a k such that $G_k(I) \neq I_k$, picking an arbitrary edge $[x, y] \in G_k(I) - I_k$, and adding the edge to I_k in each iteration. The sequence of values I obtained at the end of each iteration still satisfies the conditions on the sequence $s_0, s_1, \ldots$ in Theorem 2. Applying this transformation, we get the following algorithm for computing alias information.

$(\forall i = 1 \ldots N)\ \ I_i := \{\}$
while *exists* $[n, [x, y]] \in \{[k, [x_1, y_1]] :\ k \in 1 \ldots N,$
$\quad [x_1, y_1] \in (G_k(I) - I_k)\}$ **loop** (17)
$\quad I_n$ *with* $:= [x, y]$
endloop

The main source of inefficiency in Algorithm (17) is the recomputation of the set $\{[k, [x_1, y_1]] : k \in 1 \ldots N,\ [x_1, y_1] \in (G_k(I) - I_k)\}$ in each iteration. The finite differencing transformation to Algorithm (14) was based on incrementally maintaining the invariant $Workset = \{k \in 1 \ldots N \mid I_k \neq G_k(I)\}$ with respect to changes to I. For Algorithm (17) we create the invariant

$$Workset(k) = G_k(I) - I_k \text{ for } k = 1 \ldots N \tag{18}$$

on entry to the while loop and incrementally maintain it with respect to the set element addition I_n *with* $:= [x, y]$ where $[x, y] \in (G_n(I) - I_n)$.

In order to see how Invariant 18 can be incrementally maintained, consider an arbitrary SEG node n such that $G_n(I) - I_n$ is nonempty and let $[x, y]$ be an arbitrary arc of $G_n(I) - I_n$. Let $I'_n = I_n \cup \{[x, y]\}$, $I'_k = I_k$ for $k \neq n$, $I' = [I'_1, \ldots, I'_N]$, and $Workset'(k) = G_k(I') - I'_k$ for $k = 1 \ldots N$. Also assume without loss of generality that node n is not a predecessor of itself. This can be ensured by adding a dummy header node for each loop in the input program. Then, it is easy to show that

$$Workset'(k) = \begin{cases} Workset(n) - \{[x, y]\} & \text{if } k = n \\ Workset(k) \cup (F_n(I'_n) - F_n(I_n)) & \text{if } k \in Succ(n) \\ Workset(k) & \text{otherwise} \end{cases} \tag{19}$$

```
1 ∀i = 1...N I_i := {}
2 ∀i = 1...N Workset(i) := G_i(I) − I_i
3 while exists n ∈ {k ∈ 1...N | Workset(k) ≠ {}} loop
4     [x, y] := (∋ Workset(n))  -- arbitrary element of Workset(n)
5     Workset(n) less := [x, y] -- Update Workset(n)
6     if Succ(n) ≠ {} then
7         new_edges := F_n(I_n ∪ {[x, y]}) − F_n(I_n)
8         for k ∈ Succ(n) loop
  -- Update Workset(k) where k ∈ Succ(n)
9             Workset(k)  := Workset(k) ∪ new_edges
10        endloop
11    endif
12    I_n with := [x, y]
13 endloop
```

Fig. 4. Algorithm obtained by using a second dominated convergence transformation.

Using Equation (19), we get the algorithm in Figure 4. The only remaining problem is the efficient recomputation of $F_n(I_n)$ with respect to the modification I_n *with* $:= [x, y]$ (line 7 of the algorithm in Figure 4). In order to compute the time complexity of the algorithm in Figure 4, we need to compute the cumulative cost of each of these recomputations over all iterations of the while loop. Both of these problems, i.e., the efficient recomputation of expressions with respect to small modifications, and estimation of the cost of these recomputations, were studied in [7]. In Section 7.1 we review the main results of [7] that are relevant to this paper. In Section 7.2 we extend these results and in Section 7.3 we use these results to derive our final algorithm.

7.1 Complexity of Incremental Evaluation of Expressions

For the examples given in the rest of the paper, we will use the symbols R, S, T, U, and V to denote sets, the symbols F and G to denote maps, the symbols x, y, z to denote elements of sets, and the symbols $x_1, x_2, \ldots, y_1, y_2, \ldots$ to denote input variables (parameters) of an expression.

We will use the term *modification* to be associated with both a "kind" of a modification, and the input variable to which the modification is applied. For example, in the case of modification S *with* $:= z$, the kind of modification is "set element addition" and the associated input variable is S. Similarly, in the case of modification $T := U$, the kind of modification is "set assignment" and the associated input variable is T. We will refer to a modification abstractly as δx_i where δ refers to the kind of the modification and x_i refers to the associated input variable. We will also refer to the code required to recompute an expression with respect to some modification δx_i to one of its input variables x_i as *difference* code. For the rest of the paper, whenever we say the *cost* of computing an expression, we mean the time complexity of computing the expression. We use the function *Size* to denote the size of a set or map as defined by Equation (2) in Section 2.

Definition 2 (Continuity) Let D be a set of modifications to the input variables $x_1, \ldots, x_n$ of an expression $E(x_1, \ldots, x_n)$. Let $Cost(m)$ denote the cumulative time complexity of performing an arbitrary sequence m of modifications selected from D. We say that an invariant is maintained "eagerly" with respect to a sequence of modifications if the invariant is reestablished after each modification. Expression E is said to be *continuous* with respect to set D if the cumulative cost (time complexity) of eagerly maintaining the invariant $t = E(x_1, \ldots, x_n)$, where t is a variable distinct from $x_1 \ldots x_n$, with respect to the sequence of modifications m is:

$$O(Cost(m) + \sum_{i=1}^{n} Size(x_i) + Size(E) + \sum_{i=1}^{n} Size(x_{i_{Final}}) + Size(E_{Final})),$$

where x_i, and E refer to the values before the modifications and $x_{i_{Final}}$, and E_{Final} refer to the values after the sequence m of modifications. In other words, Expression E is continuous with respect to set D if the cumulative cost of eagerly maintaining the invariant $t = E(x_1, \ldots, x_n)$ with respect to a sequence of modifications is linearly bounded by the sum of the cost of the modifications plus the sum of the sizes of the inputs plus that of the expression, before and after the modifications.

Example 2 The following examples illustrate the concept of continuity of expressions.

1. Expressions $E_1(S,T) = S \cup T$, $E_2(S,T) = S \cap T$, and $E_3(S,T) = S - T$ are continuous with respect to the set of modifications $\{S \textit{ with} := z, S \textit{ less} := z, T \textit{ with} := z, T \textit{ less} := z\}$. In general, if the cost of recomputing an expression E with respect to a single modification is bounded by the cost of the modification itself, then the expression is continuous with respect to the modification.
2. Expressions E_1, E_2, and E_3 as defined above are not continuous with respect to the modifications $\{S := U, T := U\}$
3. Expression $E_4(S,T) = S \times T$ is continuous with respect to the set of modifications $\{S \textit{ with} := z, T \textit{ with} := z\}$.
4. Expression $E_4(S,T) = S \times T$ is not continuous with respect to set of modifications $\{S \textit{ with} := z, S \textit{ less} := z\}$.

Example 3 Consider the expression $E_2(F,S) = F[S]$, where F is a map and S is a set. Recall that $F[S]$ is defined to be $\bigcup_{x \in S} F\{x\}$. Expression $F[S]$ is continuous with respect to the set of modifications $\{S \textit{ with} := z, F\{x\} \textit{ with} := y\}$. The following difference code can be executed before applying the modification S *with* $:= z$ to recompute expression $F[S]$ (assuming that the old value of expression $F[S]$ is stored in variable E_2).

```
if z ∉ S then
   for z' ∈ F{z} loop
      E2 with := z'
   endloop
endif
```
(20)

Similarly, the following difference code can be executed before applying the modification $F\{x\}$ *with* $:= y$ to recompute expression $F[S]$.

```
if x ∈ S then
    E2 with := y                                                    (21)
endif
```

The cost of Code Fragment (20) is $O(\#F\{z\})$ if $z \notin S$, and $O(1)$ otherwise. The cost of Code Fragment (21) is $O(1)$. Thus, the cumulative cost of n modifications to set S and m modifications to map F is bounded by $O(n + m + Size(F))$ where $Size(F)$ refers to the size of map F before the modifications. Thus, expression $F[S]$ is continuous with respect to the set of modifications $\{S \textit{ with} := z, F\{x\} \textit{ with} := y\}$.

Example 4 Expression $\{[x, y] \in F \mid x \notin S\}$ is continuous with respect to the set of modifications $\{F\{x\} \textit{ with} := y, S \textit{ with} := z\}$.

Example 5 Expression $F[S]$ is not continuous with respect to the set of modifications $\{S \textit{ with} := z, S \textit{ less} := z\}$.

Composition of continuous expressions does not always result in continuous expressions. For example, consider expressions $E_1(S, T) = S \cup T$, $E_2(S, T) = S \times T$, $E_3(F, U) = F[U]$. Expressions E_1 and E_2 are continuous with respect to S *with* $:= z$, and expression E_3 is continuous with respect to U *with* $:= y$. The composition of expressions E_3 and E_1, given by expression $E_3(F, E_1(S, T)) = F[S \cup T]$, is continuous with respect to the modification S *with* $:= z$. However, the composition of expressions E_3 and E_2, given by expression $E_3(F, E_2(S, T)) = F[S \times T]$, is not continuous with respect to S *with* $:= z$. In Lemma 6 we define sufficient conditions under which the composition of continuous expressions is also continuous. To state Lemma 6, we need the following concepts.

Definition 3 (Input-Bounded) Expression $E(x_1, \ldots, x_n)$ is *input-bounded* if
$Size(E) = O(\sum_{i=1}^{n} Size(x_i))$.

Definition 4 (Output-Bounded) Expression $E(x_1, \ldots, x_n)$ *absorbs* its k*-th* parameter x_k if $Size(x_k) = O(Size(E))$. Expression $E(x_1, \ldots, x_n)$ is *output-bounded* if it absorbs all of its parameters, i.e., $\forall i = 1, \ldots, n,\ Size(x_i) = O(Size(E))$.

Example 6 The following expressions illustrate the concepts of input-boundedness and output-boundedness.

1. Expression $E_1(S, T) = S \cup T$ is output-bounded because it absorbs both parameters S and T. Expression E_1 is also input-bounded.
2. Expression $E_2(S, T) = S \cap T$ is input-bounded but not output-bounded. Expression E_2 does not absorb any of its parameters S and T.
3. Expression $E_4(F, G) = F \circ G$ is neither input-bounded, nor output-bounded. Recall that $\circ$ is the map composition operator, i.e., $F \circ G = \{[x, z] : [x, y] \in G, [y', z] \in F \mid y = y'\}$.
4. Expression $E_5(F, S) = F[S]$ is input-bounded but is not output-bounded.

Lemma 6 *The following are closure rules for continuity.*

1. *If expression $E(x_1, \ldots, x_n)$ is continuous with respect to a set of modifications D, it is also continuous with respect to any nonempty subset of D.*
2. **Parameter Substitution Rule:** *If $E(x_1, \ldots, x_n)$ is continuous with respect to the set of modifications $\{\delta x_1, \delta x_2, \delta_3 x_3, \ldots, \delta_n x_n\}$, where variables x_1 and x_2 are associated with the same kind of modification δ (e.g., δ may be set element addition), then, expression*

   $$E'(y_1, x_3, \ldots, x_n) = E(y_1, y_1, x_3, \ldots, x_n)$$

   is continuous with respect to the modifications $\{\delta y_1, \delta_3 x_3, \ldots, \delta_n x_n\}$.

3. **Composition Rule:** *Let expression $E_1(x_1,\ldots,x_n)$ be continuous with respect to a set D_1 of modifications to input variables $x_1,\ldots,x_n$. Let D_2 be the set of modifications applied to a variable y_k distinct from $x_1,\ldots,x_n$ in the course of maintaining the invariant $y_k = E_1(x_1,\ldots,x_n)$ relative to modifications from set D_1. Let*

   $E_2(y_1,\ldots,y_k,\ldots,y_m)$

   be an expression such that input variables $y_1,\ldots,y_m$ are distinct from variables $x_1,\ldots,x_n$. Let D_3 be a set of modifications to input variables $y_1,\ldots,y_{k-1},y_{k+1},\ldots,y_m$. Let expression E_3 be obtained by substituting expression E_1 for input variable y_k in expression E_2, i.e.,

   $$E_3(y_1,\ldots,y_{k-1},x_1,\ldots,x_n,y_{k+1},\ldots,y_m) = E_2(y_1,\ldots,y_{k-1},E_1(x_1,\ldots,x_n),y_{k+1},\ldots,y_m).$$

   If expression E_2 is continuous with respect to the set of modifications $D_2 \cup D_3$, then expression E_3 is continuous with respect to the set of modifications $D_1 \cup D_3$, if either (1) *expression E_2 absorbs its k-th parameter, or* (2) *if expression E_1 is input-bounded.*

The proof for composition rule of Lemma 6 can be found in [18]. The following examples illustrate some applications of Lemma 6. These examples constitute substeps in the derivation of our final algorithm.

Example 7 Consider the expression $E_1(R,T) = R \cup T$, and expression $E_2(F,S) = F[S]$ where R, T and S are sets and F is a map. Let $D_1 = \{R \textit{ with} := z, T \textit{ with} := z\}$ be a set of modifications to the input variables R and T of expression E_1. Consider the invariant $S = E_1(R,T) = R \cup T$. The only modification to set S required to maintain this invariant with respect to the modifications from D_1 is $S \textit{ with} := z$. Let D_2 denote the set containing this modification, i.e., $D_2 = \{S \textit{ with} := z\}$. Let $D_3 = \{F\{x\} \textit{ with} := y\}$ be a set of modifications to variable F of expression E_2. From Example 3 we know that Expression E_2 is continuous with respect to set $D_2 \cup D_3$. From Example 2 we know that expression E_1 is continuous with respect to set D_1. The composition rule of Lemma 6 states that if either expression E_1 is input-bounded, or if expression E_2 absorbs its second parameter, then, expression $E_3(F,R,T) = E_2(F,E_1(R,T)) = F[R \cup T]$ is continuous with respect to $D_1 \cup D_3$. From Example 6, we know that expression E_1 is input-bounded. Hence, it follows that expression E_3 is continuous with respect to $D_1 \cup D_3$.

The composition rule can also be made constructive. Given the difference code for recomputing expressions E_1 and E_2 relative to modifications in D_1 and $D_2 \cup D_3$ respectively, the difference code for recomputing expression E_3 relative to modifications in $D_1 \cup D_3$ can be obtained as follows. The difference code to recompute E_3 relative to a modification in D_3 is the same as the difference code to recompute E_2 relative to the same modification. For any modification in D_1, we first apply the difference code for E_1 relative to this modification. The composition rule is applicable only if all modifications made to variable y_k while maintaining the invariant $y_k = E_1$ belong to D_2. For each such modification, the difference code to recompute E_2 is applied. For example, the new value of expression $F[R \cup T]$ can be recomputed with respect to $R \textit{ with} := z$ using the following difference code. We assume that the old value of expression $R \cup T$ is stored in a variable S, and that the old value of expression $F[R \cup T]$ is stored in variable O.

```
-- difference code for S = R ∪ T with respect to R with := z
if z ∉ S then
    -- difference code for F[S] with respect to S with := z
    for y ∈ F{z} loop
        O with := y
    endloop                                                    (22)
    S with := z
endif
```

Example 8 Expression $F[G[S]]$ (where F and G are maps and S is a set) is continuous with respect to the set of modifications $\{S\ with := x, G\{y\}\ with := z, F\{y\}\ with := z\}$. Consider the two expressions $E_1(G,S) = G[S]$ and $E_2(F,T) = F[T]$. As shown in Example 3, both E_1 and E_2 are continuous with respect to the sets of modifications $D_1 = \{S\ with := x, G\{y\}\ with := z\}$ and $D_2 = \{T\ with := x, F\{y\}\ with := z\}$ respectively. The only modification required to set T to maintain the invariant $T = G[S]$ with respect to D_1 is $T\ with := x$. Moreover, we know from Example 6 that E_1 is input-bounded. Thus, it follows from the composition rule of Lemma 6 that expression $E_2(F, E_1(G,S)) = F[G[S]]$ is continuous with respect to the set of modifications $D = \{S\ with := x, G\{y\}\ with := z, F\{y\}\ with := z\}$.

Example 9 Expression $F[F[S]]$ (or $F^2[S]$) is continuous with respect to the set of modifications $\{S\ with := x, F\{y\}\ with := z\}$. The proof follows by applying the parameter substitution rule from Lemma 6 to the expression $F[G[S]]$.

Example 10 Expression $F^k[S]$ for a fixed constant value k is continuous with respect to $\{S\ with := x, F\{y\}\ with := z\}$. It is easy to show that each of the expressions $F^i[S]$ is input-bounded for $i = 1,\ldots,k-1$. The result follows by $k-1$ repeated applications of Lemma 6.

Example 11 The boolean-valued expression $\#S = k$ for any constant k is an $O(1)$ time computable expression, and therefore continuous with respect to any set of modifications to S. Using the composition rule of Lemma 6, it follows that if expression $E(x_1,\ldots,x_l)$ is an input-bounded expression which is continuous with respect to a set D of modifications to variables $x_1,\ldots,x_l$, then the boolean-valued expression $\#E(x_1,\ldots,x_l) = k$ is also continuous with respect to set D.

7.2 Extension to Lemma 6

We now present an example that illustrates a limitation of Lemma 6 as stated in [7]. Since we run into a similar limitation in the derivation of our final algorithm, we also propose an extension to lemma 6 that overcomes this limitation.

Example 12 Consider the expression $E = R \cup$ ($\texttt{if}$ $\#S = 1$ $\texttt{then}$ T $\texttt{else}$ U $\texttt{endif}$). Consider the set D_1 of modifications $\{S\ with := z, T\ with := z, U\ with := z\}$. Expression ($\texttt{if}$ $\#S = 1$ $\texttt{then}$ T $\texttt{else}$ U $\texttt{endif}$) is both input-bounded and continuous with respect to D_1. Now we will introduce a set D_2 of modifications applied to variable V in the process of maintaining the invariant $V =$ ($\texttt{if}$ $\#S = 1$ $\texttt{then}$ T $\texttt{else}$ U $\texttt{endif}$) with respect to modifications drawn from D_1. The modification $S\ with := z$ could cause the value of predicate $\#S = 1$ to shift from *false* to *true* if S was empty before the modification, or shift from *true* to *false* if $\#S = 1$ before the modification and S did not already contain z. The corresponding modifications required to variable V are $V := T$ or $V := U$. The modifications $T\ with := z$ and $U\ with := z$ require the modification $V\ with := z$ to variable V. Thus, the set D_2 of modifications to V is given by $\{V\ with := z, V := T, V := U\}$. The expression $R \cup V$ is not continuous with respect to the set $\{V\ with := z, V := T, V := U\}$. For example, if we alternate the modifications $V := T$ and $V := U$, then the time spent in recomputing $R \cup V$ each time is $O(\#T + \#U)$, and the cumulative cost of recomputing $R \cup V$ over m such modifications cannot be bounded by the sum of the costs of the modifications ($O(1)$ time each) plus the initial and final sizes of the input plus that of the output. Consequently, the composition rule of Lemma 6 is not applicable over here.

The problem in applying the composition rule arises because of the updates required when predicate $\#S = 1$ changes from *false* to *true*, or from *true* to *false*. However, if all modifications have the form $S\ with := z$, then the value of predicate $\#S = 1$ may change at most twice, once from *false* to *true*, and once from *true* to *false*. Consequently, the modifications $V := T$, or $V := U$ can be applied to variable V at most once each. Given this fact, it can be shown that E is indeed continuous with respect to D_1. However, if set D_1 additionally includes the modification $S\ less := z$, then the number of times that predicate $\#S = 1$ can change value (on application of an arbitrary sequence of modifications from D_1) is not bounded by a constant, and in this case expression E is not continuous with respect to D_1. Based on this intuition, we propose the following extension to the composition rule of Lemma 6.

Lemma 7 (Extension to the composition rule of Lemma 6) *Let*

$$E_1(x_1, \ldots, x_n) = \texttt{if } E_{11}(x_1, \ldots, x_n) \texttt{ then } E_{12}(x_1, \ldots, x_n) \texttt{ else } E_{13}(x_1, \ldots, x_n) \texttt{ endif}$$

where E_{11} is a boolean-valued predicate, E_{12} and E_{13} are input-bounded expressions. Let D_1 be a set of modifications to input variables $x_1, \ldots, x_n$. Let D_1 be such that, for any variable x_i, the modifications to x_i in D_1 may either only cause the size of x_i to increase, or only to decrease. Let E_{11}, E_{12}, and E_{13} be continuous with respect to D_1. Let c_1 be some constant bound on the number of times that the value of predicate E_{11} can change on application of an arbitrary sequence of modifications from D_1.

Let $E_2(y_1, \ldots, y_k, \ldots, y_m)$ be an expression such that variables $y_1, \ldots, y_m$ are distinct from variables $x_1, \ldots, x_n$. Let D_3 be a set of modifications to input variables $y_1, \ldots, y_{k-1}$, $y_{k+1}, \ldots, y_m$. Let expressions E_{21} and E_{22} be obtained by substituting variable y_k in expression E_2 by expressions E_{12} and E_{13} respectively, and expression E_3 be obtained by substituting variable y_k in expression E_2 by expression E_1, i.e.,

$$\begin{aligned}
E_{21}(y_1, \ldots, y_{k-1}, x_1, \ldots, x_n, y_{k+1}, \ldots, y_m) &= \\
&E_2(y_1, \ldots, y_{k-1}, E_{12}(x_1, \ldots, x_n), y_{k+1}, \ldots, y_m), \\
E_{22}(y_1, \ldots, y_{k-1}, x_1, \ldots, x_n, y_{k+1}, \ldots, y_m) &= \\
&E_2(y_1, \ldots, y_{k-1}, E_{13}(x_1, \ldots, x_n), y_{k+1}, \ldots, y_m), \text{ and,} \\
E_3(y_1, \ldots, y_{k-1}, x_1, \ldots, x_n, y_{k+1}, \ldots, y_m) &= \\
&E_2(y_1, \ldots, y_{k-1}, E_1(x_1, \ldots, x_n), y_{k+1}, \ldots, y_m).
\end{aligned}$$

Let the cost of recomputing expression E_2 with respect to the modification $y_k := E'$ be $O(Size(y_{k_i}) + Size(y_{k_f}))$, where y_{k_i} denotes the value of variable y_k before the modification and y_{k_f} denotes the value of variable y_k after the modification. Let

$$|Size(E_2(y_1, \ldots, y'_k, \ldots, y_m)) - Size(E_2(y_1, \ldots, y_k, \ldots, y_m))| = O(Size(y_k) + Size(y'_k)).$$

If expressions E_{21} and E_{22} are both continuous with respect to $D_1 \cup D_3$, then expression E_3 is also continuous with respect to $D_1 \cup D_3$.

The proof of Lemma 7 may be found in [18].

Example 13 From Lemma 7, it follows that expression

$$E(R, S, T, U) = R \cup (\texttt{if } \#S = 1 \texttt{ then } T \texttt{ else } U \texttt{ endif})$$

is continuous with respect to $\{S\ with := z,\ \ T\ with := z,\ \ U\ with := z\}$.

7.3 An Improved Alias Analysis Algorithm

We now return to the derivation of the alias analysis algorithm. Recall that the main source of inefficiency in the algorithm in Figure 4 in Section 7 was the incremental computation of $F_n(I_n)$ with respect to the modification $I_n\ with := [x, y]$. We now show that expression $F_n(I_n)$ is continuous with respect to the set of modifications $D = \{I_n\ with := [x, y]\}$.

Lemma 8 *Expression $F_n(I_n)$, given by Equation (5), is continuous with respect to the set of modifications $D = \{I_n\ with := [x, y]\}$.*

```
-- Incrementally compute the changes inc_P_i and inc_Q_{j+1} to P_i and Q_{j+1}
-- using the code in Example 10.
inc_P_i = (I_n ∪ {[x,y]})^i[{p}] − I_n^i[{p}]
inc_Q_{j+1} = (I_n ∪ {[x,y]})^{j+1}[{q}] − I_n^{j+1}[{q}]
-- Incrementally compute changes to E_2 with respect to I_n with := [x,y]
-- and P_i := P_i ∪ inc_P_i.
inc_E_2 := if x ∉ P_i and x ∉ inc_P_i then{[x,y]} else {} endif
dec_E_2 := ∪_{x_1 ∈ inc_P_i}{[x_1,y_1] : y_1 ∈ I_n{x_1}}
-- Incrementally compute changes to E_1.
if #P_i = 0 and #inc_P_i = 0 then
    inc_E_1 := {}
elseif #P_i = 0 and #inc_P_i = 1 then
    inc_E_1 := E_2 ∪ inc_E_2 − dec_E_2
elseif #P_i = 0 and #inc_P_i > 1 then
    inc_E_1 := I_n ∪ {[x,y]}
elseif #P_i = 1 and #inc_P_i = 0 then
    inc_E_1 := inc_E_2
elseif #P_i = 1 and #inc_P_i > 0 then
    inc_E_1 := I_n ∪ {[x,y]} − E_2
else
    inc_E_1 := {[x,y]}
endif
-- Incrementally compute changes to E_3
inc_E_3 := P_i × inc_Q_{j+1} ∪ inc_P_i × Q_{j+1} ∪ inc_P_i × inc_Q_{j+1}
-- Incrementally compute changes to E_4
inc_E_4 := inc_E_1 ∪ inc_E_3
```

Fig. 5. Outline of the code used to incrementally evaluate $F_n(I_n)$ with respect to the modification I_n *with* $:= [x, y]$.

The proof of Lemma 8 may be found in [18].

The code for efficiently recomputing $F_n(I_n)$ with respect to I_n *with* $:= [x, y]$ can be derived as follows. From the proof of Lemma 8, it can be determined that the following invariants need to be maintained.

1. $P_i = I_n^i[\{p\}]$
2. $Q_{j+1} = I_n^{j+1}[\{q\}]$
3. $E_2 = \{[x_1, x_2] \in I_n \mid x_1 \notin P_i\}$
4. $E_1 =$ **if** $\#P_i = 0$ **then** $\{\}$ **elseif** $\#P_i = 1$ **then** E_2 **else** I_n **endif**
5. $E_3 = P_i \times Q_{j+1}$
6. $E_4\ (= F_n(I_n)) = E_1 \cup E_3$

The outline of the code to incrementally compute the changes to E_4 with respect to I_n *with* $:= [x, y]$ is given in Figure 5.

Time Complexity of the Algorithm

Let us now compute the time complexity of the algorithm in Figure 4 in which Line 7 is efficiently implemented using the code outlined in Figure 5. There is one remaining source of inefficiency in the algorithm in Figure 4, which is the repeated computation of expression $\{k \in 1 \ldots N \mid Workset(k) \neq \{\}\}$ on Line 3 for each iteration of the while loop. This can be easily eliminated using another simple finite differencing transformation in which we maintain the invariant

$$Workset' = \{k \in 1 \ldots N \mid Workset(k) \neq \{\}\}. \tag{23}$$

```
1 ∀i = 1...N  I_i := {}
2 ∀i = 1...N  Workset(i) := G_i(I) − I_i
3 Workset' := {k ∈ 1...N | Workset(k) ≠ {}}
4 while exists n ∈ Workset' loop
5     [x, y] := ∋ Workset(n)  -- arbitrary element of Workset(n)
6     if #Workset(n) = 1 then
7         Workset' less := n
8     endif
9     Workset(n) less := [x, y] -- Update Workset(n)
10    if Succ(n) ≠ {} then
11        new_edges := F_n(I_n ∪ {[x, y]}) − F_n(I_n) -- computed using Figure 5
12        for k ∈ Succ(n) loop
13            if #new_edges ≠ 0 then
14                Workset' with := k
15            endif
   -- Update Workset(k) where k ∈ Succ(n)
16            Workset(k)  :=  Workset(k)  ∪  new_edges
17        endloop
18    endif
19    I_n with := [x, y]
20 endloop
```

Fig. 6. The final $O(N^3)$ time algorithm for computing may–alias information.

This invariant needs to be updated on execution of Lines 5 and 9 of the algorithm of Figure 4. The final algorithm which incorporates this transformation is given in Figure 6.

Let $I_{n_{Final}}$ denote the final value of I_n on termination of the algorithm, i.e., the set of edges in the alias graph computed just before node n.

Lemma 9 *The while loop in Figure* 6 (*Lines* 4–20) *gets executed* $\Sigma_{n=1...N} I_{n_{Final}}$ *times.*

Lemma 10 *The time complexity of the Algorithm in Figure* 6 *is* $O(\Sigma_{n=1...N} I_{n_{Final}})$.

The proofs can be found in [18].

Our final Algorithm in Figure 6 is linear in the size of the output for all programs whose SEG's in-degree and out-degree is bounded by a constant. Thus, for this class of programs, our algorithm is asymptotically optimal.

For other programs, we can show that even though the algorithm is not guaranteed to be linear in the size of the output, it is still guaranteed to be $O(N^3)$ where N denotes the number of nodes in the transformed SEG, i.e., the SEG which has been transformed to have a maximum in-degree and out-degree of 2. Note that N is proportional to the number of edges in the original SEG. For programs in which the in-degree and out-degree of the SEG is not bounded by a constant, the time complexity of our algorithm is still $O(\Sigma_{i=1...N} I_{n_{Final}})$, but the size of the relevant output may be smaller, since we do not care about the values $I_{n_{Final}}$ where n is a dummy SEG node. However, $I_{n_{Final}} = O(V^2)$, where V is the number of distinct named objects in the program. Thus, the time complexity of the algorithm is $O(N \times V^2)$. Since $V = O(N)$, the time complexity of the algorithm is $O(N^3)$ where N denotes the number of edges in the CFG of the input program.

8 Conclusion

In this paper, we have presented a new $O(N^3)$ time intraprocedural flow sensitive may–alias analysis algorithm where N denotes the number of edges in the CFG of the program. The improvement in time complexity is obtained without deterioration of space complexity.

However, the main contribution of the paper is not the alias analysis algorithm per se but the techniques that are used to derive the algorithm and the fact that our derivation technique leads to a simplified yet precise analysis of time complexity. We believe that the techniques used in this paper are of independent interest for algorithm designers. Another example of a problem where these techniques are applicable is the problem of Escape Analysis for Java programs [12]. In [12], an $O(N^5)$ time algorithm for doing Escape Analysis is presented. This algorithm uses Kildall's workset strategy to compute the least fixed point. The techniques used in our paper can be directly applied to obtain a new $O(N^3)$ time Escape Analysis algorithm.

The algorithm presented in our paper computes may–alias information independently at each program point. One approach for improving the algorithm could be to exploit the fact that alias graphs computed at successive program points share a lot of common structure. The use of persistent data structures [15] to store alias graphs would significantly reduce the space requirements of the algorithm in practice. It would be worthwhile to study how techniques such as finite differencing can be used together with persistent data structures.

References

1. A. Aho, J. Hopcroft, and J. Ullman: *Design and Analysis of Computer Algorithms.* Addison-Wesley, Reading, MA, 1974.
2. A. V. Aho, R. Sethi, and J. D. Ullman: *Compilers: Principles, Techniques, and Tools.* Addison-Wesley, 1988.
3. G. Birkhoff: *Lattice Theory.* American Mathematical Society, Providence, 1966.
4. B. Bloom: *Ready Simulation, Bisimulation, and the Semantics of CCS-like Languages.* Ph.D. thesis, Massachusets Institute of Technology, 1989.
5. M. Burke: An interval-based approach to exhaustive and incremental interprocedural data-flow analysis. *ACM Transactions on Programming Languages and Systems*, 12(3):341–395, July 1990.
6. J. Cai: Fixed point computation and transformational programming. Technical Report DCS-TR-217, The State University of New Jersey, Rutgers. Ph.D. Thesis, 1987.
7. J. Cai and R. Paige: Binding performance at language design time. In *Proc. Fourteenth ACM Symp. on Principles of Programming Languages*, 85–97, Jan. 1987.
8. J. Cai and R. Paige: Program derivation by fixed point computation. *Science of Computer Programming*, 11:3, 197–261, 1989.
9. D. R. Chase, M. Wegman, and F. K. Zadeck: Analysis of pointers and structures. In *SIGPLAN'90 Conference on Programming Language Design and Implementation*, 296–310, 1990.
10. J. D. Choi, M. Burke, and P. Carini: Automatic construction of sparse data flow evaluation graphs. In *18th annual ACM symposium on Principles of Programming Languages*, 55–66, 1991.
11. J. D. Choi, M. Burke, and P. Carini: Efficient flow-sensitive interprocedural computation of pointer-induced aliases and side-effects. In *20th SIGACT-SIGPLAN ACM Symposium on the Principles of Programming Languages*, 232–245, 1993.
12. J. D. Choi, M. Gupta, M. Serrano, V. C. Sreedhar, and S. Midkiff: Escape analysis for Java. In *Conference on Object-Oriented Programming Systems, Languages, and Applications*, Nov. 1999.
13. P. Cousot: Asynchronous iterative methods for solving a fixed point system of monotone equations in a complete lattice. Res. rep. R.R. 88, Laboratoire IMAG, Université scientifique et médicale de Grenoble, Grenoble, France, 15 pages, Sept. 1977.
14. R. Dewar, A. Grand, S. Liu, and J. Schwartz: Programming by refinement, as exemplified by the SETL representation sublanguage. *TOPLAS*, 1(1):27–49, July 1979.

15. J. R. Driscoll, N. Sarnak, D. D. Sleator, and R. E. Tarjan: Making data structures persistent. *Journal of Computer and System Sciences*, 38(1), Feb. 1989.
16. J. Earley: High level iterators and a method for automatically designing data structure representation. *J. of Computer Languages*, 1(4):321–342, 1976.
17. D. Goyal: *A Language-Theoretic Approach to Algorithms.* Ph.D. thesis, Computer Science Dept., New York University, January 2000. http://cs.nyu.edu/deepak/ThinThesis.ps.
18. D. Goyal: Transformational derivation of an improved alias analysis algorithm. *Higher Order And Symbolic Computation*, 18(1-2):15–49, 2005.
19. M. Hind: Pointer analysis: Haven't we solved this problem yet? In *2001 ACM SIGPLAN-SIGSOFT Workshop on Program Analysis for Software Tools and Engineering (PASTE'01)*, Snowbird, Utah, June 2001.
20. M. Hind, M. Burke, P. Carini, and J. D. Choi: Interprocedural pointer alias analysis. *ACM TOPLAS*, 21(4):848–894, July 1999.
21. S. Horwitz, P. Pfeiffer, and T. Reps: Dependence analysis for pointer variables. In *Programming Language Design and Implementation*, 28–40, 1989.
22. J. B. Kam and J. D. Ullman: Monotone data flow analysis frameworks. *Acta Informatica*, 7:305–317, 1977.
23. G. A. Kildall: A unified approach to global program optimization. In *ACM Symp. on Principles of Prog. Lang.*, 194–206, 1973.
24. W. Landi: Undecidability of static analysis. *ACM Letters on Programming Languages and Systems*, 1(4):323–337, Dec. 1992.
25. J. R. Larus and P. N. Hilfinger: Detecting conflicts between structure accesses. In *Programming Language Design and Implementation*, 21–34, 1988.
26. J.-L. Lassez, V. L. Nguyen, and L. Sonenberg: Fixed point theorems and semantics: A folk tale. *Information Processing Letters*, 14(3):112–116, 1982.
27. R. Paige: *Formal Differentiation: A Program Synthesis Technique.* UMI Research Press, 1981. Revision of Ph.D. thesis, NYU, June 1979.
28. R. Paige and S. Koenig: Finite differencing of computable expressions. *ACM Trans. on Programming Languages and Systems*, 4(3):401–454, 1982.
29. R. Paige, R. Tarjan, and R. Bonic: A linear time solution to the single function coarsest partition problem. *Theoretical Computer Science*, 40(1):67–84, Sept. 1985.
30. G. Ramalingam: The undecidability of aliasing. *ACM Transactions on Programming Languages and Systems*, 16(6):1467–1471, Nov. 1994.
31. E. Schonberg, J. Schwartz, and M. Sharir: An automatic technique for selection of data representations in SETL programs. *ACM TOPLAS*, 3(2):126–143, Apr. 1981.
32. J. Schwartz: *On Programming: An Interim Report on the SETL Project, Installments I and II.* New York University, New York, 1974.
33. J. Schwartz, R. Dewar, E. Dubinsky, and E. Schonberg: *Programming with Sets: An Introduction to SETL.* Springer-Verlag, New York, 1986.
34. B. Steensgaard. Points-to analysis in almost linear time: In *23rd SIGACT-SIGPLAN ACM Symposium on the Principles of Programming Languages*, 32–41, 1996.
35. P. Suppes: *Axiomatic Set Theory.* Dover, 1972.
36. A. Tarski: A lattice-theoretical fixpoint theorem and its application. *Pacific J. of Mathematics*, 5:285–309, 1955.

Dynamic Programming via Static Incrementalization

Yanhong A. Liu and Scott D. Stoller*

State University of New York at Stony Brook, Stony Brook, NY 11794, USA.
{liu,stoller}@cs.sunysb.edu

Summary. Dynamic programming is an important algorithm design technique. It is used for problems whose solutions involve recursively solving subproblems that share subsubproblems. While a straightforward recursive program solves common subsubproblems repeatedly, a dynamic programming algorithm solves every subsubproblem just once, saves the result, and reuses it when the subsubproblem is encountered again. This can reduce the time complexity from exponential to polynomial. This paper describes a systematic method for transforming programs written as straightforward recursions into programs that use dynamic programming. The method extends the original program to cache all possibly computed values, incrementalizes the extended program with respect to an input increment to use and maintain all cached results, prunes out cached results that are not used in the incremental computation, and uses the resulting incremental program to form an optimized new program. Incrementalization statically exploits semantics of both control structures and data structures and maintains as invariants equalities characterizing cached results. It provides the basis of a general method for achieving drastic program speedups. Compared with previous methods that perform memoization or tabulation, the method based on incrementalization is more powerful and systematic. It has been implemented in a prototype system CACHET and applied to numerous problems and succeeded on all of them.

Keywords: caching, dependence analysis, dynamic programming, incremental computation, incrementalization, memoization, program optimization, program transformation, static analysis, tabulation.

1 Introduction

Dynamic programming is an important technique for designing efficient algorithms [2,15,53]. It is used for problems whose solutions involve recursively solving subproblems that share subsubproblems. While a straightforward recursive program solves common subsubproblems repeatedly, a dynamic programming algorithm solves every subsubproblem just once, saves the result in a table, and reuses the result when the subsubproblem is encountered again. This can reduce the time complexity from exponential to polynomial. The technique is generally applicable to all problems that can be solved efficiently by memoizing results of subproblems [4,5].

Given a straightforward recursion, there are two traditional ways to achieve the effect of dynamic programming [15]: memoization [41] and tabulation [5].

Memoization uses a mechanism that is separate from the original program to save the result of each function call or reduction as the program executes [1,19,20,25,27,28,41,42,

* This work is supported in part by ONR under grants N00014-99-1-0132 and N00014-99-1-0358 and by NSF under grants CCR-9711253 and CCR-9876058. This article is a revised and extended version of a paper that appeared in *Proceedings of the 8th European Symposium on Programming*, Amsterdam, The Netherlands, March 1999.

O. Danvy et al. (eds.), Automatic Program Development – A Tribute to Robert Paige, 71–92.

46,50,52]. The idea is to keep a separate table of solutions to subproblems, modify recursive calls to first look up in the table, and then, if the subproblem has been computed, use the saved result, otherwise, compute it and save the result in the table. This method has two advantages. First, the original recursive program needs virtually no change. The underlying interpretation mechanism takes care of table filling and lookup. Second, only values needed by the original program are actually computed, which is optimal in a sense. Memoization has two disadvantages. First, the mechanism for table filling and lookup has an interpretive overhead. Second, no general strategy for table management is efficient for all problems.

Tabulation statically determines what shape of table is needed to store the values of all possibly needed subcomputations, introduces appropriate data structures for the table, and computes the table entries in a bottom-up fashion so that the solution to a superproblem is computed using available solutions to subproblems [5, 11–14, 24, 46–49]. This overcomes both disadvantages of memoization. First, table filling and lookup are compiled into the resulting program, so no separate mechanism is needed for the execution. Second, strategies for table filling and lookup can be specialized to be efficient for particular problems. However, tabulation has two drawbacks. First, it usually requires a thorough understanding of the problem and a complete manual rewrite of the program [15]. Second, to statically ensure that all values possibly needed are computed and stored, a table that is larger than necessary is often used; it may also include solutions to subproblems not actually needed in the original computation.

This paper presents a powerful method that statically analyzes and transforms straightforward recursive programs to efficiently cache and use the results of needed subproblems at appropriate program points in appropriate data structures. The method has three steps: (1) extend the original program to cache all possibly computed values, (2) incrementalize the extended program, with respect to an input increment, to use and maintain all cached results, (3) prune out cached results that are not used in the incremental computation, and use the resulting incremental program to form an optimized program. The method overcomes both drawbacks of tabulation. First, it consists of static program analyses and transformations that are general and systematic. Second, it stores only values that are necessary for the optimization; it also shows exactly when and where subproblems not in the original computation have to be included.

Our method is based on a number of static analyses and transformations studied previously by others [6,9,21,43,48,56,57,63] and ourselves [30,37,38,40] and improves them. Each of the caching, incrementalization, and pruning steps is simple, automatable, and efficient and has been implemented in a prototype system, CACHET. The system has been used in optimizing many programs written as straightforward recursions, including all dynamic programming problems found in [2, 15, 53], most in semiautomatic mode and some in fully automatic mode. Performance measurements confirm drastic asymptotic speedups.

The rest of the paper is organized as follows. Section 2 formulates the problem. Sections 3, 4, and 5 describe the three steps. Section 6 summarizes and discuses related issues. Section 7 presents the experimentation and performance measurements. Section 8 compares with related work and concludes.

2 Formulating the Problem

Straightforward solutions to many combinatorics and optimization problems can be written as simple recursions [15,53]. For example, the matrix-chain multiplication problem [15, pages 302–314] computes the minimum number of scalar multiplications needed by any parenthesization in multiplying a chain of n matrices, where matrix i has dimensions $p_{i-1} \times p_i$. This can be computed as $m(1,n)$, where $m(i,j)$ computes the minimum number of scalar multiplications for multiplying matrices i through j and can be defined as: for $i \le j$,

$$m(i,j) = \begin{cases} 0 & \text{if } i = j \\ \min_{i \le k \le j-1}\{m(i,k) + m(k+1,j) + p_{i-1} * p_k * p_j\} & \text{otherwise} \end{cases}$$

The longest-common-subsequence problem [15, pages 314–320] computes the length $c(n,m)$ of the longest common subsequence of two sequences $\langle x_1, x_2, ..., x_n\rangle$ and $\langle y_1, y_2, ..., y_m\rangle$, where $c(i,j)$ can be defined as: for $i, j \geq 0$,

$$c(i,j) = \begin{cases} 0 & \text{if } i = 0 \text{ or } j = 0 \\ c(i-1, j-1) + 1 & \text{if } i \neq 0 \text{ and } j \neq 0 \text{ and } x_i = y_j \\ \max(c(i, j-1), c(i-1, j)) & \text{otherwise} \end{cases}$$

Both of these examples are literally copied from the textbook by Cormen, Leiserson, and Rivest [15].

These recursive functions can be written straightforwardly in the following first-order, call-by-value functional programming language. A program is a function f_0 defined by a set of mutually recursive functions of the form

$$f(v_1, ..., v_n) \triangleq e$$

where an expression e is given by the grammar

$e ::= v$	variable
$\mid\ c(e_1, ..., e_n)$	constructor application
$\mid\ p(e_1, ..., e_n)$	primitive function application
$\mid\ f(e_1, ..., e_n)$	function application
$\mid$ **if** e_1 **then** e_2 **else** e_3	conditional expression
$\mid$ **let** $v = e_1$ **in** e_2	binding expression

We include arrays as variables and use them for indexed access such as x_i and p_j above. For convenience, we allow global variables, i.e., variables that do not change across function calls, to be implicit parameters to functions; such variables can be identified easily for our language even if they are given as explicit parameters. For a conditional expression whose condition depends on a global variable, we assume that both branches may be executed without divergence or runtime errors regardless of the value of the condition, which holds for the large class of combinatorics and optimization problems we handle.

Figure 1 gives programs for the examples above. Invariants about an input are not part of a program but are written explicitly to be used by the transformations. Clearly, x and y are implicit parameters to c, and p is an implicit parameter to m and $msub$. Condition $x[i] = y[j]$ in c depends on global variables x and y. These examples do not use data constructors, but our previous papers contain a number of examples that use them [37,38,40] and our method handles them.

<table>
<tr><td colspan="2">

$c(i,j)$ where $i, j \geq 0$
$\triangleq$ **if** $i = 0 \lor j = 0$ **then** 0
else if $x[i] = y[j]$ **then** $c(i-1, j-1) + 1$
else $\max(c(i, j-1), c(i-1, j))$

</td></tr>
<tr><td>

$m(i,j)$ where $i \leq j$
$\triangleq$ **if** $i = j$ **then** 0
else $msub(i,j,i)$

</td><td>

$msub(i,j,k)$ where $i \leq k \leq j-1$
$\triangleq$ **let** $s = m(i,k) + m(k+1,j) + p[i-1] * p[k] * p[j]$ **in**
if $k+1 = j$ **then** s
else $\min(s, msub(i,j,k+1))$

</td></tr>
</table>

Fig. 1. Example programs.

These straightforward programs repeatedly solve common subproblems and take exponential time. For example, $m(i,j)$ computes $m(i,k)$ for all k from i to $j-1$ and computes $m(k,j)$ for all k from $i+1$ to j. We transform them into dynamic programming algorithms that perform efficient caching and take polynomial time.

We use an asymptotic cost model for measuring time complexity. Assuming that all primitive functions take constant time, we need to consider only values of function applications as candidates for caching. Caching takes extra space, which reflects the well-known trade-off between time and space. Our primary goal is to improve the asymptotic running time of the program. Our secondary goal is to save space by caching only values useful for achieving the primary goal.

Caching requires appropriate data structures. In Step 1, we cache all possibly computed results in a recursive tree following the structure of recursive calls. Each node of the tree is a tuple that bundles recursive subtrees with the return value of the current call. We use <> to denote a tuple, and we use selectors $1st$, $2nd$, $3rd$, etc. to select the first, second, third, etc. elements of a tuple.

In Step 2, cached values are used and maintained in efficiently computing function calls on slightly incremented inputs. We use an infix operation $\oplus$ to denote an input increment operation, also called an input change (or update) operation. It combines a previous input $x = \langle x_1, ..., x_n \rangle$ and an increment parameter $y = \langle y_1, ..., y_m \rangle$ to form an incremented input $x' = \langle x'_1, ..., x'_n \rangle = x \oplus y$, where each x'_i is some function of x_j's and y_k's. An input increment operation we use for program optimization always has a corresponding decrement operation $prev$ such that for all x, y, and x', if $x' = x \oplus y$ then $x = prev(x')$. Note that y might or might not be used. For example, an input increment operation to function m in Figure 1 could be $\langle x'_1, x'_2 \rangle = \langle x_1, x_2 + 1 \rangle$ or $\langle x'_1, x'_2 \rangle = \langle x_1 - 1, x_2 \rangle$, and the corresponding decrement operations are $\langle x_1, x_2 \rangle = \langle x'_1, x'_2 - 1 \rangle$ and $\langle x_1, x_2 \rangle = \langle x'_1 + 1, x'_2 \rangle$, respectively. An input increment to a function that takes a list could be $x' = cons(y, x)$, and the corresponding decrement operation is $x = cdr(x')$.

In Step 3, cached values that are not used for an incremental computation are pruned away, yielding functions that cache, use, and maintain only useful values. Finally, the resulting incremental program is used to form an optimized program. The optimized program computes in an incremental fashion with step $\oplus$, caching and reusing results of subcomputations as needed, and thus avoids repeatedly solving common subproblems.

For a function f in an original program, $\overline{f}$ denotes the function that caches all possibly computed values of f, and $\widehat{f}$ denotes the pruned function that caches only useful values. We use x to denote an unincremented input and use r, $\overline{r}$, and $\widehat{r}$ to denote the return values of $f(x)$, $\overline{f}(x)$, and $\widehat{f}(x)$, respectively. For any function f, we use f' to denote the incremental function that computes $f(x')$, where $x' = x \oplus y$, using cached results about x such as $f(x)$. So, f' may take parameter x', as well as extra parameters each corresponding to a cached result. Figure 2 summarizes the notation.

Function	Return Value	Denoted as	Incremental Function
f	original value	r	f'
$\overline{f}$	all possibly computed values	$\overline{r}$	$\overline{f}'$
$\widehat{f}$	useful values	$\widehat{r}$	$\widehat{f}'$

Fig. 2. Notation.

3 Step 1: Caching All Possibly Computed Values

Consider a function f_0 defined by a set of recursive functions. Program f_0 may use global variables, such as x and y in function $c(i, j)$. A *possibly computed value* is the value of a function call that is computed for some but not necessarily all values of the global variables. For example, function $c(i, j)$ computes the value of $c(i-1, j-1)$ only when $x[i] = y[j]$. Such values occur exactly in branches of conditional expressions whose conditions depend on any global variable.

We construct a program $\overline{f_0}$ that caches all possibly computed values in f_0. For example, we extend $c(i, j)$ to always compute the value of $c(i-1, j-1)$ regardless of whether $x[i] = y[j]$. We first apply a simple *hoisting transformation* to lift function calls out of conditional expressions whose conditions depend on global variables. We then apply an *extension transformation* to cache all intermediate results, i.e., values of all function calls, in the return value.

3.1 Hoisting Transformation

Hoisting transformation $\mathcal{H}st$ identifies conditional expressions whose condition depends on any global variable and then applies the transformation

$$\begin{aligned}\mathcal{H}st[\![\textbf{if } e_1 \textbf{ then } e_2 \textbf{ else } e_3]\!] = \; & \textbf{let } v_2 = e_2 \textbf{ in} \\ & \textbf{let } v_3 = e_3 \textbf{ in} \\ & \textbf{if } e_1 \textbf{ then } v_2 \textbf{ else } v_3\end{aligned}$$

For example, the hoisting transformation leaves m and $msub$ unchanged and transforms c into

$$\begin{aligned}c(i,j) \triangleq \; & \textbf{if } i=0 \vee j=0 \textbf{ then } 0 \\ & \textbf{else let } u_1 = c(i-1,j-1)+1 \textbf{ in} \\ & \qquad \textbf{let } u_2 = \max(c(i,j-1), c(i-1,j)) \textbf{ in} \\ & \qquad \textbf{if } x[i]=y[j] \textbf{ then } u_1 \textbf{ else } u_2\end{aligned}$$

$\mathcal{H}st$ simply lifts up the entire subexpressions in the two branches, not just the function calls in them. Administrative simplification performed at the end of the extension transformation will unwind bindings for computations that are used at most once in subsequent computations; thus computations other than function calls will be put down into the appropriate branches then. $\mathcal{H}st$ is simple and efficient. The resulting program has essentially the same size as the original program, so $\mathcal{H}st$ does not increase the running time of the extension transformation or the running times of the later incrementalization and pruning.

If we apply the hoisting transformation on arbitrary conditional expressions, the resulting program may run slower, become nonterminating, or have errors introduced, since the transformed program may perform certain computations not performed in some branches of the original program. By applying the hoisting transformation only on conditional expressions whose conditions depend on global variables, our assumption in Section 2 eliminates the last two problems. The first problem is discussed in Section 6.

3.2 Extension Transformation

For each hoisted function definition $f(v_1, ..., v_n) \triangleq e$, we construct a function definition

$$\overline{f}(v_1, ..., v_n) \triangleq \mathcal{E}xt[\![e]\!]$$

where $\mathcal{E}xt[\![e]\!]$, defined in [38], extends an expression e to return a nested tuple that contains the values of all function calls made in computing e, i.e., it examines subexpressions of e in applicative order, introduces bindings that name the results of function calls, builds up tuples of these values together with the values of the original subexpressions, and passes these values from subcomputations to enclosing computations. The first component of a tuple corresponds to an original return value. Next, administrative simplifications clean up the resulting program. This yields a program $\overline{f}_0$ that embeds values of all possibly computed function calls in its return value. For the hoisted programs m and c, the extension transformation produces the following functions:

$$\begin{aligned}\overline{m}(i,j) \triangleq \; & \textbf{if } i=j \textbf{ then } <0> \textbf{ else } \overline{msub}(i,j,i) \\ \overline{msub}(i,j,k) \triangleq \; & \textbf{let } v_1 = \overline{m}(i,k) \textbf{ in} \\ & \textbf{let } v_2 = \overline{m}(k+1,j) \textbf{ in} \\ & \textbf{let } s = 1st(v_1) + 1st(v_2) + p[i-1]*p[k]*p[j] \textbf{ in} \\ & \textbf{if } k+1=j \textbf{ then } <s, v_1, v_2> \\ & \textbf{else let } v = \overline{msub}(i,j,k+1) \textbf{ in} \\ & \qquad <\min(s, 1st(v)), v_1, v_2, v>\end{aligned}$$

$$
\begin{array}{l}
\overline{c}(i,j) \triangleq \textbf{ if } i=0 \ \vee j=0 \textbf{ then } <0> \\
\quad\quad\quad\quad \textbf{else let } v_1 = \overline{c}(i-1,j-1) \textbf{ in} \\
\quad\quad\quad\quad\quad\quad \textbf{let } v_2 = \overline{c}(i,j-1) \textbf{ in} \\
\quad\quad\quad\quad\quad\quad \textbf{let } v_3 = \overline{c}(i-1,j) \textbf{ in} \\
\quad\quad\quad\quad\quad\quad \textbf{if } x[i] = y[j] \textbf{ then } <1st(v_1)+1, v_1, v_2, v_3> \\
\quad\quad\quad\quad\quad\quad \textbf{else } <\max(1st(v_2), 1st(v_3)), v_1, v_2, v_3>
\end{array}
$$

We have $m(i,j) = 1st(\overline{m}(i,j))$ and $c(i,j) = 1st(\overline{c}(i,j))$.

4 Step 2: Static Incrementalization

The essence of our method is to transform a program to use and maintain cached values efficiently as the computation proceeds. This is done by incrementalizing $\overline{f}_0$ with respect to an input increment operation $\oplus$, i.e., we transform $\overline{f}_0(x \oplus y)$ to use the cached value of $\overline{f}_0(x)$ rather than compute from scratch.

An *input increment operation* $\oplus$ describes a minimal update to the input parameters. We first describe a general method for determining $\oplus$. We then give a method, called *static incrementalization*, that constructs an incremental version $\overline{f}\,'$ for each function $\overline{f}$ in the extended program and allows an incremental function to have multiple parameters that represent cached values.

4.1 Determining Input Increment Operation

An input increment reflects how a computation proceeds. In general, a function may have multiple ways of proceeding, depending on the particular computations involved. There is no general method for identifying all of them or the most appropriate ones. Here we propose a method that can systematically identify a general class of them. The idea is to use a *minimal* input change that is in the *opposite* direction of change compared to arguments of recursive calls. Using the opposite direction of change yields an increment; using a minimal change allows maximum reuse, i.e., maximum incrementality.

Consider a recursively defined function f_0. Formulas for the possible arguments of recursive calls to f_0 in computing $f_0(x)$ can be determined statically. For example, for function $c(i,j)$, recursive calls to c have the set of possible arguments $S_c = \{\langle i-1, j-1\rangle, \langle i, j-1\rangle, \langle i-1, j\rangle\}$, and for function $m(i,j)$, recursive calls to m have the set of possible arguments $S_m = \{\langle i,k\rangle, \langle k+1, j\rangle \mid i \le k \le j-1\}$. The latter is simplified from $S_m = \{\langle a,c\rangle, \langle c+1, b\rangle \mid a \le c \le b-1,\ a=i,\ b=j\}$ where a,b,c are fresh variables that correspond to i,j,k in *msub*; the equalities are based on arguments of the function calls involved (in this case calls to *msub*); the inequalities are obtained from inequalities on these arguments (in the where-clause of *msub*). The simplification here, as well as the manipulations below, can be done automatically using Omega [51], a system for manipulating linear constraints over integer variables, Presburger formulas, and integer tuple relations and sets.

Represent the arguments of recursive calls so that the differences between them and x are explicit. For function c, S_c is already in this form, and for function m, S_m is rewritten as $\{\langle i, j-l\rangle, \langle i+l, j\rangle \mid 1 \le l \le j-i\}$. Then, extract minimal differences that cover all of these recursive calls. The partial ordering on differences is: a difference involving fewer parameters is smaller; a difference in one parameter with smaller magnitude is smaller; other differences are incomparable. A set of differences *covers* a recursive call if the argument to the call can be obtained by repeated application of the given differences. So, we first compute the set of minimal differences and then remove from it each element that is covered by the remaining elements. For function c, we obtain $\{\langle i, j-1\rangle, \langle i-1, j\rangle\}$, and for function m, we obtain $\{\langle i, j-1\rangle, \langle i+1, j\rangle\}$. Elements of this set represent decrement operations *prev*. Finally, take the opposite of each decrement operation to obtain an increment operation $\oplus$, introducing a parameter y if needed (e.g., for increments that use data constructions). For function c, we obtain $\langle i, j+1\rangle$ and $\langle i+1, j\rangle$, and for function m, we obtain $\langle i, j+1\rangle$ and $\langle i-1, j\rangle$. Although finding input increment operations is theoretically hard in general (and a decrement operation might not have an inverse, in which case our algorithm does not apply), it is usually straightforward.

Typically, a function that involves repeatedly solving common subproblems contains multiple recursive calls to itself. If there are multiple input increment operations, then any one may be used to incrementalize the program and then form an optimized program; the rest may be used to further incrementalize the resulting optimized program, if it still involves repeatedly solving common subproblems. For example, for program c, either $\langle i, j+1\rangle$ or $\langle i+1, j\rangle$ leads to a final optimized program that takes polynomial time; the resulting program does not contain multiple recursive calls that solve common subproblems. For program m, either $\langle i-1, j\rangle$ or $\langle i, j+1\rangle$ leads to an optimized program that is an exponential factor faster, but the program still contains multiple recursive calls that solve common subproblems and takes exponential time; incrementalizing that program again under the other increment operation leads to a final optimized program that takes polynomial time. In other words, both $\langle i-1, j\rangle$ and $\langle i, j+1\rangle$ need to be used, and they may be used in either order.

4.2 Static Incrementalization

Given a program $\overline{f}_0$ and an input increment operation $\oplus$, incrementalization symbolically transforms $\overline{f}_0(x')$ for $x' = x \oplus y$ to replace subcomputations with retrievals of their values from the value $\overline{r}$ of $\overline{f}_0(x)$. This exploits equality reasoning, based on control and data structures of the program and properties of primitive operations. The resulting program $\overline{f}_0'$ uses $\overline{r}$ or parts of $\overline{r}$ as additional arguments, called *cache arguments*, and satisfies: if $\overline{f}_0(x) = \overline{r}$ and $\overline{f}_0(x') = \overline{r}\,'$, then $\overline{f}_0'(x', \overline{r}) = \overline{r}\,'$.

Note. In previous papers, we defined $\overline{f}_0'$ slightly differently: if $\overline{f}_0(x) = \overline{r}$ and $\overline{f}_0(x \oplus y) = \overline{r}\,'$, then $\overline{f}_0'(x, y, \overline{r}) = \overline{r}\,'$. This difference is insubstantial since when incrementalization is used for program optimization, as we do here, both x and y (if it is used) can be obtained from x'.□

The idea is to establish the strongest invariants we can, especially those about cache arguments, for each function at all calls of it and maximize the usage of the invariants. At the end, unused candidate cache arguments are eliminated. Reducing running time corresponds to maximizing uses of invariants; reducing space corresponds to maintaining weakest invariants that suffice for all uses. It is important that the methods for establishing and using invariants are not only powerful but also systematic so that they are automatable. The algorithm is described below. Its use is illustrated afterwards using the running examples.

The algorithm starts with transforming $\overline{f}_0(x')$ for $x' = x \oplus y$ and $\overline{f}_0(x) = \overline{r}$ and first uses the decrement operation to establish an invariant about function arguments. More precisely, it starts with transforming $\overline{f}_0(x')$ with invariant $\overline{f}_0(prev(x')) = \overline{r}$, where $\overline{r}$ is a candidate cache argument. It may use other invariants about x' if given. Invariants given or formed from the enclosing conditions and bindings are called *context*. The algorithm transforms function applications recursively. There are four cases at a function application $f(e_1', ..., e_n')$.

(i) If $f(e_1', ..., e_n')$ specializes, by definition of f, under its context to a base case, i.e., an expression with no recursive calls, then replace it with the specialized expression.

   `Example.` For function application $f(e)$ with definition $f(x) \triangleq$ **if** $x \leq 0$ **then** 0 **else** $g(x)$ and context $e = 0$, we specialize $f(e)$ to 0.

(ii) Otherwise, if $f(e_1', ..., e_n')$ equals a retrieval from a cache argument based on an invariant about the cache argument that holds at $f(e_1', ..., e_n')$, then replace it with the retrieval.

   `Example.` If invariant $f(e) = 2nd(\overline{r})$ holds at function application $f(e)$, then we replace $f(e)$ with $2nd(\overline{r})$.

(iii) Otherwise, if an incremental version f' of f has been introduced, then replace $f(e_1', ..., e_n')$ with a call to f' if the invariants associated with f' can be maintained; if some invariants cannot be maintained, then eliminate them and retransform from where f' was introduced. Maintaining invariants includes maintaining both the invariants about a cache argument, which have the form of a function application equaling a retrieval from a cache argument, and the other usual invariants.

`Example.` After introducing $\overline{f_0}'(x',\overline{r})$ to compute $\overline{f_0}(x')$ with invariant $\overline{f_0}(prev(x')) = \overline{r}$, we replace $\overline{f_0}(e)$ by $\overline{f_0}'(e, 3rd(\overline{r}))$ if we have $\overline{f_0}(prev(e)) = 3rd(\overline{r})$.

`Example.` After introducing $f'(x, r_1, r_2)$ to compute $f(x)$ with invariants $f(x-1) = r_1$, $f(x-2) = r_2$, and $x > 0$, if we encounter a function application $f(e)$ at which $f(e-1) = e_{r1}$ holds for some e_{r1} but neither $f(e-2) = e_{r2}$ for any e_{r2} nor $e > 0$ can be inferred, then we retransform from where f' was introduced and introduce $f'(x, r_1)$ with only invariant $f(x-1) = r_1$.

(iv) Otherwise, introduce an incremental version f' of f and replace $f(e'_1, ..., e'_n)$ with a call to f', as described below.

In general, the replacement in Case (i) is also done, repeatedly, if the specialized expression contains only recursive calls whose arguments are closer to, and will equal after a bounded number of such replacements, arguments for base cases or arguments for which retrievals can be done. Since a bounded number of invariants are used at a function application, as described below, the retransformation in Case (iii) happens at most a bounded number of times, so the algorithm always terminates; in the worst case, no invariants can be maintained and used, and f' is the same as f. Since f is just one function in the given program; the final program that uses f' after Step 3 might or might not be faster than the original program. There is no way to tell in general which is the case, but for the class of dynamic programming problems we consider, it can be conservatively determined that the original programs always contain repeated recursive calls and thus take exponential time, and the final programs proceed in nested linear fashion and thus take polynomial time.

Case (iv) To introduce an incremental version f' of f at $f(e'_1, ..., e'_n)$, we (iv.1) determine candidate invariants associated with f' based on the invariants that hold at $f(e'_1, ..., e'_n)$ and (iv.2) obtain a definition of f' based on the definition of f and the candidate invariants. Finally, we (iv.3) replace $f(e'_1, ..., e'_n)$ with a call to f'.

Case (iv.1) To determine candidate invariants associated with f', let *Inv* be the set of invariants about a cache argument or in the context that hold at $f(e'_1, ..., e'_n)$. Invariants about a cache argument are of the form $g_i(e_{i1}, ..., e_{in_i}) = e_{ri}$, where e_{ri} is either a candidate cache argument in the enclosing environment or a selector applied to such an argument. Invariants in the context are of a form given, e.g., $i \le j$ for $m(i, j)$, or of the form $e = true$, $e = false$, or $v = e$ obtained from enclosing conditions or bindings. For simplicity, we assume that all bound variables are renamed so that they are distinct.

`Example.` As an example for Step (iv.1), consider function application $\overline{msub}(i', j', i')$, and assume that invariants about a cache argument, $\overline{msub}(i', j'-1, i') = \overline{r}$ and $\overline{m}(i', j'-1) = \overline{r}$, and invariants in the context, $i' \le j'$, $i' \ne j'$, and $i' \ne j'-1$, hold at the application.

We are introducing $f'(x''_1, ..., x''_n, ...)$ to compute $f(x''_1, ..., x''_n)$ for $x''_1 = e'_1, ..., x''_n = e'_n$, where $x''_1, ..., x''_n$ are fresh variables, and the second ... in the parameters of f' denote the cache arguments to be determined. So we deduce invariants about $x''_1, ..., x''_n$ based on *Inv* and $x''_1 = e'_1, ..., x''_n = e'_n$.

`Example.` For the example above, we let $\overline{msub}'$ compute $\overline{msub}(i'', j'', k'')$ for $i'' = i'$, $j'' = j'$, and $k'' = i'$, and deduce invariants about i'', j'', and k''.

The deduction has four steps.

(1) Use equations $e'_1 = x''_1, ..., e'_n = x''_n$ to try to eliminate all variables in *Inv* other than those in e_{ri}'s. This can be done automatically using Omega [51].

`Example.` For the example above, we give the following formula to Omega:

$$\exists i', j' : \overline{msub}(i', j'-1, i') = \overline{r},\ \ \overline{m}(i', j'-1) = \overline{r},\ \ i' \le j',\ \ i' \ne j',\ \ i' \ne j'-1, \\ i' = i'',\ \ j' = j'',\ \ i' = k''$$

and we obtain the following result:

$\overline{msub}(i'', j''-1, i'') = \overline{r},\ \ \overline{m}(i'', j''-1) = \overline{r},\ \ i'' \le j''-2,\ \ k'' = i''$

(2) Remove resulting invariants that still use variables in *Inv* other than those in e_{ri}'s.

`Example.` For the example above, this has no effect.

(3) Use equations relating $x_1'', ..., x_n''$ to add additional forms of other invariants. This is done as follows: if $x_j'' = x_k''$ or $x_k'' = x_j''$ is a resulting equation, and i is another resulting invariant that involves x_j'', then for each invariant i' that can be obtained by replacing some occurrences of x_j'' in i with x_k'', add i' to the resulting set of invariants.

`Example.` For the example above, this yields

$\overline{msub}(i'', j''-1, i'') = \overline{r},\ \overline{msub}(i'', j''-1, k'') = \overline{r},\ \overline{m}(i'', j''-1) = \overline{r},\ \ i'' \le j''-2,\ \ k'' = i'',$
$\overline{msub}(k'', j''-1, i'') = \overline{r},\ \overline{msub}(k'', j''-1, k'') = \overline{r},\ \overline{m}(k'', j''-1) = \overline{r},\ k'' \le j''-2,$

(4) For each invariant about a cache argument, replace its right side with a fresh variable.

`Example.` For the above example, six fresh variables, $\overline{r}_1$ to $\overline{r}_6$, are used to replace the right sides of the invariants about a cache argument.

We call the resulting invariants *candidate invariants*; each of them either uses only variables $x_1'', ..., x_n''$ or is of the form $g_i(e_{i1}'', ..., e_{in_i}'') = r_i$, where $e_{i1}'', ..., e_{in_i}''$ use only variables $x_1'', ..., x_n''$ and r_i is a fresh variable. They are now associated with f', which has arguments $x_1'', ..., x_n''$ and candidate cache arguments r_i's.

Given invariants *Inv* and equations $x_1'' = e_1', ..., x_n'' = e_n'$, the set of strongest invariants about $x_1'', ..., x_n''$, expressed using no other variables in *Inv* except those in e_{ri}'s, are in general uncomputable. However, the deduction using Omega in (1) allows us to obtain such invariants automatically when only Presburger arithmetic is involved, which is the case for all the dynamic programming problems we consider. The removal in (2) allows us to fall back to weaker invariants in the general cases. The additional forms in (3) allow us to, when some invariants deduced can not be maintained at other calls to f, keep the strongest subset of invariants that can be obtained based on direct equalities.

Case (iv.2) To obtain a definition of f', first unfold $f(x_1'', ..., x_n'')$. Then exploit control structures, i.e., conditionals in $f(x_1'', ..., x_n'')$ and $g_i(e_{i1}'', ..., e_{in_i}'')$'s, and data structures, i.e., components in r_i's, together with other candidate invariants associated with f'. Exploiting data structures allows us to use not only a cached result as a whole but also components of it. Exploiting control structures helps us obtain different forms of cached results under different conditions.

(1) To exploit conditionals in $f(x_1'', ..., x_n'')$, in the unfolded expression, move function applications into branches of the conditionals whenever possible, preserving control dependencies incurred by the order of conditional tests and data dependencies incurred by the bindings. This allows transformations of function applications to use as many conditions in their contexts as possible. This is done by repeatedly applying the following transformation in applicative, i.e., leftmost and innermost first, order to the unfolded expression:

For any expression $t(e_1, ..., e_k)$,
being $c(e_1, ..., e_k)$, $p(e_1, ..., e_k)$, $f(e_1, ..., e_k)$, **if** e_1 **then** e_2 **else** e_3, or **let** $v = e_1$ **in** e_2:
if subexpression e_i is **if** e_{i1} **then** e_{i2} **else** e_{i3}
where if t is a conditional, $i \ne 2, 3$, and
if t is a binding expression, $i \ne 2$ or e_{i1} does not depend on v,
then transform $t(e_1, ..., e_k)$ to
if e_{i1} **then** $t(e_1, ..., e_{i-1}, e_{i2}, e_{i+1}, ..., e_k)$ **else** $t(e_1, ..., e_{i-1}, e_{i3}, e_{i+1}, ..., e_k)$.

`Example.` If the unfolded expression is **let** $v = h(e)$ **in if** e_1 **then** e_2 **else** e_3, where e_1 does not depend on v, then we transform it to **if** e_1 **then let** $v = h(e)$ **in** e_2 **else let** $v = h(e)$ **in** e_3.

This transformation preserves the semantics. It may increase the code size, but it does not increase the running time of the resulting program.

(2) To exploit the conditionals in $g_i(e''_{i1},...,e''_{in_i})$'s, first choose, among $g_i(e''_{i1},...,e''_{in_i})$'s that are applications of f, one whose arguments differ minimally from $x''_1,...,x''_n$, denote it $f(e''_1,...,e''_n)$, and call it the *corresponding previous application*. If the corresponding previous application is found, then introduce in the expression obtained from (1) conditions that appear in $f(e''_1,...,e''_n)$, and put function applications inside both branches that follow such a condition. Again, this allows transformations of function applications to use as many conditions in their contexts as possible and, in particular, to use different forms of cached values from $f(e''_1,...,e''_n)$ under different conditions. This is done by applying the following transformation in outermost-first order to the conditionals in the expression obtained from (1):

For each branch e of a conditional that contains a function application:
let e' be the condition of the leftmost and outermost conditional in $f(e''_1,...,e''_n)$ such that the context of e does not imply e' and does not imply $\neg e'$;
if e' uses only variables defined in the context of e and takes constant time to compute, and the two branches in $f(e''_1,...,e''_n)$ that are conditioned on e' contain different function applications in some component
then transform e to **if** e' **then** e **else** e.

`Example.` If a branch e of a conditional is $...h(x-1)...$, the corresponding previous application $f(x-1)$ by definition equals **if** $x-1=0$ **then** $<0>$ **else** $<e_1,e_2>$, and the context of e does not imply whether $x-1=0$ or not, then transform e to **if** $x-1=0$ **then** e **else** e.

Exploiting conditionals in the corresponding previous application $f(e''_1,...,e''_n)$ is a heuristic. In general, one may exploit conditionals in all $g_i(e''_{i1},...,e''_{in_i})$'s; afterwards, conditionals whose two branches are the same are optimized by eliminating the condition and merging the two branches. We use this heuristic here since it simplifies the transformations and is sufficient for all examples we have seen. The rationale is that the arguments $e''_1,...,e''_n$ differ minimally from $x''_1,...,x''_n$, so the values of $f(e''_1,...,e''_n)$ under various conditions are most likely to be reused in computing $f(x''_1,...,x''_n)$.

(3) To exploit each component in a candidate cache argument r_i where there is an invariant $g_i(e''_{i1},...,e''_{in_i}) = r_i$, for each branch in the transformed expression from (2), specialize $g_i(e''_{i1},...,e''_{in_i})$ under the context of that branch. This may yield additional invariants that are equalities between function applications and components of r_i. It does not change the resulting expression from (2).

`Example.` Continuing the above example, if $f(x-1)=r$ is an invariant at e, and if e_2 is $h(x-1)$, then specializing $f(x-1)$ under $x-1\neq 0$ yields $h(x-1)=2nd(r)$. This invariant will enable one to replace $h(x-1)$ in e in the else-branch with $2nd(r)$.

After these control structures and data structures are exploited, we perform the following transformations on the expression from (2) in applicative order: we simplify subexpressions using algebraic properties and transform function applications recursively based on the four cases described.

Case (iv.3) After we finish transforming the expression for defining f', we eliminate dead code. Finally, after we obtain a definition of f', replace the function application $f(e'_1,...,e'_n)$ with a call to f' with arguments $e'_1,...,e'_n$ and cache arguments e_{ir}'s for the invariants used.

After an application of f, other than the initial application $\overline{f}_0(x')$, is replaced by an application of f', if f' is not recursively defined, then we unfold the application of f' and repeat transformations (1) to (3) in (iv.2) on the enclosing expression that will become the body of the enclosing function. This enables, in defining the enclosing function, more exploitation of control structures and data structures based on the conditionals and binding expressions in the unfolded application of f'. This may increase the code size but not the running time of the resulting program.

4.3 Longest Common Subsequence

Incrementalize $\overline{c}$ under $\langle i',j'\rangle = \langle i+1,j\rangle$. We start with $\overline{c}(i',j')$, with cache argument $\overline{r}$ and invariant $\overline{c}(prev(i',j')) = \overline{c}(i'-1,j') = \overline{r}$; the invariants $i',j'>0$ may also be included but do not affect any transformation below, so for brevity, we omit them. This is case (iv), so we introduce incremental version $\overline{c}'$ to compute $\overline{c}(i',j')$. Unfolding the definition of $\overline{c}$ and exploiting control structures according to (1) and (2) in (iv.2), we obtain the code below, where the annotations on the right are explained in the two paragraphs that follow. In particular, according to (2) in (iv.2), the false branch of $\overline{c}(i',j')$ is duplicated and put inside both branches of the additional condition $i'-1=0 \ \vee j'=0$, which is copied from the condition in the corresponding previous application $\overline{c}(i'-1,j')$; for convenience, the three function applications bound to v_1 through v_3 are not put inside branches that follow condition $x[i']=y[j']$, since their transformations are not affected, and simplification at the end can take them back out.

$\overline{c}(i',j')$ — with invariant $\overline{c}(i'-1,j')=\overline{r}$

$=$

if $i'=0 \ \vee j'=0$ **then** $<0>$
else if $i'-1=0 \ \vee j'=0$ **then** — context includes: $i'-1=0$
  let $v_1=\overline{c}(i'-1,j'-1)$ **in** — $=<0>$
  let $v_2=\overline{c}(i',j'-1)$ **in** — $=\overline{c}'(i',j'-1,\overline{c}(i'-1,j'-1))=\overline{c}'(i',j'-1,<0>)$
  let $v_3=\overline{c}(i'-1,j')$ **in** — $=<0>$
  if $x[i']=y[j']$ **then** $<1st(v_1)+1,v_1,v_2,v_3>$
  else $<\max(1st(v_2),1st(v_3)),v_1,v_2,v_3>$
else — context includes: $i'\neq 0,\ i'-1\neq 0,\ j'\neq 0$
  let $v_1=\overline{c}(i'-1,j'-1)$ **in** — $=3rd(\overline{r})$
  let $v_2=\overline{c}(i',j'-1)$ **in** — $=\overline{c}'(i',j'-1,\overline{c}(i'-1,j'-1))=\overline{c}'(i',j'-1,3rd(\overline{r}))$
  let $v_3=\overline{c}(i'-1,j')$ **in** — $=\overline{r}$
  if $x[i']=y[j']$ **then** $<1st(v_1)+1,v_1,v_2,v_3>$
  else $<\max(1st(v_2),1st(v_3)),v_1,v_2,v_3>$

In the second branch (lines 2–7), $i'-1=0$ is true, since $j'=0$ would imply that the first branch is taken. Therefore, the first and third calls fall in case (i) and specialize to $<0>$. The second call falls in case (iii) and equals a recursive call to $\overline{c}'$ with arguments $i',j'-1$ and cache argument $<0>$, since we have a corresponding invariant $\overline{c}(i'-1,j'-1)=<0>$ from specialization. Additional simplification unwinds bindings for v_1 and v_3, simplifies $1st(<0>)+1$ to 1, and simplifies $\max(1st(v_2),1st(<0>))$ to $1st(v_2)$.

In the third branch (lines 8–13), condition $i'-1=0 \vee j'=0$ is false; under this condition, the corresponding previous application $\overline{c}(i'-1,j')$ by definition of $\overline{c}$ equals its second branch where $\overline{c}(i'-1,j'-1)$ is bound to v_2, and thus the invariant $\overline{c}(i'-1,j')=\overline{r}$ implies $\overline{c}(i'-1,j'-1)=3rd(\overline{r})$. Therefore, in this third branch, the first call falls in case (ii) and equals $3rd(\overline{r})$. The second call falls in case (iii) and equals a recursive call to $\overline{c}'$ with arguments $i',j'-1$ and cache argument $3rd(\overline{r})$ since we have a corresponding invariant $\overline{c}(i'-1,j'-1)=3rd(\overline{r})$. The third call falls in case (ii) and equals $\overline{r}$. We obtain

$\overline{c}'(i',j',\overline{r}) \triangleq$ **if** $i'=0 \ \vee j'=0$ **then** $<0>$
  else if $i'-1=0$ **then**
    let $v_2=\overline{c}'(i',j'-1,<0>)$ **in**
    if $x[i']=y[j']$ **then** $<1,<0>,v_2,<0>>$
    else $<1st(v_2),<0>,v_2,<0>>$
  else let $v_1=3rd(\overline{r})$ **in**
    let $v_2=\overline{c}'(i',j'-1,3rd(\overline{r}))$ **in**
    let $v_3=\overline{r}$ **in**
    if $x[i']=y[j']$ **then** $<1st(v_1)+1,v_1,v_2,v_3>$
    else $<\max(1st(v_2),1st(v_3)),v_1,v_2,v_3>$

If $\overline{r} = \overline{c}(i'-1,j')$, then $\overline{c}'(i',j',\overline{r}) = \overline{c}(i',j')$, and $\overline{c}'$ takes time and space linear in j', for caching and maintaining a linear list. It is easy to see that $\overline{c}'$ takes linear time, since $\overline{c}'(i',j',...)$ only calls $\overline{c}'(i',j'-1,...)$ recursively, and $j'=0$ is a base case. It is also easy to see that $\overline{c}'$ takes linear space, since each call to $\overline{c}'$ creates a tuple of length no more than 4.

4.4 Matrix-Chain Multiplication

Incrementalize $\overline{m}$ under $\langle i',j'\rangle = \langle i,j+1\rangle$. We start with $\overline{m}(i',j')$, with cache argument $\overline{r}$ and invariants $\overline{m}(i',j'-1) = \overline{r}$ and $i' \leq j'$. This is case (iv), so we introduce incremental version $\overline{m}'$ to compute $\overline{m}(i',j')$. Unfolding $\overline{m}$, exploiting control structures according to (1) and (2) in (iv.2), and specializing the second branch according to case (i), we obtain the code below.

$$\begin{array}{rl} \overline{m}(i',j') = & \textbf{if } i'=j' \textbf{ then } <0> \\ & \textbf{else if } i'=j'-1 \textbf{ then } < p[i'-1]*p[i']*p[j'], <0>, <0>> \\ & \textbf{else } \overline{msub}(i',j',i') \end{array} \tag{1}$$

In the third branch, condition $i' = j'-1$ is false; under this condition, $\overline{m}(i',j'-1)$ by definition of $\overline{m}$ equals $\overline{msub}(i',j'-1,i')$, and thus the invariant $\overline{m}(i',j'-1) = \overline{r}$ implies $\overline{msub}(i',j'-1,i') = \overline{r}$. The call $\overline{msub}(i',j',i')$ falls in case (iv). We introduce $\overline{msub}'$ to compute $\overline{msub}(i'',j'',k'')$ for $i''=i', j''=j', k''=i'$, with collected invariants

$$\overline{msub}(i',j'-1,i') = \overline{r}, \quad \overline{m}(i',j'-1) = \overline{r}, \quad i' \leq j', \quad i' \neq j', \quad i' \neq j'-1 \tag{2}$$

where the first two are about cache arguments; the third is given; and the last two are from the enclosing conditionals, in concise form, rather than, e.g., $(i'=j') = \mathit{false}$, for ease of reading. According to (iv.1), we express these invariants as invariants on i'',j'',k'' using Omega, and introduce fresh variables $\overline{r}_i$ for candidate cache arguments. We obtain candidate invariants associated with $\overline{msub}'$:

$$\begin{array}{ll} \overline{msub}(i'',j''-1,k'') = \overline{r}_1,\ \overline{m}(i'',j''-1) = \overline{r}_2, & i'' \leq j''-2,\ k''=i'', \\ \overline{msub}(i'',j''-1,i'') = \overline{r}_3, & k'' \leq j''-2, \\ \overline{msub}(k'',j''-1,k'') = \overline{r}_4,\ \overline{m}(k'',j''-1) = \overline{r}_5, & \\ \overline{msub}(k'',j''-1,i'') = \overline{r}_6, & \end{array} \tag{3}$$

The arguments of $\overline{msub}(i'',j''-1,k'')$ have a minimum difference from the arguments of $\overline{msub}(i'',j'',k'')$, and thus $\overline{msub}(i'',j''-1,k'')$ is the corresponding previous application according to (2) in (iv.2).

Unfolding $\overline{msub}(i'',j'',k'')$ and exploiting control structures according to (1) and (2) in (iv.2), we obtain the code below, where the annotations on the right are explained in the four paragraphs that follow. In particular, according to (1) in (iv.2), the code for v_1 and v_2 is duplicated and moved into both branches that follow the condition $k''+1=j''$. According to (2) in (iv.2), in the else-branch, the code for v is duplicated and put inside both branches that follow the additional condition $k''+1=j''-1$, which is copied from the condition in the corresponding previous application $\overline{msub}(i'',j''-1,k'')$; for convenience, the code for v_1 and v_2 is not put inside branches that follow $k''+1=j''-1$, since their transformations are not affected, and simplification at the end can take them back out.

$\overline{msub}(i'',j'',k'')$ — start with invariants in (3),
$=$ — later with $k''=i''$ eliminated

$$\begin{array}{l} \textbf{if } k''+1=j'' \textbf{ then} \\ \quad \textbf{let } v_1 = \overline{m}(i'',k'') \textbf{ in} \\ \quad \textbf{let } v_2 = \overline{m}(k''+1,j'') \textbf{ in} \\ \quad \textbf{let } s = 1st(v_1)+1st(v_2)+p[i''-1]*p[k'']*p[j''] \textbf{ in} \\ \quad < s, v_1, v_2 > \end{array}$$

$$
\begin{array}{ll}
\textbf{else} & \text{context includes: } k''+1 \neq j'' \\
\quad \textbf{let } v_1 = \overline{m}(i'',k'') \textbf{ in} & = <0> \text{ or } 2nd(\overline{r}_1), \text{where the former uses } k''=i'' \\
\quad \textbf{let } v_2 = \overline{m}(k''+1,j'') \textbf{ in} & = \overline{m}'(k''+1,j'',\overline{m}(k''+1,j''-1)) \\
 & = \overline{m}'(k''+1,j'',3rd(\overline{r}_1)) \\
\quad \textbf{let } s = 1st(v_1)+1st(v_2)+p[i''-1]*p[k'']*p[j''] \textbf{ in} & \\
\quad \textbf{if } k''+1 = j''-1 \textbf{ then} & \text{context includes: } k''+1 = j''-1 \\
\qquad \textbf{let } v = \overline{msub}(i'',j'',k''+1) \textbf{ in} & = (4) \text{ below} \\
\qquad < \min(s,1st(v)), v_1, v_2, v > & \\
\quad \textbf{else} & \text{context includes: } k''+1 \neq j'', k''+1 \neq j''-1 \\
\qquad \textbf{let } v = \overline{msub}(i'',j'',k''+1) \textbf{ in} & = \overline{msub}'(i'',j'',k''+1,\overline{msub}(i'',j''-1,k''+1), \\
\qquad < \min(s,1st(v)), v_1, v_2, v > & \qquad \overline{m}(i'',j''-1), \overline{msub}(i'',j''-1,i'')) \\
 & = \overline{msub}'(i'',j'',k''+1,4th(\overline{r}_1),\overline{r}_2,\overline{r}_3)
\end{array}
$$

The first branch gets simplified away, since we have invariant $k'' \leq j''-2$.

In the else-branch, the corresponding previous application $\overline{msub}(i'',j''-1,k'')$ by definition of $\overline{msub}$ has $\overline{m}(i'',k'')$ bound to v_1 and $\overline{m}(k''+1,j''-1)$ bound to v_2, and thus the invariant $\overline{msub}(i'',j''-1,k'') = \overline{r}_1$ implies $\overline{m}(i'',k'') = 2nd(\overline{r}_1)$ and $\overline{m}(k''+1,j''-1) = 3rd(\overline{r}_1)$. The first call $\overline{m}(i'',k'')$ falls in case (i), since we have invariant $k'' = i''$, and equals $<0>$. The second call falls in case (iii) and equals a recursive call to $\overline{m}'$ with arguments $k''+1, j''$ and cache argument $3rd(\overline{r}_1)$, since we have a corresponding invariant $\overline{m}(k''+1,j''-1) = 3rd(\overline{r}_1)$.

In the branch where $k''+1 = j''-1$ is true, the call to $\overline{msub}$ falls in case (i) and equals

$$
\begin{array}{l}
\textbf{let } v_1 = \overline{m}(i'',j''-1) \textbf{ in let } v_2 = \overline{m}(j'',j'') \textbf{ in} \\
\textbf{let } s = 1st(v_1)+1st(v_2)+p[i''-1]*p[k''+1]*p[j''] \textbf{ in } < s, v_1, v_2 >
\end{array}
\qquad (4)
$$

which then equals $<1st(\overline{r}_2)+p[i''-1]*p[k''+1]*p[j''], \overline{r}_2, <0>>$, because the first call equals $\overline{r}_2$, and the second call equals $<0>$.

In the last branch, the call to $\overline{msub}$ falls in case (iii). However, the arguments of this call do not satisfy the invariants corresponding to $k'' = i''$ or corresponding to those on the third and fourth lines of the candidate invariants in (3). Specifically, the invariant corresponding to $k'' = i''$ is $k''+1 = i''$, which is false because the context includes $k'' = i''$; the others can not be maintained because we can not infer that $\overline{msub}(k''+1,j''-1,k''+1)$, $\overline{m}(k''+1,j''-1)$, or $\overline{msub}(k''+1,j''-1,i'')$ equals a retrieval from any cache argument $\overline{r}_1$ to $\overline{r}_6$. So we delete these invariants and retransform $\overline{msub}$. Everything remains the same except that $\overline{m}(i'',k'')$ does not fall in case (i) any more; it falls in case (ii) and equals $2nd(\overline{r}_1)$. We replace this last call to $\overline{msub}$ by a recursive call to $\overline{msub}'$ with arguments $i'',j'',k''+1$ and cache arguments $4th(\overline{r}_1)$, $\overline{r}_2$, $\overline{r}_3$ since we have corresponding invariants $\overline{msub}(i'',j''-1,k''+1) = 4th(\overline{r}_1)$, $\overline{m}(i'',j''-1) = \overline{r}_2$, $\overline{msub}(i'',j''-1,i'') = \overline{r}_3$.

We eliminate unused candidate cache argument $\overline{r}_3$, and we replace the original call $\overline{msub}(i',j',i')$ in (1) with $\overline{msub}'(i',j',i',\overline{r},\overline{r})$ according to (iv.3). We obtain

$$
\begin{array}{ll}
\overline{m}'(i',j',\overline{r}) \triangleq & \textbf{if } i' = j' \textbf{ then } <0> \\
 & \textbf{else if } i' = j'-1 \textbf{ then } < p[i'-1]*p[i']*p[j'], <0>, <0>> \\
 & \textbf{else } \overline{msub}'(i',j',i',\overline{r},\overline{r})
\end{array}
$$

$$
\begin{array}{ll}
\overline{msub}'(i'',j'',k'',\overline{r}_1,\overline{r}_2) \triangleq & \\
 & \textbf{let } v_1 = 2nd(\overline{r}_1) \textbf{ in} \\
 & \textbf{let } v_2 = \overline{m}'(k''+1,j'',3rd(\overline{r}_1)) \textbf{ in} \\
 & \textbf{let } s = 1st(v_1)+1st(v_2)+p[i''-1]*p[k'']*p[j''] \textbf{ in} \\
 & \textbf{if } k''+1 = j''-1 \textbf{ then} \\
 & \qquad \textbf{let } v = <1st(\overline{r}_2)+p[i''-1]*p[k''+1]*p[j''], \overline{r}_2, <0>> \textbf{ in} \\
 & \qquad < \min(s,1st(v)), v_1, v_2, v > \\
 & \textbf{else let } v = \overline{msub}'(i'',j'',k''+1,4th(\overline{r}_1),\overline{r}_2) \textbf{ in} \\
 & \qquad < \min(s,1st(v)), v_1, v_2, v >
\end{array}
$$

Note that for the six invariants about cache arguments, $\overline{r}_1$ to $\overline{r}_6$, in (3), $\overline{r}_4$ to $\overline{r}_6$ can not be maintained at the recursive call and are weakened away; $\overline{r}_3$ can be maintained but is not used and hence is eliminated; and $\overline{r}_1$ and $\overline{r}_2$ can be maintained and are used and hence are kept.

For the resulting program $\overline{m}'$, if $\overline{r} = \overline{m}(i', j'-1)$, then $\overline{m}'(i', j', \overline{r}) = \overline{m}(i', j')$, and $\overline{m}'$ is an exponential factor faster. Function $\overline{m}'$ still takes exponential time due to repeated recursive calls to $\overline{m}'$, since each $\overline{m}'(i', j', ...)$ calls $\overline{m}'(k', j', ...)$ for all k from $i'+1$ to $j'-1$. Incrementalizing the resulting optimized program $\overline{m}(i,j)$ obtained from Step 3 under $\langle i', j' \rangle = \langle i-1, j \rangle$ yields a quadratic-time incremental program that involves no repeated recursive calls.

5 Step 3: Pruning Unnecessary Values

The first component of $\overline{f_0}'(x', \overline{r})$ is the return value of $f_0(x')$. Components of $\overline{r}$ that are not useful for computing this value need not be cached and maintained. We prune the programs $\overline{f_0}$ and $\overline{f_0}'$ to obtain a program $\widehat{f_0}$ that caches only the useful values and a program $\widehat{f_0}'$ that uses and maintains only the useful values. Finally, we form an optimized program that computes f_0 by using the base cases in $\widehat{f_0}$ and by repeatedly using the incremental version $\widehat{f_0}'$.

5.1 Pruning

Pruning requires a dependence analysis that can precisely describe substructures of trees [38]. We use an analysis based on regular tree grammars [31, 35]. We have designed and implemented a simple algorithm that uses regular tree grammar based constraints to efficiently produce precise analysis results [33, 35]. Pruning can save space and time and reduce code size.

For example, in program $\overline{c}'$, only the third component of $\overline{r}$ is useful. Pruning the second and fourth components of $\overline{c}$ and $\overline{c}'$, which makes the third component become the second component, and doing a few simplifications, which transform $1st(\overline{c})$ back to c and unwind bindings for v_1 and v_3, we obtain $\widehat{c}$ and $\widehat{c}'$ below:

$$
\begin{array}{ll}
\widehat{c}(i,j) \triangleq & \textbf{if } i=0 \lor j=0 \textbf{ then } <0> \\
& \textbf{else let } v_2 = \widehat{c}(i, j-1) \textbf{ in} \\
& \quad \textbf{if } x[i] = y[j] \textbf{ then } < c(i-1, j-1)) + 1, v_2 > \\
& \quad \textbf{else } < \max(1st(v_2), c(i-1, j))), v_2 > \\
\widehat{c}'(i', j', \widehat{r}) \triangleq & \textbf{if } i'=0 \lor j'=0 \textbf{ then } <0> \\
& \textbf{else if } i'-1=0 \textbf{ then} \\
& \quad \textbf{let } v_2 = \widehat{c}'(i', j'-1, <0>) \textbf{ in} \\
& \quad \textbf{if } x[i'] = y[j'] \textbf{ then } <1, v_2> \\
& \quad \textbf{else } < 1st(v_2), v_2 > \\
& \textbf{else let } v_2 = \widehat{c}'(i', j'-1, 2nd(\widehat{r})) \textbf{ in} \\
& \quad \textbf{if } x[i'] = y[j'] \textbf{ then } < 1st(2nd(\widehat{r})) + 1, v_2 > \\
& \quad \textbf{else } < \max(1st(v_2), 1st(\widehat{r})), v_2 >
\end{array}
$$

It is easy to see that $\widehat{c}'$, like $\overline{c}'$, also takes time and space linear in j, but each call to $\widehat{c}'$ creates a tuple of length no more than 2, compared to 4 for $\overline{c}'$.

Pruning leaves programs $\overline{m}$ and $\overline{m}'$ unchanged. We obtain the same programs $\widehat{m}$ and $\widehat{m}'$, respectively.

The first components of these functions remain unchanged, so we have $m(i,j) = 1st(\widehat{m}(i,j))$ and $c(i,j) = 1st(\widehat{c}(i,j))$.

5.2 Forming Optimized Programs

We redefine functions f_0 and $\widehat{f_0}$ and use function $\widehat{f_0}'$:

$$f_0(x) \triangleq 1st(\widehat{f_0}(x))$$
$$\widehat{f_0}(x) \triangleq \textbf{if } base\_cond(x) \textbf{ then } base\_val(x) \textbf{ else let } \widehat{r} = \widehat{f_0}(prev(x)) \textbf{ in } \widehat{f_0}'(x, \widehat{r})$$

where *base_cond* is the base-case condition, and *base_val* is the corresponding value, both copied from the original definition of $\widehat{f_0}$ obtained by pruning. This new definition of $\widehat{f_0}$ is called the optimized $\widehat{f_0}$. In general, there may be multiple base cases, and we just list them all; to be conservative, we may include here all cases not containing multiple recursive calls.

For examples c and m, we obtain directly

$$c(i,j) \triangleq 1st(\widehat{c}\,(i,j))$$
$$\widehat{c}\,(i,j) \triangleq \textbf{if } i=0 \ \vee j=0 \textbf{ then } <0> \textbf{ else let } \widehat{r} = \widehat{c}\,(i-1,j) \textbf{ in } \widehat{c}\,'(i,j,\widehat{r})$$

$$m(i,j) \triangleq 1st(\widehat{m}(i,j))$$
$$\widehat{m}(i,j) \triangleq \textbf{if } i=j \textbf{ then } <0> \textbf{ else let } \widehat{r} = \widehat{m}(i,j-1) \textbf{ in } \widehat{m}\,'(i,j,\widehat{r})$$

It is easy to see that $\widehat{c}\,(i,j)$ takes $O(i*j)$ time, since it only calls $\widehat{c}\,(i-1,j)$ recursively, and $i=0$ is a base case; each call to $\widehat{c}$ calls $\widehat{c}\,'$, and $\widehat{c}\,'$ takes $O(j)$ time. Thus, for $c(n,m)$, while the original program takes $O(2^{n+m})$ time, the optimized program takes $O(n*m)$ time. For $m(1,n)$, while the original program takes $O(n*3^n)$ time, the optimized program takes $O(n^2*2^n)$ time. Incrementalizing the optimized program again under the increment to the other parameter allows us to obtain a final optimized program that takes $O(n^3)$ time; the time complexity of the final optimized program is also easy to analyze.

The precise exponential complexities are not as easy to see, but for the purpose of our optimization, it is sufficient to know that they are exponential due to repeated recursive calls.

6 Summary and Discussion

To summarize, we emphasize that it is the incremental computation under step $\oplus$ that determines appropriate values to cache so as to avoid repeated subcomputations. It yields a kind of regularity, in particular linearity, that we think is important for efficient computation.

Correctness. The transformations for caching, incrementalization, and pruning together preserve semantics in the sense that if the original $\widehat{f_0}(x')$ terminates with a value, then the incremental program $\widehat{f_0}'(x',\widehat{r})$, given $\widehat{r} = \widehat{f_0}(prev(x'))$, terminates with the same value and is asymptotically at least as fast. This is because each transformation preserves semantics except that unfolding function definitions and eliminating unused values may make the resulting program terminate more often. Possible problems associated with hoisting causing $\widehat{f_0}$ to compute values not computed in the original program f_0 are avoided as described in Section 3.1 and below. Forming optimized programs is straightforward for all the problems we have encountered, and it is easy to see that the resulting programs are correct, but a rigorous and general correctness argument for this needs further research. Overall, these transformations together preserve semantics in the sense that if the original program terminates with a value, then the optimized program terminates with the same value.

Mechanization. Our method for dynamic programming is composed completely of static program analyses and transformations and is systematic. It is based on a general approach for program optimization—incrementalization—which helps it to be systematic. The analyses and transformations used for caching and pruning are fully automatic and highly efficient [35, 38]. The analyses and transformations for incrementalization are fully automatic modulo the simplifications and equality reasoning used for establishing and using invariants. Such simplifications and equalities needed for all the problems we have encountered involve only Presburger arithmetic [51] and simple facts about recursive data structures and thus can be fully automated; for the same reason, determining input increment operations can also be fully automated. Also, as we have seen, forming optimized programs is straightforward to automate. Characterizing the exact class of problems to which these automated techniques apply needs further study.

Monovariance. Although our static incrementalization allows only one incremental version for each original function, i.e., is monovariant, it is still powerful enough to incrementalize all examples in [37,38,40], including various list manipulations, matrix computations, attribute evaluation, and graph problems. In general, while monovariant analyses and transformations are usually simpler and more efficient, they might not be sufficiently powerful when a function is used in multiple contexts for different roles. To overcome this potential problem, we propose to introduce a new function for each function composition that appears in the original program. This is based on the observation that different roles of a function are usually played in its composition with other functions. This method does not involve creating copies of existing functions, as is usually done but often causes code blowup and needs additional heuristics. We believe that the method based on static incrementalization can achieve dynamic programming for all problems whose solutions involve recursively solving subproblems that overlap, but a formal justification awaits more rigorous study.

Space usage and data structures. In our method, only values that are necessary for the incrementalization are stored, in appropriate data structures. For the longest-common-subsequence example, only a linear list is needed, whereas in standard textbooks, a quadratic two-dimensional array is used, and an additional optimization is needed to reduce it to a one-dimensional array [15]. For the matrix-chain multiplication example, our optimized program uses a list of lists that forms a triangle shape, rather than a two-dimensional array of square shape. It is nontrivial to see that recursive data structures give the same asymptotic speedup as arrays for some examples. Our recent work on transforming recursion into iteration can help eliminate the linear stack space used [36]. There are dynamic programming problems, e.g., 0-1 knapsack, for which the use of an array, with constant-time access to elements, helps achieve desired asymptotic speedups. Such situations become evident when doing incrementalization and can be accommodated easily, as is described in [39]. Although we present the optimizations for a functional language, the underlying principle is general and has been applied to programs that use loops and arrays [30,34].

Auxiliary information. Some values computed in a hoisted program might not be computed by the original program and are therefore called *auxiliary information* [37]. Both incrementalization and pruning produce programs that are at least as fast as the given program, but caching auxiliary information may result in a slower program on certain inputs. We can determine statically whether such information is cached in the final program. If so, we can use time and space analysis [32,55,60–62] to determine conservatively whether it is worthwhile to use and maintain such information. The trade-off between time and space is an open problem for future study.

Additional properties. Many dynamic programming algorithms can be further improved by exploiting additional properties, such as greedy properties [7], of the given problems. For example, Hu and Shing [22,23] give an $O(n * \log n)$-time algorithm for the matrix-chain multiplication problem. Our method is not aimed at discovering such properties. Nevertheless, it can help preserve such properties once they are added. For example, for the paragraph-formatting problem [15,18], we can derive a quadratic-time algorithm that uses dynamic programming; if the original program contains a simple extra conditional that follows from a kind of thinning property, our derived dynamic programming program uses it as well and takes linear time with a factor of line width. How to systematically discover and use such additional properties in general is a subject for future study.

7 Implementation and Experimentation Results

All three steps—caching, incrementalization, and pruning—have been implemented in a prototype system, CACHET, using the Synthesizer Generator [54]. Incrementalization as currently implemented is semiautomatic [29] and is being automated [64]. Determining input increment operations and forming optimized programs are currently done manually, but both are straightforward for all the problems we have encountered.

Figure 3 summarizes some of the examples derived, most of them semiautomatically and some automatically. The second column shows whether more than one cache argument is needed in an incremental program. The third column shows whether the incremental program computes values not necessarily computed by the original program. For the last two examples, the letter "a" in the third column shows that cached values are stored in arrays. The last two columns compare the asymptotic running times of the original programs and the optimized programs. For Fibonacci function, n is the input number, rather than the size of the input, and the running time is the numbers of additions performed. The matrix-chain multiplication, optimal binary search tree, and optimal polygon triangulation problems have similar control structures for recursive calls, which is reflected in the running times; yet, the optimal costs for these problems are computed in different ways, and our general method handles all of them in the same systematic manner. Paragraph formatting 2 [18] includes a conditional that reflects a greedy property, as described in Section 6.

Examples	multiple cache arg	aux info	original program's running time	optimized prog's running time
Fibonacci function [46]			$O(2^n)$	$O(n)$
binomial coefficients [46]			$O(2^n)$	$O(n*k)$
longest common subsequence [15]		√	$O(2^{n+m})$	$O(n*m)$
matrix-chain multiplication [15]	√		$O(n*3^n)$	$O(n^3)$
string editing distance [53]			$O(3^{n+m})$	$O(n*m)$
dag path sequence [6]		√	$O(2^n)$	$O(n^2)$
optimal polygon triangulation [15]	√		$O(n*3^n)$	$O(n^3)$
optimal binary search tree [2]	√		$O(n*3^n)$	$O(n^3)$
paragraph formatting [15]	√		$O(n*2^n)$	$O(n^2)$
paragraph formatting 2	√		$O(n*2^n)$	$O(n*width)$
0-1 knapsack [15]		√a	$O(2^n)$	$O(n*weight)$
context-free-grammar parsing [2]	√	√a	$O(n*(2*size+1)^n)$	$O(n^3*size)$

Fig. 3. Summary of examples.

To measure and compare the actual running times, we translated some of the programs into Java. The translation is straightforward, where tuples are implemented using class `Vector`. Other languages could be used, but Java is particularly good for testing whether caching additional information could incur a significant overhead in running time on small input, since operations on objects of the `Vector` class in Java are relatively expensive compared with operations on constructed data in languages such as ML, Scheme, or C.

Figure 4 presents the running times of the straightforward programs and the optimized programs for some examples; results for other examples are similar. Stars indicate running times longer than 48 hours. These measurements were taken for programs compiled with Sun JDK 1.0.2 and running on a Sun Ultra 10 with 300 MHz UltraSPARC-IIi CPU and 128 MB main memory. One can see that the optimized programs run much faster than the straightforward programs even on small inputs.

8 Related Work and Conclusion

Dynamic programming was first formulated by Bellman [4], where "programming" refers to the use of tabular solution method, and has been studied extensively since [59]. Bird [5], de Moor [17], and others have studied it in the context of program transformation. While some work addresses the derivation of recursive equations, including the original work by Bellman [4] and the later work by Smith [58], our work addresses the derivation of efficient programs that use tabulation. Previous methods for this problem either apply to specific

input	binomial coefficients		longest comm. subseq.		matrix-chain multipl.		paragraph formatting	
size	original	optim.	original	optim.	original	optim.	original	optim.
10	0	0	10	5	42	16	10	5
15	3	1	185	7	7453	23	111	7
20	305	2	36250	9	3282564	75	3712	10
25	4924	3	1454555	16	*******	180	123360	16
40	2436874	6	*******	60	*******	752	******	68
60	*******	11	*******	113	*******	2617	******	157
80	*******	15	*******	211	*******	6187	******	325
100	*******	20	*******	383	*******	11846	******	714

Fig. 4. Running times of original programs and optimized programs (in milliseconds).

subclasses of problems [11, 13, 14, 24, 47, 49] or give general frameworks or strategies rather than precise derivation algorithms [3, 5, 6, 8, 9, 16, 17, 46, 48, 56, 57, 63]. Our work is based on the general principle of incrementalization [37, 45] and consists of precise program analyses and transformations.

In particular, tupling [9, 47, 48] aims to compute multiple values together in an efficient way. It is improved to be automatic on subclasses of problems [11] and to work on more general forms [13]. It is also extended to store lists of values [49], but such lists are generated in a fixed way, which is not the most appropriate way for many programs. A special form of tupling can eliminate multiple data traversals for many functions [24]. A method specialized for introducing arrays was proposed for tabulation [12], but as our method has shown, arrays are not essential for the speedup of many programs. Also, that method relies on explicit coding of the appropriate increment operation in the original function; it is not clear how to code the original function when multiple increment operations are needed, such as for the matrix chain multiplication problem. To summarize, no previous method can perform all the powerful optimizations our method can. Each of our examples is nontrivial and requires advanced algorithm design discipline to derive even by hand.

Compared with our previous work for incrementalizing functional programs [37, 38, 40], this work contains several substantial improvements. First, our previous work addresses the systematic derivation of an incremental program f' given both program f and operation $\oplus$. This paper describes a systematic method for identifying an appropriate operation $\oplus$ given a function f and forming an optimized version of f using the derived incremental program f'. Second, since it is difficult to introduce appropriate cache arguments, our previous method allows at most one cache argument for each incremental function. This paper allows multiple cache arguments, without which many programs could not be incrementalized, e.g., the matrix-chain multiplication program. Third, our previous method introduces incremental functions using an on-line strategy, i.e., on-the-fly during the transformation, so it may attempt to introduce an unbounded number of new functions and thus not terminate. The algorithm in this paper introduces one incremental function for each function in the original program, i.e., it is monovariant; even though it is theoretically more limited, it is simpler, always terminates, and is able to incrementalize all previous examples. Finally, based on the idea of cache-and-prune [38], the method in this paper uses hoisting to extend the set of intermediate results [38] to include a kind of auxiliary information [37] that is sufficient for dynamic programming. This method is simpler than our previous general method for discovering auxiliary information [37]. Additionally, we now use a more precise and efficient dependence analysis for pruning [35].

Finite differencing [43–45] is based on the same underlying principle as incremental computation. Fifteen years ago, Paige explicitly asked whether finite differencing can be generalized to handle dynamic programming [43]; it is clear that he perceived an important connection. However, finite differencing has been formulated for set expressions in while loops [45], which can be obtained from fixed-point specifications [10], while straightforward solutions to dynamic programming problems are usually formulated as recursive functions, so it has been difficult to establish the exact connection. A major open problem is how to transform general recursive functions into set expressions extended with fixed-point operations [10].

Overall, being able to incrementalize complicated recursion in a general and systematic way is a substantial improvement complementing previous methods for incrementalizing loops [30, 45]. Our new method based on static incrementalization is both powerful and automatable. Based on our existing implementation, we believe that a complete system will perform incrementalization efficiently.

Acknowledgments

The authors would like to thank the anonymous referees for many helpful comments and suggestions.

References

1. M. Abadi, B. Lampson, and J.-J. Lévy: Analysis and caching of dependencies. In *Proceedings of the 1996 ACM Sigplan International Conference on Functional Programming*, 83–91. ACM, New York, 1996.
2. A. V. Aho, J. E. Hopcroft, and J. D. Ullman: *The Design and Analysis of Computer Algorithms.* Addison-Wesley, Reading, MA, 1974.
3. F. L. Bauer, B. Möller, H. Partsch, and P. Pepper: Formal program construction by transformations—Computer-aided, intuition-guided programming. *IEEE Trans. Softw. Eng.*, 15(2):165–180, Feb. 1989.
4. R. E. Bellman: *Dynamic Programming.* Princeton University Press, Princeton, NJ, 1957.
5. R. S. Bird: Tabulation techniques for recursive programs. *ACM Comput. Surv.*, 12(4):403–417, Dec. 1980.
6. R. S. Bird: The promotion and accumulation strategies in transformational programming. *ACM Trans. Program. Lang. Syst.*, 6(4):487–504, Oct. 1984.
7. R. S. Bird and O. de Moor: From dynamic programming to greedy algorithms. In B. Möller, H. Partsch, and S. Schuman, editors, *Formal Program Development*, volume 755 of *Lecture Notes in Computer Science*, 43–61. Springer, Berlin, 1993.
8. E. A. Boiten: Improving recursive functions by inverting the order of evaluation. *Sci. Comput. Program.*, 18(2):139–179, Apr. 1992.
9. R. M. Burstall and J. Darlington: A transformation system for developing recursive programs. *J. ACM*, 24(1):44–67, Jan. 1977.
10. J. Cai and R. Paige: Program derivation by fixed point computation. *Science of Computer Programming*, 11:3, 197–261, 1989.
11. W.-N. Chin: Towards an automated tupling strategy. In *Proceedings of the ACM Sigplan Symposium on Partial Evaluation and Semantics-Based Program Manipulation*, 119–132. ACM, New York, 1993.
12. W.-N. Chin and M. Hagiya: A bounds inference method for vector-based memoization. In ICFP 1997 [26], 176–187, 1997.
13. W.-N. Chin and S.-C. Khoo: Tupling functions with multiple recursion parameters. In P. Cousot, M. Falaschi, G. Filè, and A. Rauzy, editors, *Proceedings of the 3rd International Workshop on Static Analysis*, volume 724 of *Lecture Notes in Computer Science*, 124–140. Springer, Berlin, Sept. 1993.
14. N. H. Cohen: Eliminating redundant recursive calls. *ACM Trans. Program. Lang. Syst.*, 5(3):265–299, July 1983.
15. T. H. Cormen, C. E. Leiserson, and R. L. Rivest: *Introduction to Algorithms.* The MIT Press/McGraw-Hill, 1990.

16. S. Curtis: Dynamic programming: A different perspective. In R. Bird and L. Meertens, editors, *Algorithmic Languages and Calculi*, 1–23. Chapman & Hall, London, UK, 1997.
17. O. de Moor: A generic program for sequential decision processes. In M. Hermenegildo and D. S. Swierstra, editors, *Programming Languages: Implementations, Logics, and Programs*, volume 982 of *Lecture Notes in Computer Science*, 1–23. Springer, Berlin, 1995.
18. O. de Moor and J. Gibbons: Bridging the algorithm gap: A linear-time functional program for paragraph formatting. Technical Report CMS-TR-97-03, School of Computing and Mathematical Sciences, Oxford Brookes University, July 1997.
19. J. Field and T. Teitelbaum: Incremental reduction in the lambda calculus. In *Proceedings of the 1990 ACM Conference on LISP and Functional Programming*, 307–322. ACM, New York, 1990.
20. D. P. Friedman, D. S. Wise, and M. Wand: Recursive programming through table look-up. In *Proceedings of the 1976 ACM Symposium on Symbolic and Algebraic Computation*, 85–89. ACM, New York, 1976.
21. Y. Futamura and K. Nogi: Generalized partial evaluation. In B. Bjørner, A. P. Ershov, and N. D. Jones, editors, *Partial Evaluation and Mixed Computation*, 133–151. North-Holland, Amsterdam, 1988.
22. T. C. Hu and M. T. Shing: Computation of matrix chain products. Part I. *SIAM J. Comput.*, 11(2):362–373, 1982.
23. T. C. Hu and M. T. Shing: Computation of matrix chain products. Part II. *SIAM J. Comput.*, 13(2):228–251, 1984.
24. Z. Hu, H. Iwasaki, M. Takeichi, and A. Takano: Tupling calculation eliminates multiple data traversals. In ICFP 1997 [26], 164–175, 1997.
25. J. Hughes: Lazy memo-functions. In *Proceedings of the 2nd Conference on Functional Programming Languages and Computer Architecture*, volume 201 of *Lecture Notes in Computer Science*, 129–146. Springer, Berlin, Sept. 1985.
26. *Proceedings of the 1997 ACM Sigplan International Conference on Functional Programming*. ACM, New York, 1997.
27. R. M. Keller and M. R. Sleep: Applicative caching. *ACM Trans. Program. Lang. Syst.*, 8(1):88–108, Jan. 1986.
28. H. Khoshnevisan: Efficient memo-table management strategies. *Acta Informatica*, 28(1):43–81, 1990.
29. Y. A. Liu: CACHET: An interactive, incremental-attribution-based program transformation system for deriving incremental programs. In *Proceedings of the 10th IEEE Knowledge-Based Software Engineering Conference*, 19–26. IEEE CS Press, Los Alamitos, CA, 1995.
30. Y. A. Liu: Principled strength reduction. In R. Bird and L. Meertens, editors, *Algorithmic Languages and Calculi*, 357–381. Chapman & Hall, London, UK, 1997.
31. Y. A. Liu: Dependence analysis for recursive data. In *Proceedings of the IEEE 1998 International Conference on Computer Languages*, 206–215. IEEE CS Press, Los Alamitos, CA, 1998.
32. Y. A. Liu and G. Gómez: Automatic accurate cost-bound analysis for high-level languages. *IEEE Transctions on Computers*, 50(12):1295–1309, Dec. 2001.
33. Y. A. Liu, N. Li, and S. D. Stoller: Solving regular tree grammar based constraints. In *Proceedings of the 8th International Static Analysis Symposium*, volume 2126 of *Lecture Notes in Computer Science*, 213–233. Springer, Berlin, 2001.
34. Y. A. Liu and S. D. Stoller: Loop optimization for aggregate array computations. In *Proceedings of the IEEE 1998 International Conference on Computer Languages*, 262–271. IEEE CS Press, Los Alamitos, CA, 1998.
35. Y. A. Liu and S. D. Stoller: Eliminating dead code on recursive data. In *Proceedings of the 6th International Static Analysis Symposium*, volume 1694 of *Lecture Notes in Computer Science*, 211–231. Springer, Berlin, 1999.

36. Y. A. Liu and S. D. Stoller: From recursion to iteration: what are the optimizations? In *Proceedings of the ACM Sigplan 2000 Workshop on Partial Evaluation and Semantics-Based Program Manipulation*, 73–82. ACM, New York, 2000.
37. Y. A. Liu, S. D. Stoller, and T. Teitelbaum: Discovering auxiliary information for incremental computation. In *Conference Record of the* 23*rd Annual ACM Symposium on Principles of Programming Languages*, 157–170. ACM, New York, 1996.
38. Y. A. Liu, S. D. Stoller, and T. Teitelbaum: Static caching for incremental computation. *ACM Trans. Program. Lang. Syst.*, 20(3):546–585, May 1998.
39. Y. A. Liu and S. D. Stoller. Program optimization using indexed and recursive data structures. In *Proceedings of the ACM Sigplan 2002 Workshop on Partial Evaluation and Semantics-Based Program Manipulation*, 108–118. ACM, New York, 2002.
40. Y. A. Liu and T. Teitelbaum: Systematic derivation of incremental programs. *Sci. Comput. Program.*, 24(1):1–39, Feb. 1995.
41. D. Michie: "memo" functions and machine learning. *Nature*, 218:19–22, Apr. 1968.
42. D. J. Mostow and D. Cohen: Automating program speedup by deciding what to cache. In *Proceedings of the* 9*th International Joint Conference on Artificial Intelligence*, 165–172. Morgan Kaufmann Publishers, San Francisco, CA, Aug. 1985.
43. R. Paige: Programming with invariants. *IEEE Software*, 3(1):56–69, Jan. 1986.
44. R. Paige: Symbolic finite differencing—Part I. In N. D. Jones, editor, *Proceedings of the* 3*rd European Symposium on Programming*, volume 432 of *Lecture Notes in Computer Science*, 36–56. Springer, Berlin, 1990.
45. R. Paige and S. Koenig: Finite differencing of computable expressions. *ACM Trans. Program. Lang. Syst.*, 4(3):402–454, July 1982.
46. H. A. Partsch: *Specification and Transformation of Programs—A Formal Approach to Software Development.* Springer, Berlin, 1990.
47. A. Pettorossi: A powerful strategy for deriving efficient programs by transformation. In *Conference Record of the 1984 ACM Symposium on LISP and Functional Programming.* ACM, New York, 1984.
48. A. Pettorossi and M. Proietti: Rules and strategies for transforming functional and logic programs. *ACM Comput. Surv.*, 28(2):360–414, June 1996.
49. A. Pettorossi and M. Proietti: Program derivation via list introduction. In R. Bird and L. Meertens, editors, *Algorithmic Languages and Calculi.* Chapman & Hall, London, UK, 1997.
50. W. Pugh: An improved cache replacement strategy for function caching. In *Proceedings of the 1988 ACM Conference on LISP and Functional Programming*, 269–276. ACM, New York, 1988.
51. W. Pugh: The Omega Test: A fast and practical integer programming algorithm for dependence analysis. *Commun. ACM*, 31(8):102–114, Aug. 1992.
52. W. Pugh and T. Teitelbaum: Incremental computation via function caching. In *Conference Record of the* 16*th Annual ACM Symposium on Principles of Programming Languages*, 315–328. ACM, New York, 1989.
53. P. W. Purdom and C. A. Brown: *The Analysis of Algorithms.* Holt, Rinehart, and Winston, 1985.
54. T. Reps and T. Teitelbaum: *The Synthesizer Generator: A System for Constructing Language-Based Editors.* Springer-Verlag, New York, 1988.
55. M. Rosendahl: Automatic complexity analysis. In *Proceedings of the* 4*th International Conference on Functional Programming Languages and Computer Architecture*, 144–156. ACM, New York, 1989.
56. W. L. Scherlis: Program improvement by internal specialization. In *Conference Record of the* 8*th Annual ACM Symposium on Principles of Programming Languages*, 41–49. ACM, New York, 1981.
57. D. R. Smith: KIDS: A semiautomatic program development system. *IEEE Trans. Softw. Eng.*, 16(9):1024–1043, Sept. 1990.

58. D. R. Smith: Structure and design of problem reduction generators. In B. Möller, editor, *Constructing Programs from Specifications*, 91–124. North-Holland, Amsterdam, 1991.
59. M. Sniedovich: *Dynamic Programming*. Marcel Dekker, New York, 1992.
60. L. Unnikrishnan, S. D. Stoller, and Y. A. Liu: Automatic accurate stack space and heap space analysis for high-level languages. Technical Report TR 538, Computer Science Department, Indiana University, Apr. 2000.
61. L. Unnikrishnan, S. D. Stoller, and Y. A. Liu: Automatic accurate live memory analysis for garbage-collected languages. In *Proceedings of the ACM Sigplan 2001 Workshop on Languages, Compilers, and Tools for Embedded Systems*, 102–111. ACM, New York, 2001.
62. B. Wegbreit: Mechanical program analysis. *Commun. ACM*, 18(9):528–538, Sept. 1975.
63. B. Wegbreit: Goal-directed program transformation. *IEEE Trans. Softw. Eng.*, SE-2(2):69–80, June 1976.
64. Y. Zhang and Y. A. Liu: Automating derivation of incremental programs. In *Proceedings of the 1998 ACM Sigplan International Conference on Functional Programming*, page 350. ACM, New York, 1998.

Automatic Program Generation from Formal Specifications using APTS

Elizabeth I. Leonard and Constance L. Heitmeyer*

Center for High Assurance Computer Systems, Naval Research Laboratory, Code 5546, Washington, DC 20375, USA.
{leonard,heitmeyer}@itd.nrl.navy.mil

Summary. A promising trend in software practice is the increasing adoption of model-driven design. In this approach, a developer first constructs an abstract specification of the required program behavior in a language, such as Statecharts, Stateflow, or LUSTRE, and then uses a code generator to automatically translate the specification into an executable program. This approach has major advantages over more traditional approaches. First, because a specification is more concise, it is usually more understandable than code, and hence manual inspections can detect more errors in specifications than in code. Second, a specification is also more amenable to both user validation (e.g., via simulation) and formal verification, which together provide high confidence that the specification captures the desired behavior. Finally, the automatic generation of source code usually produces software with fewer errors than handcrafted code. This paper describes a case study in which Bob Paige's program transformation system APTS was used to produce code generators that construct C source code from a requirements specification in the SCR (Software Cost Reduction) tabular notation. Two different code generation strategies were explored. The first strategy uses rewrite rules to transform the parse tree of an SCR specification into a parse tree for the corresponding C code. The second strategy associates a relation with each node of the specification parse tree. Each member of this relation acts as an attribute, holding the C code corresponding to the tree at the associated node; the root of the tree has the entire C program as its member of the relation. This paper describes the two code generators, how each was used to synthesize code for two example SCR requirements specifications, and lessons learned about APTS from the case study.

Keywords: automatic program derivation, formal specifications, code generation strategies, program transformation, program validation.

1 Introduction

In developing complex software, a formal specification of the system's required behavior can be extremely useful. Such a specification can be (1) formally verified to show that it satisfies critical properties and (2) validated using simulation to show that it captures the intended system behavior. Additionally, because specifications contain much less detail than programs, users find specifications easier to understand than code and hence find more errors in specifications than in code.

Using both inspections and software analysis tools, users can develop high confidence that a formal specification is correct. Unfortunately, high assurance in the correctness of the specification does not guarantee that the system implementation is correct, since the implementation is often separately developed with no formal link to the specification. In such cases, some confidence in the correctness of the actual code can be achieved by testing, but that confidence is only as good as the tests that are applied. One way to transfer high confidence in the specification to the implementation is to automatically generate the program

* This research was funded by the Office of Naval Research.

O. Danvy et al. (eds.), Automatic Program Development – A Tribute to Robert Paige, 93–113.

code from the specification, thus eliminating errors introduced by hand-coding. Such an approach is a promising and growing trend in current software practice: Increasingly, software developers are writing specifications in languages such as Statecharts [14], LUSTRE [12], and Mathwork's Stateflow [8] and then using automatic code generators to transform these specifications into executable code.

A program transformation system such as Cai and Paige's APTS [6, 37, 38] can be extremely useful in developing an automatic code generator. APTS, which is implemented in SETL2 [45], is an improved version of Paige's earlier RAPTS system [35]. It includes a syntax analyzer, which uses a given grammar to parse a specification; a relational database, which accumulates information about the specification needed during code generation; and a transformation engine. Code can be generated either in the relational database or via the transformation engine. APTS also includes optimization techniques, such as *finite differencing* [35, 36], a technique for optimizing code by replacing frequently repeated calculations with less expensive incremental updates. Cai and Paige used APTS [6] to implement translators from SETL2 [45] and SQ2+ [5] to C.

Unlike Statecharts, Stateflow, and LUSTRE, which were developed to capture software designs, SCR (Software Cost Reduction) is a language for specifying software *requirements*, i.e., the required externally visible behavior of a software system. The SCR language and method have been applied successfully by many organizations in industry and government (e.g., Bell Laboratories [23], Grumman [33], Lockheed [10], the Naval Research Laboratory [19, 29], Ontario Hydro [39], and Rockwell Aviation [34]) to develop and analyze specifications of practical systems, including flight control systems [10, 34], weapons systems [19], space systems [9], and cryptographic devices [29]. The SCR toolset [18] provides a user-friendly interface for writing requirements specifications in the SCR tabular notation and several analysis tools, including a consistency checker [21], simulator [20], model checker [19], theorem prover [3], and invariant generator [24, 28]. A context-free grammar is the underlying communication medium for the different tools. By applying the SCR tools, a user can develop high confidence that a specification is a correct statement of the required behavior.

Once an SCR specification has been formally verified and validated via simulation, a logical next step is to automatically construct executable code from the SCR specification. This paper describes a case study in which APTS was used to generate C code from SCR requirements specifications. Using APTS, two code generation strategies were investigated. The first strategy, which uses only the APTS relational database, applies rewrite rules to transform code in the source language into target language code. The source language used for these experiments was the SCR specification language and the target language was C. The transformations modify the parse tree of the requirements specification, replacing each node containing SCR code with a new node containing the corresponding C code. The second strategy, which used both the APTS relational database and its transformation engine, treats the code to be generated as a synthesized attribute of the parse tree of the SCR specification. A relation is developed that associates target language code with each node in the parse tree of the source language program. In the relational model, the code associated with a given node is formed by combining the code for the node's children with additional code specific for the node and the result is stored in a relation. In both strategies, auxiliary information (e.g., variable dependencies) is stored in relations. Frequently a node of the parse tree is the key for the relation. In these cases, the relations are analogous to the attributes in an attribute grammar [30].

The use in APTS of a relational database to store attribute information has important advantages over traditional attribute grammars. In APTS, information can be passed directly from one node to any other node via pattern matching in the relational database, while in an attribute grammar, information must flow along a path through the tree. Although any attribute grammar can be augmented with auxiliary data structures to hold inherited attributes for lookup, the relational database used by APTS is an integral part of the system, holding all attribute information. Also, APTS allows relations to be defined over domains other than nodes of the parse tree. For example, the tree domain allows attribute information to be assigned to all nodes representing the same term. Such a tree relation can be used to assign the same attribute value to all occurrences of a variable at the same time. In contrast, in a traditional attribute grammar, this information must be passed around the tree.

Our study used APTS because its flexible framework allowed us to experiment with the two different code generation strategies described above. Because both approaches required common information about an SCR specification, we were able to reuse a large portion of the first generator in the implementation of the second. Similar experiments could have been conducted with other systems, such as REFINE [41] or the Synthesizer Generator [42], but both rely on attribute grammars and thus suffer from the restrictions on information flow and the absence of the built-in auxiliary data structures described above. An additional advantage of APTS over the Synthesizer Generator is that program transformations in APTS are automatic rather than interactive as in the Synthesizer Generator and may be conditioned on information stored in the relations. However, the primary advantage of APTS over other systems is its built-in optimization capabilities. In our case study, we planned to make use of the APTS finite differencing engine, but the lack of adequate documentation for the finite differencing engine prevented us from doing so.

This paper is organized as follows. Section 2 reviews APTS and SCR and describes the C code that can be generated from SCR specifications. Section 3 describes in more detail the two strategies for generating C code described above, one based on rewrite rules and the second based on the accumulating relation approach. Section 4 describes the results of applying the code generators to an SCR requirements specification of a cryptographic device [29]. It also compares the two strategies, discusses what we learned by implementing them in APTS, and describes our recent research on generating optimized code from SCR specifications [43]. Section 5 discusses related work. Finally, Section 6 presents some conclusions and describes our plans for future work.

2 Background

2.1 APTS

From a given grammar, the APTS syntax analyzer builds parse trees for input files. Each node of the parse tree forms the root of a subtree corresponding to the SCR syntax for that portion of the specification. The grammar used to specify SCR in APTS is described in [31].

The relational database is built using a set of inference rules. These rules usually consist of three parts: a pattern to match a portion of the parse tree, a set of conditions (logical combinations of relations) under which a new member of a relation is added to the database, and a result portion stating which new member to add to a relation in the database. Pattern matching variables, whose names in APTS begin with a dot, are used to correlate portions of the pattern matching condition with the relations appearing in the rule's condition and result portions or to correlate information contained in different relations in the rule. For example, the following rule states that if an expression composed of two subexpressions separated by a ">" is found in the parse tree such that the two subexpressions are members of the **arithexpr** relation, then the matched expression (denoted by `loc()`) is added to the relation **boolexpr**:

```
match(%expr, .x > .y%)| arithexpr(.x) and arithexpr(.y)
  -> boolexpr(loc());\vspace*{-6pt}
```

Relations are predicates whose arguments can be taken from several possible domains. Frequently, nodes in the parse tree serve as keys for relations. Thus, these relations may be viewed as attributes of the parse tree. Since there are no restrictions on which relations can be in the condition of an inference rule for a relation, both synthesized and inherited attributes can be defined in APTS in terms of relations. However, not every relation in the implementation need be a parse tree attribute. For example, in SCR specifications, the value of a variable may depend on the current value of another variable. During execution of the corresponding C code, the values of variables must be updated in an order that respects these dependencies. Thus, this dependency information is necessary for code generation. In the implementations, the dependency relation between variables is implemented as a relation defined over pairs of strings (the variable names). Relations in APTS are grouped together into *transcripts*. To execute a transcript, the inference engine partially instantiates all applicable rules from the transcript and then nondeterministically tries to complete them until there are no more rules that can be instantiated.

APTS also allows user-defined SETL2 routines to be used to add relations to the database. The user specifies an interface in APTS for each SETL2 routine, stating which relations the routine receives as input and the relations it will produce upon execution. The SCR-to-C code generators use a SETL2 routine to calculate the order in which variables are to be updated based on the information in the dependency relation.

In addition to using the relational database, the code generator based on rewrite rules also uses the APTS transformation engine. It uses a set of rewrite rules to transform the parse tree, replacing SCR language constructs with the corresponding C code. These rewrite rules match a pattern in the tree and if the conditions for the rule are met, the tree is transformed into the given replacement tree. For example, the rule `equal` below states that if an expression consisting of two subexpressions separated by "=" is found in the parse tree, then that piece of the tree is rewritten in C-style, with the "=" replaced by "==". Rewrite rules can be read as matching a piece of the parse tree and if the given conditions are met, the matched part of the parse tree is replaced by the tree given inside the rewrite.

```
equal: match(%expr, .x = .y%) |true ->
  rewrite(%expr, .x == .y%);
```

In APTS, rewrite rules are collected into groups called *closures*. When a closure is applied to the parse tree, the transformation engine works bottom up on the tree, nondeterministically applying rules from the closure to the tree until no more rules can be applied. In our implementation, we apply closures in a certain order to guarantee that certain rules will be applied before others.

2.2 SCR Specifications

Originally formulated to document the requirements of the flight program of the US Navy's A-7 aircraft [22], the SCR requirements method is designed to detect and correct errors during the requirements phase of software development [17, 21]. In SCR, the required behavior of a software system is defined in terms of *monitored* and *controlled variables*, which represent quantities in the system environment that the system monitors and controls. A set of *assumptions* describes the constraints imposed on the monitored and controlled quantities by physical laws and the system environment, and a relation on the monitored and controlled variables describes how the system is required to change the values of the controlled quantities in response to changes in the values of the monitored quantities. A set of *assertions* describes properties, such as security and safety properties, that the specification is expected to satisfy.

To specify the required behavior of a software system in a practical and efficient manner, the A-7 requirements document introduced two kinds of predicates—conditions and events—and two kinds of auxiliary variables—mode classes and terms. Conditions and events are defined in terms of the system state, where a *system state* is a function that maps each *state variable* (a monitored, controlled, or auxiliary variable) to a type-correct value. A *condition* is a predicate defined on a single system state, while an *event* is a predicate defined on two system states that denotes some change in the values of the state variables between those states. An event "occurs" if it evaluates to true for a given pair of consecutive states. A *monitored event* occurs when the value of a monitored variable changes. A *conditioned event*, which has the form "`@T(c) WHEN` d", occurs if an event occurs (i.e., condition c becomes true) when a specified condition d is true. A *mode class* may be viewed as a state machine, whose states are called *modes* and whose transitions are triggered by events. A *term* is a state variable, defined in terms of monitored variables, mode classes, or other terms. Mode classes and terms capture history—the changes that occurred in the values of the monitored variables—and help make the specification more concise.

SCR specifications include two kinds of tables: condition tables and event tables. Each defines the value of a dependent variable (a controlled or auxiliary variable) by means of a mathematical function. Usually, a condition table defines a variable as a function of a mode and a *condition*, and an event table defines a variable as a function of a mode and an *event*.

The purpose of the SCR requirements model [21] is to provide a precise semantics for the notation used in SCR requirements specifications. The model defines a conditioned event "@T(c) `WHEN` d" as

$$@\mathrm{T}(c)\ \texttt{WHEN}\ d \;=\; \neg c \,\wedge\, c' \,\wedge\, d \qquad (1)$$

where c and d are conditions, and the unprimed c denotes c in the old state and the primed c denotes c in the new state. The model also defines the functions that can be derived from the SCR tables. In the SCR model, a software system Σ is represented as a state machine $\Sigma = (S, S_0, E^m, T)$, where S is a set of states, $S_0 \subseteq S$ is the initial state set, E^m is the set of monitored events, and T is the transform describing the allowed state transitions. To compute the new state, the transform T composes the functions derived from the condition and event tables. For T to be well defined, no circular dependencies are allowed in the definitions of the new state variable values. To achieve this, the model requires the new state dependencies (i.e., dependencies among the new values of the state variables) to be a partial order of the state variables.

While an SCR specification is represented as a collection of tables, the underlying communication medium between the SCR tools is a context-free grammar. An abstract grammar for SCR can be found in [31]. This grammar focuses on the basic syntax of the constructs, omitting the precedence rules needed for unambiguous parsing. The abstract grammar is similar to the grammars used by our APTS implementations to parse SCR specifications. However, those grammars contain modifications, including precedence rules, necessary for unambiguous parsing.

The syntax of the language is best illustrated by an example. Below is the specification of a simplified version of a control system for safety injection (SIS) in a nuclear power plant [7]. (The line numbers are not part of the actual specification. They are included for ease of reference.) The SIS system monitors water pressure and if the pressure is too low, the system injects coolant into the reactor core. There are three monitored variables in this specification (lines 9–11).[1] The first is `mWaterPres`, which represents the actual value of water pressure. The other two monitored variables are switches—`mBlock`, a switch that overrides safety injection, and `mReset`, a switch that resets the system after blockage. An assumption of the specification is that the water pressure will not change by more than 10 units at a time (lines 18–21). A mode class `mcPressure`, with three possible values `TooLow`, `Permitted`, and `High`, associates the pressure with the appropriate range (lines 16–17). At any given time, the system must be in one and only one of these modes. The term variable `tOverridden` (lines 14–15) is *true* if safety injection is blocked, and *false* otherwise. The specification contains one controlled variable, `cSafety_Injection`, which represents a switch indicating whether the safety injection is on or off (lines 12–13). The value of each dependent variable is defined by a table function. Event tables define the value of the mode class `mcPressure` (lines 24–39) and the term variable `tOverridden` (lines 40–47). A condition table defines the

[1] By convention, the names of monitored variables begin with "m", of controlled variables begin with "c", of terms begin with "t", and of mode classes begin with "mc". The names of user-defined types begin with "y", and the names of types associated with mode classes begin with "type_".

value of the controlled variable `cSafety_Injection` (lines 48–60). The tabular format of the function definitions on lines 24–60 can be found in [31].

```
1   spec Safety_Injection_System
2   type definitions
3     ySwitch: enum in {Off, On};
4     type_mcPressure: enum in {TooLow, Permitted,High};
5     yWPres: integer in [0, 2000];
6   constant definitions
7     Low=900:integer;
8     Permit=1000:integer;
9   monitored variables
10    mWaterPres: yWPres, initially 0;
11    mBlock, mReset: ySwitch, initially Off;
12  controlled variables
13    cSafety_Injection: ySwitch, initially On;
14  term variables
15    tOverridden: boolean, initially false;
16  mode classes
17    mcPressure:  type_mcPressure, initially TooLow;
18  assumptions
19     A1: (mWaterPres' >= mWaterPres AND mWaterPres'-
20         mWaterPres <=10) OR (mWaterPres' < mWaterPres
21         AND mWaterPres - mWaterPres' <= 10);
22  assertions
23  function definitions
24  var mcPressure :=
25    case mcPressure
26     [] TooLow
27        ev
28           [] @T(mWaterPres >= Low) -> Permitted
29        ve
30     [] Permitted
31        ev
32           [] @T(mWaterPres >= Permit) -> High
33           [] @T(mWaterPres <  Low) -> TooLow
34        ve
35     [] High
36        ev
37           [] @T(mWaterPres < Permit) -> Permitted
38        ve
39    esac
40  var tOverridden  :=
41    ev
42      [] (@T(mBlock=On) WHEN (mReset=Off AND
43         NOT(mcPressure = High))) -> true
44      [] (@T(mReset=On) WHEN NOT(mcPressure = High))
45         OR @T(mcPressure = High)
46         OR @T(NOT(mcPressure = High))-> false
47    ve
48  var cSafety_Injection ==
49    case mcPressure
50     [] TooLow
51        if
52          [] tOverridden -> Off
53          [] NOT tOverridden -> On
54        fi
55     [] Permitted, High
56        if
57          []  false -> On
58          []  true -> Off
59        fi
60    esac
```

2.3 C Code Generated From SCR Specifications

An SCR specification contains several sections. Code must be generated from each section. Additionally, code is needed to drive the reactive program. In practice, this code would be replaced by the device driver software for the system. This section describes the C code generated from an SCR specification. The format of the C code to be generated is given as part of the grammar file in APTS. Our synthesizers produce code that is very closely related to the SCR specification. This makes correspondence between the specification and the code easy to observe.

Some pieces of code are generated for every specification. At the beginning of each generated code file are two file pointers, `infile` and `outfile`, which will be associated with the input and output files that drive the reactive program. Also included are input and output routines for boolean values (represented internally by the integer constants false = 0 and true = 1).

For every SCR specification, the corresponding C code contains a header file "scr-header.h". This header file is exactly the same for every specification and is not generated by APTS. This file contains definitions of formats for reading and writing strings and integers. For example, the string output format macro is defined as `# define strformout "%s\n"`. Defining these format routines in a header file that is not generated by APTS is necessary because APTS treats the `"%"` as a special character that cannot appear in relations or rewrite rules. The file also contains definitions used by the generated C code. For example, `boolean` and `integer` are defined as additional names for the C type `int`; and `false, true, AND, OR`, and `NOT` are defined as corresponding C code values and operators. None of these definitions is strictly necessary, but they are included to make the C code resemble the SCR specification.

Type Definitions

Both SCR and C support enumerated types. However, unlike C, SCR allows overloading of value names in enumerated types. To handle this soundly in our encoding, we simply prepend the type name to each enumerated value. For example, the value `Off` of the enumerated type `ySwitch` in the SCR specification is transformed into `ySwitch_Off` in the C type definition.

```
/* type definitions and range declarations */
   enum ySwitch { ySwitch_Off , ySwitch_On } ;
   typedef enum ySwitch ySwitch ;\vspace*{-4pt}
```

Each enumerated type also requires special input and output routines to convert the value names used in the specification to the corresponding value names used in the C code, and vice versa.

The user-defined range types in SCR have no counterpart in C. In the generated C code, the name of the range type becomes an alias for integer, and a check function is created to correspond to the range of the type. Each time a variable with a range type is assigned a value, the corresponding check function is called to ensure that the value is within the specified range. The SIS example contains one range type, `yWPres`, with the range [0,2000]. Below is the C code corresponding to this range type.

```
  # define yWPres int
   void check_yWPres (char * name, int value) { if ((value
     < 0) OR (value > 2000)) {printf(" value out of
     range : "); printf (strformout, name); } }
```

Constant Definitions

The generated C code for an SCR constant definition is a straightforward rearrangement of the SCR definition. Below is the C code generated for the constant definitions of the SIS example.

```
/* constant definitions */
    const integer Low = 900 ;
    const integer Permit = 1000 ;
```

Variable Declarations

In an SCR specification, x represents the value of variable x in the old state, and x′ represents the value of x in the new state. To refer to both the old and new values of the variable x, the generated C code represents each variable x in the SCR specification by two variables, x and prime_x. Initial values, if given, are defined by constants. (In SCR specifications, the initial value of a dependent variable can often be derived from the initial values of variables upon which the variable depends.) The name given to these constants is constructed by prepending init_val_ to the name of the first variable in the declaration. For example, the initial value for mBlock and mReset is named init_val_mBlock. Below is the C code corresponding to the declaration of the monitored variables. The other variable declarations may be transformed into C code in a similar way.

```
/* monitored variables */
  yWPres mWaterPres; yWPres prime_mWaterPres;
     const yWPres init_val_mWaterPres = 0;
  ySwitch mBlock, mReset; ySwitch prime_mBlock,
  prime_mReset;
     const ySwitch init_val_mBlock = ySwitch_Off;
```

Assumptions and Assertions

In SCR, assumptions and assertions are predicates describing relationships between the variables. These logical formulas may refer to both the old and new state values of the variables and can use a full range of logical operators. Event expressions may also appear in predicates and are expanded using definition (1). Each assumption or assertion in the specification is transformed into an evaluation function which returns true if the predicate is true and false otherwise. Additionally, two functions, check_assumptions and check_assertions, which call these functions and produce an error message if a predicate is false, are generated if there are assumptions and assertions in the SCR specification. The violation of an assumption indicates that the input does not obey the assumed environmental constraints. If an assertion is violated, then the specification does not satisfy a property that it was expected to satisfy. The code we generate for assumptions and assertions is not a necessary part of an implementation of the SCR specification. We generate the evaluation and check functions to provide information on violations of expected behaviors for use in simulation. In the SIS example, an evaluation function is generated for the assumption A1 along with the function check_assumptions which calls the evaluation function. There are no assertions in SIS, so the function check_assertions is not generated.

```
/* assumptions */
 boolean eval_A1( ) {return((prime_mWaterPres >=
   mWaterPres AND prime_mWaterPres - mWaterPres <= 10 ) OR
   ( prime_mWaterPres < mWaterPres AND mWaterPres -
   prime_mWaterPres <= 10 ) ) ; } ;
 void check_assumptions ( ) {
   if ( eval_A1( ) == false ) printf (" A1 violated \n ");
 }
 /* assertions */
```

Function Definitions

As stated in Section 2.2, each dependent variable in an SCR specification is associated with a function. This function is defined by either a condition or an event table, describing how the variable's value is updated when a monitored variable changes. For each SCR table function, the C code contains a corresponding update function in which the successful branches assign the newly calculated value to the primed version of the variable. Each branch of the SCR **case** statement in a condition table becomes a C **if** statement conditioned on the value of the primed version of the mode class variable. Each SCR **if** statement is transformed into a C **if-else** statement. Below is the C update function corresponding to the condition table for `cSafety_Injection`. Note that no code is generated for the false branch (line 57) in the table for `cSafety_Injection`.[2] The tables for `mcPressure` and `tOverridden` are transformed into C code in a similar way.

```
void update_cSafety_Injection ( ) {
    if (( prime_mcPressure == type_mcPressure_TooLow )) {
        if ( prime_tOverridden ) {
          prime_cSafety_Injection = ySwitch_Off ;
          fprintf ( outfile , " cSafety_Injection = " ) ;
        put_ySwitch ( prime_cSafety_Injection ) ; }
      else if ( NOT prime_tOverridden ) {
        prime_cSafety_Injection = ySwitch_On ;
        fprintf ( outfile , " cSafety_Injection = " ) ;
      put_ySwitch ( prime_cSafety_Injection ) ; }
    } ;
    if ( ( prime_mcPressure == type_mcPressure_Permitted )
      OR ( prime_mcPressure == type_mcPressure_High ) ) {
      if ( true ) {
        prime_cSafety_Injection = ySwitch_Off ;
        fprintf ( outfile , " cSafety_Injection = " ) ;
      put_ySwitch ( prime_cSafety_Injection ) ; }
    } ;
}
```

Execution Code

In addition to generating code from the specification, we also generate code which executes the specified state machine. The generated code simulates input and output using text files. Input is from a file which lists monitored events, each specified by the name of a monitored variable and a value to be assigned to that variable. The execution model is similar to the execution model of SCR systems used in the translation of SCR into Promela (the language of the SPIN model checker) by Bharadwaj and Heitmeyer [4] and can be described (in pseudocode) as follows.

```
<open files>
state = 0;
<initialize new state variables>;
<check assumptions and assertions>;
while (<infile contains another monitored event> ) {
    state = state+1;
    <copy new state variables to old state variables>;
    <update new state variable corresponding to monitored
       event>;
    <update new state dependent variables in dependency
       order>;
    <check assumptions and assertions>;
}
<close files>
```

Separate functions are generated for performing the initialization, copying the variables, and updating the dependent variables. Note that the dependent variables are updated in an order consistent with the partial order describing the new state dependencies as discussed in Section 2.2. The previously generated `check_assumptions` and `check_assertions` functions

[2] The entry `false -> On` in the function defining `cSafety_Injection` is an artifact of the tabular format. It means that `cSafety_Injection` is never equal to `On` when the mode is `High` or `Permitted` and would therefore correspond to dead code.

are also called by this main routine. All SCR specifications generate a similar main routine; the only differences are in the names of the update functions for the dependent variables and the updating of the monitored variables in response to monitored events.

3 Generating C Code from SCR Specifications

This section describes our implementation of the two strategies for code generation. Both strategies use many of the same relations in their generation of code. Section 3.1 describes these relations. Sections 3.2 and 3.3 describe the strategies, the first using rewrite rules and the second using accumulating relations.

3.1 Relations Common to Both Strategies

To generate C code from an SCR specification, each code generator makes extensive use of the APTS relational database. Relations are defined to compute and store information needed to generate code. In the implementations, some relations have rules that are conditioned on other relations not holding for a node. Thus, relations that appear negated in the conditions of rules need to be fully calculated before the rules that contain those negations can be applied. To accomplish this, the relations are organized into groups called *transcripts* that can be calculated in the same pass. The transcripts are executed in an order that respects the dependencies of relations in one group on relations in another group. Both code generation strategies require similar information to be stored, i.e., the variable dependencies, information about the types of expressions, and several pieces of code that need to be calculated and stored in the nodes before the C code is generated.

Both strategies need information about variable dependencies. The order in which the dependent variables are updated depends on the new state dependencies. This dependency information is calculated by the SCR toolset and then converted into an APTS relation **depend**.[3] In the SIS example, the rules for **depend** are as follows:

```
true -> depend(mcPressure, mWaterPres);
true -> depend(tOverridden, mBlock);
true -> depend(tOverridden, mReset);
true -> depend(tOverridden, mcPressure);
true -> depend(cSafety_Injection, mcPressure);
true -> depend(cSafety_Injection, tOverridden);
```

These rules can be read as `mcPressure` depends on `mWaterPres`, `tOverridden` depends on `mBlock`, and so on. This relation is passed to a SETL2 routine that constructs a topological sort of the variables with respect to the dependency constraints. The results of this SETL2 routine are stored in a relation **followedby** that holds the ordering of the variables.

Several relations are used to check that the input specification is a valid SCR specification. These relations are necessary because in the parsing grammar, expressions, events, and predicates are condensed into one category. The relations mark the nodes containing each of these separate types of expression. For example, `@T(mBlock = on)` is a member of the **eventexpr** relation, as is `@T(mBlock =on) WHEN (mReset=Off AND NOT (McPressure=High))`. An additional relation marks the nodes containing primes. This **primeexpr** relation includes `mWaterPres'`, `mWaterPres' - mWaterPres`, and `mWaterPres' - mWaterPres <= 10`. Using the information in these relations, checks are done to determine whether expressions including primes, events, and predicates are used only as allowed in the SCR language.

Other relations store information needed to generate the code. For example, each variable in the generated code has a corresponding primed variable. In the relational database, a

[3] The conversion is currently done by hand but would be easy to automate.

relation **primename**, associated with a variable in the specification, holds the name to be used for the primed version of the variable.

```
match(%declist, .x,.y%)| true
   -> primename(.x,concat('prime_',str(.x)));

match(%declist, .x %) | isavar(rchild(.x))
   -> primename(rchild(.x),concat('prime_',str(rchild(.x))));
```

Most APTS relational database rules first match a specific construct in the parse tree, in this case, a list of variables that is part of a declaration. (Recall that, in APTS, pattern matching variables have names beginning with a dot.) In the first rule above for **primename**, if the list has the form of an identifier .x followed by a comma and list of identifiers .y, then we convert the node .x to a string and prepend the string "prime_" to it and associate the resulting string with the tree at .x. (**primename** is a relation between trees and strings, meaning that the string is associated with every instance of the identifier .x in the tree, not just the instance of .x in the node matched by the rule.) The elements in the list .y are assigned their prime names by repeated applications of the rules. The first rule handles all multiple element lists but not single element lists. In the second rule, the match condition states that .x must be a declist. This match condition will match any declist with any number of elements. In the second rule, we only wish to match declists with a single element. The condition that the rightmost child of .x be a variable ensures that the second rule is only triggered by a single element list. (If the list has multiple elements, the right child of .x will be a declist, not a variable.) The actual identifier is the child of node .x and this is what is associated with the new string. Note that when matching more complex patterns, as in the first rule, APTS is able to associate the pattern matching variables with the children of the matched node (e.g., .x is the leftmost child), but when the pattern consists of just a single pattern matching variable, as in the second rule, the pattern matching variable is associated with the matched node rather than with one of its children, even in the case when there is only one child. This makes it necessary to use rchild(.x) to refer to the child in the second rule.

For enumerated type variables, relations hold the names of the relevant input and output routines, as described in Section 2.3. Another relation marks the nodes containing obviously dead code, e.g., branches labeled by `false`, `never`, `@T(true)`, or `@T(false)`. In the SIS example, the branch `[] false -> On` on line 57 is marked as dead code by the rule below. No code is generated for such branches. In the rule, `loc()` refers to the node matched by the pattern matching portion of the rule.

```
match(%if_stmt_body, [] false -> .y%) |true
  -> deadcode(loc());
```

Relations are also used to hold some pieces of the C code. This is done when the code needs to be calculated at one point in the parse tree and generated somewhere else in the tree. The code for these functions is generated during one of the earlier phases of the relational database creation and then passed to the transformation engine or used in the calculation of the code accumulating relation. For example, the code for the C functions that execute the specification must be placed at the end of the generated code. These functions must contain code for each variable in the specification, and thus the code must be calculated in the variable declarations portion of the parse tree because that is where the variables are actually listed. As another example, each value in the list of values for an enumerated type requires code for reading and writing that value (because of the previously described conversion between the names used in the specification and the names used in the code). For example, the **inputcode** relation will contain a pair composed of the variable value, `On`, and the code, `if(strcmp(compname = "On") ==0) return(ySwitch_on); else{printf("not a valid input value\n"); return(-1);}`. This code is stored in relations because both the input and output routines must be generated at the node in the parse tree where the entire type definition occurs, not where the value itself occurs.

3.2 Code Generation Using Rewrite Rules

The generation of C code from an SCR specification using APTS rewrite rules is performed in three steps. First, a grammar is created that combines the language of the specification and the form of the corresponding C code. Second, relations are defined that capture the information in the specification. Finally, a set of transformations is defined that replaces the SCR specification with the corresponding C code.

The transformation-based code generator uses a grammar that combines both the form of an acceptable SCR specification and the form of the C code corresponding to the specification. A combined grammar is used so that during transformation, when the tree is a combination of SCR and C code, it is still a valid program. (This is a design decision that we made. During transformation APTS does not check that the parse tree remains valid, so the combined grammar is not strictly required by APTS.) The grammar used gives a parse tree where the nodes alternate between general structure and language-specific structure. For example, consider the following general structure rule from the grammar:

```
case_ev = scr_case_ev | c_case_ev;
```

and its corresponding language-specific rules:

```
scr_case_ev = 'case' id case_branch_evs 'esac'; c_case_ev =
case_branch_evs ;
```

A node denoting a case statement in an event table, `case_ev`, has only one child which is either a node denoting an SCR event table case statement, `scr_case_ev`, or a node denoting the C code corresponding to the event table case statement, `c_case_ev`. The SCR event table case statement is delimited by the keywords '`case`' and '`esac`' and includes the name of the identifier whose value the branches are conditioned on. It is also defined in terms of the general structure node `case_branch_evs`, which denotes the branches of the case statement. The C language-specific node is also defined in terms of `case_branch_evs`, which, in turn, is defined in terms of language-specific nodes denoting the branches. When the parse tree is initially created from an SCR specification, it contains general structure nodes alternating with SCR structure nodes. During the transformation process, SCR nodes are replaced by C nodes, so that at all times during the transformation, we have a valid parse tree.

Once an input SCR specification is parsed using this grammar, the relational database inference engine is called. In addition to the relations described in the preceding section, this code generator needs a new relation checking that the input is a valid SCR specification. This is necessary because the layered nature of the grammar allows an input file containing a mix of SCR and C code to be accepted by APTS as being syntactically valid. We check that only SCR nodes are used in alternation with the general structure nodes.

After the input specification has been parsed and all necessary relations have been calculated, the SCR specification is transformed into C code. The translation is done in several stages and the order of these stages matters because the transformations change the parse tree and thus may cause matches for later transformations to fail.

The first stage of transformation eliminates some of the dead code in the specification. The dead code on line 57 of the SIS example would be removed by the rewrite rule `if2` below. The match condition matches lists of if-statements where the first member .y is an if-statement and the second member .x is a list of if-statements. Recall that the node corresponding to line 57 has already been added to the **deadcode** relation during the building of the relational database, so the condition `deadcode(.y)` will be true. The complete list is replaced in the parse tree by the second component, eliminating the dead code branch.

```
if2: match(%scr_if_stmt_bodies,  .y .x%) | deadcode(.y)->
  rewrite(%scr_if_stmt_bodies, .x%);
```

The second step replaces enumerated values appearing in constant definitions and variable declarations with their new, type-specific names. For example, the rule below replaces `TooLow` on line 17 of the SIS example with `type_mcPressure_TooLow`. The relation **newname** contains the type-specific name associated with the identifier.

```
enum3: match(%var_decl, .x : .t, initially .y;%) |
   newname(.y,.z) ->
   rewrite(%var_decl, .x : .t, initially .z;%);
```

The third stage of the transformation converts most of the SCR language into C code. For example, the rule below replaces the type definition on line 5 of SIS example with the corresponding C code given in Section 2.3. The relation `rangefun` contains the name to use for the range checking function in the variable .s.

```
typebody2: match(%scr_type_body, .x : integer in
  [.y,.z];%) | rangefun(.x,.s) -> rewrite(%c_type_body,
  # define .x int void .s (char * name, int value) {if
  ((value < .y) OR (value > .z)) {printf("value out of
  range:"); printf(strformout, name);}}%);
```

After this step is complete, the enumerated values remaining in expressions are replaced with their type-specific counterparts. Following this, the event operators are replaced with equivalent logical expressions. Finally, each primed expression is replaced with the name of the corresponding primed variable if the expression is a variable. If the primed expression is an enumerated value or an integer, then the prime is eliminated. The following are some of the rewrite rules used to perform these transformations.

```
equal: match(%expr, .x = .y%) |true ->
  rewrite(%expr, .x == .y%);
event1: match(%expr, @T(.x)%) |true ->
  rewrite(%expr, ((.x)') AND NOT(.x) %);
prime2: match(%expr, (.x)'%) | primename(.x,.y) ->
  rewrite(%expr, .y%);
prime11: match(%expr, (.x == .y)'%) | true ->
  rewrite(%expr, (.x)' == (.y)'%);
prime18: match(%expr, (.x)'%) | enumval(.x) or isint(.x)
  or const(.x) -> rewrite(%expr, .x%);
```

Each rule above replaces an SCR expression with an equivalent C expression. For example, for the SCR event expression `@T(x=5)`, the C code is generated as follows. First, the node containing this expression is rewritten as `@T(x==5)` using the `equal` rule. Then, using `event1`, the expression is transformed into `((x==5)') AND NOT(x==5)`. Next, `prime11` and `prime18` are used to place the prime in the correct location, rewriting the expression first as `((x)'==(5)') AND NOT(x==5)` and then as `((x)'==5) AND NOT(x==5)`. Finally, using `prime2`, the expression is rewritten as `(prime_x==5) AND NOT(x==5)`.

3.3 Code Generation Using an Accumulating Relation

An alternative form of code generation relies solely on relations. Instead of transforming the source code into the target language, the target language code is accumulated in a relation. This approach keeps the two languages separate and preserves the original parse tree. On the negative side, it requires a great deal of additional calculation of relations. The relations used in this method are all but one of those used in the transformation-based method as well as several additional relations to hold the generated code and a relation to calculate the primed version of any expression.

Because the purely relational framework keeps the grammars for SCR and C separate, the parse tree contains only the productions for SCR language constructs. There is no need for the alternating style used in the parse tree for transformation-based code generation. Additionally, there is no need for the relation that checks that the input specification is pure SCR (rather than a mix of SCR and C). Because the grammars for SCR and C are separated, the input is only accepted by the parser if it is a valid specification in the SCR grammar. Although it is no longer combined with the SCR grammar, the C grammar is still included in the APTS grammar specification because it is used to structure the C code kept in the accumulating relation.

In this framework, additional relations perform the work done by the transformations in the other approach. A relation **prime** is used to calculate the primed C version of each expression so that if the primed form is needed during code generation, it will be available. The following are some of the rules for calculating **prime**. The first rule states that the prime of an identifier is the corresponding primed identifier, stored in relation **primename**. The second states that the prime of an integer or a constant is just that integer or constant. Finally, the third rule states that the prime of an equality expression involving two values is an equality expression of the primes of those two values. Note that in this inference rule the "=" used by SCR is replaced by the "==" used by C.

```
match(%expr, .x%) | primename(.x,.y) ->
  prime(loc(),%expr,.y%);

match(%expr, .x%) | isint(.x) or const(.x) ->
  prime(loc(),%expr,.x%);

match(%expr, .x = .y%) | prime(.x,.primex) and
  prime(.y,.primey)
  -> prime(loc(),%expr,.primex == .primey%);
```

All generated code is also held in relations. The execution code is developed in relations as previously described. The code generated by rewrite rules in the transformational approach is, in this approach, also placed in a relation. For each node, the accumulating relation stores the C code corresponding to the portion of the SCR specification represented by the subtree rooted at that node. This code is determined by the structure of the tree at that node and the code associated with the node's children. The code for the entire program is associated with the root of the tree.

The relational database rules listed below perform the same functions as the rewrite rules `if2`, `enum3`, and `typebody2` described in Section 3.2. The first rule eliminates the same dead code as rewrite rule `if2` by keeping as the code for the matched node only the code associated with the pattern variable .x. The second rule uses the type-specific name of an enumerated value as the C code to be generated for all expressions equivalent to that value, including the variable declarations handled in the transformation-based code generator by the rewrite rule `enum3`. The third relational database rule handles the same situation as rewrite rule `typebody2`, associating an SCR range type definition with the corresponding C code in relation **c_code**.

```
match(%if_stmt_bodies,  .y .x%) | deadcode(.y) and
  c_code(.x,.codex)and not(deadcode(loc()))->
  c_code(loc(),%c_if_stmt_bodies, .codex%);

match(%expr, .x%) | enumval(.x) and newname(.x,.y) ->
  c_code(loc(),%expr,.y%) ;

match(%type_body, .x : integer in [.y,.z];%) |
  rangefun(.x,.s)-> c_code(loc(),%c_type_body,
  # define .x int void .s (char * name, int value)
  {if ((value < .y) OR (value > .z)) {printf("value
  out of range:");printf(strformout, name);}}%);
```

Below are several rules for generating C code for event expressions and other simpler expressions. The first two rules state that variables and integers in the SCR specification are not changed in the C code. The third rule calculates the C code for an expression that checks the equivalence of two expressions by combining the previously calculated code for each of the subexpressions. The last rule calculates the C code for the "at true" event, converting it into an equivalent logical expression. Note that this rule assumes that the code corresponding to the prime value of the expression has already been calculated and is stored in `.primex`.

```
match(%expr, .x%) | not(enumval(.x)) and isavar(.x)
  -> c_code(loc(),%expr,.x%);

match(%expr, .x%) | isint(.x) -> c_code(loc(),%expr,.x%);

match(%expr, .x = .y%) | c_code(.x,.codex) and
  c_code(.y,.codey) ->
  c_code(loc(),%expr, .codex == .codey%) ;

match(%expr, @T(.x)%) |prime(.x,.primex) and
  c_code(.x,.codex) ->
  c_code(loc(),%expr, (.primex) AND NOT(.codex) %) ;
```

Consider again the example `@T(x=5)`. Using the relational database rules above and those given earlier for the relation **prime**, we can calculate the relations **c_code** and **prime** for each of the subcomponents. For example, using these rules, `prime(5) = 5`, `prime(x) = prime_x`, `c_code(5) = 5`, and `c_code(x) = x`. Using these as a basis, we determine the values of the relation for `x=5`, namely, `prime(x=5) = prime_x ==5` and `c_code(x=5) = x==5`. Now, we can calculate the value of `c_code` for the full expression: `c_code(@T(x=5)) = (prime_x==5) AND NOT (x==5)`.

4 Discussion

Given an SCR specification, the transformation-based and relation-based strategies generate exactly the same code. For the SIS example, which generated 293 lines of C code, the transformation-based strategy required four minutes and the relation-based strategy required 20 minutes.[4] We also used both generators to generate code from the SCR requirements specification for a cryptographic device [29]. This latter specification contains 36 variables and 20 function definitions and is 658 lines long. The translation of the SCR tabular specification into the SCR grammar used by APTS was done by hand, although such a translation could be automated. In the second example, the code generator using rewrite rules required approximately 12 hours to generate 2028 lines of C code. The generator that accumulated code in a relation required far longer—more than 72 hours. We observed that the version using rewrite rules spent most of the time building the relational database. Clearly, a speedup of the APTS relational database inference engine would greatly improve the execution times for both code generators.

For our purposes, APTS was a useful tool for exploring different code generation strategies since it supports both the relational and the transformation-based approaches. However, the prohibitively lengthy times that APTS requires to generate code makes the use of APTS in a production-quality system impractical. Paige planned a number of improvements to APTS [38], which were expected to increase the translation rate by a factor of 6000. These improvements included translating the SETL2 code of APTS into C (for an expected speedup factor of 30) and using partial evaluation to convert the APTS interpreters into compilers (for an expected speedup factor of 10). If these two improvements were made, the times for generating code for SIS would be in seconds instead of minutes and for the cryptographic device would be in minutes instead of hours.

[4] Execution times are for a Sun Ultra 450 with 2 UltraSPARC-II 296 MHz CPUs and 2 GB memory, running Solaris 5.6.

While the transformation-based approach generates code more quickly in APTS, the relation-based strategy is more straightforward. Though many of the rewrite rules are easily understood because they relate directly to the inference rules used by the relation-based strategy, the necessary execution ordering of the rewrite rules is less intuitive. Note too that a purely transformational strategy was impossible because code sometimes needed to be calculated at one point in the parse tree and generated at a different point in the tree. In the generator using rewrite rules, this was done by placing the code in a relation in the relational database and passing it to the transformational engine.

Changing the target language from C to some other language (or modifying the C code to be generated) would require approximately the same amount of work for both generators. Most relations used by both strategies refer only to the SCR specification and thus would not change. For both code generators, the grammar used by APTS would require modification. With the transformation-based strategy, the C language structures in the interleaved grammar would be replaced by the language structures of the new target language. With the relation-based strategy (or if the translation-based strategy did not use an interleaved grammar), changing the grammar is even easier. Since the new language need not be interleaved with the SCR grammar, it can simply be added to the grammar file in place of the C grammar. Finally, the transformations or the accumulating relation must be modified. In both cases, the actual conditions for the rules (rewrite or inference) remain the same. What changes is the result of the rewrite rule or the value stored in the accumulating relation.

It should also be noted that code can be generated for incomplete specifications. In particular, code can be generated for partial specifications in which all of the dependent variables have not yet been defined by a table function. Because the code generated for each function definition is independent of the code generated for any other function definition, it is possible to generate code separately for each function in the specification. However, the variable declarations, type definitions, and constant definitions for any variables, types, and constants used in the table need to be included in the partial specification in order for code to be properly generated for the table.

The generated C code performs very well. The code for the communications device processed an input file with 17 monitored events in less than one second. We have no handwritten C code to which the generated code can be compared, but the SCR toolset has a simulator [20] that produces Java code to simulate the behavior of the state machine defined by the specification. Our C code runs faster than the simulator's Java code, but a fair comparison of the two is difficult. Java code is generally slower than C code and the simulator also uses a GUI interface, slowing the running time even more. The simulator has one advantage over our C code; it has been optimized to update a variable only when at least one of the variables on which it depends has been changed.

A major contribution of Paige's research is finite differencing. Although the APTS reference manual [37] states that APTS contains techniques for optimizing the generated code, how to use these techniques within APTS is not documented. However, some preliminary work on how code generated from SCR specifications could be optimized has been done. One serious source of inefficiency in the generated code is that each new input requires an update to every variable in the program. One obvious way to reduce this inefficiency, which is used by the SCR simulator, is to use the variable dependencies (computed automatically by the SCR toolset) to determine which variables could potentially change value when a given input variable changes value and eliminate updates to the remaining variables. Information from invariants may also be used to further optimize the variable updates [27].

Other ways to reduce inefficiency are to identify parts of the specification that lead to dead code and to redundant code and omit code generation for those parts. Our implementations currently only eliminate only the obviously dead code – branches labeled by `false` and `never`. State invariants constructed using the algorithms described in [24,28] can be used to identify parts of the specification that lead to dead code or to redundant code [26]. For example, [24] shows that `tOverridden` $=$ *true* $\rightarrow$ `mcPressure` $\neq$ `High` is a state invariant of the SIS specification. This invariant implies that `tOverridden` cannot change from *true* to *false* if `mcPressure` $=$ `High` in the old state. Hence, the disjunct on line 46 of the SIS specification `@T(NOT(mcPressure = High))` can be ignored because it produces dead code. Similarly, in the specification of an automobile cruise control system, one of the automatically generated state invariants is $M =$ `Inactive` $\rightarrow$ `IgnOn` [24]. Hence, the event `@T(Lever=const)` WHEN ($M =$ `Inactive` AND `EngRunning` AND NOT `Brake` AND `IgnOn`) may be replaced by the equivalent event `@T(Lever=const)` WHEN ($M =$ `Inactive` AND `EngRunning` AND NOT `Brake`).

Recent Work

Since the publication of our original research on this problem in 2003 [18], we have continued to investigate the problem of generating code from SCR specifications. In 2004, we published preliminary results on using invariants to optimize specifications before code generation [25]. Such optimizations eliminate parts of the original specification that either lead to dead code or unneeded code and, as a result, reduce the amount of code that is generated from the specification and consequently increase the efficiency of the code. These optimizations are applied to a specification prior to running one of the APTS-based translators described in this paper.

More recently, an optimized code generator for SCR specifications has been designed and implemented [43]. This new code generator did not build directly upon either of our APTS-based translators because we were interested in generating code quickly and in performing optimizations. Instead, the new code generator, written in Python, constructs an abstract syntax tree (AST) from the SCR specification, performs optimizations on the AST, and then generates code from the optimized AST. The code generator applies three types of optimization to the AST, namely, input slicing, simplification, and output slicing. The first optimization technique, input slicing, uses the update dependencies of variables to eliminate updates to variables that cannot change for a given input. The second technique, simplification, use substitutions to reduce the complexity of expressions and to remove nodes from the AST when those nodes cannot execute. Simplification derives substitutions from several sources, for example, from invariants, such as the assumptions and assertions of the specification, or from invariants automatically generated using the method described in [24, 28]. It also derives substitutions based on the known value of variables on paths in the AST and conditions that are known to be true along paths in the AST. The third technique, output slicing, uses dataflow analysis to determine which variables are live at each node of the AST. Any node which computes a value for a variable which is not live in the poststate of that node can be eliminated. This new code generator is considerably faster than our APTS-based code generators; for example, in a few seconds, it was able to generate unoptimized code from a specification containing 1114 tables.

5 Related Work

Generating code from specifications is not a new idea. The APTS translators for SETL2 and SQ2+ [6] can be used to translate specifications in these high-level languages into C. Like APTS, META-AMPHION [32], REFINE [41], and KIDS [44] can be used to design translators from high-level declarative specifications into executable programs. Moreover, several commercial tools generate code from specifications. For example, Statemate [15] generates C or Ada code from Statechart specifications, Telelogic's SCADE generates C or Ada code from LUSTRE [13], and the Stateflow Coder generates C code from Stateflow specifications. Also, C++ code generation for specifications written in RSML is discussed in [16,48], C++ code generated from Charon specifications is described in [2], and generation of Java code from Input/Output Automata is discussed in [46]. In [47], C-like imperative code is generated from specifications given in E-FRP (Event-Driven Functional Reactive Programming) and transformations are applied to optimize the code. As in SCR, E-FRP variables only change value in response to events. Interest has been expressed in translating E-FRP into SCR [47].

Our two strategies share similarities with previous code generation methods. Both store additional information in relations. Many (but not all) of these relations are defined over nodes in the parse tree, making those relations similar to the attributes used in attribute grammar systems [30], such as the Synthesizer Generator [42]. Like an attribute grammar, our relational approach treats the target code as a synthesized attribute of the parse tree for the specification.

Our use of APTS rewrite rules to generate code is similar to Cai and Paige's [6] use of APTS to translate SETL2 and SQ2+ into C. Their translations used rewrite rules to generate the code, just as our implementation did. One difference between our work and theirs is that they also made use of APTS built-in finite differencing and dominated convergence optimizations, while we did not.

Our transformational strategy is also similar to the HATS transformational programming system [49] in its use of tree rewriting rules. Both HATS and our transformational strategy use rewrite rules that modify the actual tree to hold the changed code, and both require that the transformations always produce valid trees. Both also condition the rewrite rules on the matching of patterns in the trees. One difference is that our APTS-based transformational strategy also allows the relations to hold additional information and to be used as conditions for matching in rewrite rules. Another difference between HATS and our approach is that HATS may sometimes use problem-specific transformations, which our transformational system does not currently support.

Two other systems, Twig [1] and iburg [11], produce code generators that modify the parse tree. Unlike APTS, which makes many passes over the parse tree, these code generator generators work by making only two passes over the parse tree. The first pass finds a set of minimal cost patterns that cover the tree. The second pass executes the semantic actions associated with these patterns. Twig and iburg do not replace the code in the tree with target language code as our transformational system does. Instead, a pattern is matched, code associated with the pattern is generated to a file, and then the tree is reduced using the rewrite rule for the pattern. This process is repeated until the whole tree has been reduced.

As noted in Section 2.3, the code generated by our code generators uses an execution model similar to the execution model of SCR systems used in the translation of SCR into Promela [4]. Both use two sets of variables, one for the old state values and one for the new state values. Both encode the function tables as conditional statements in the target language and both execute the code for the functions in an order determined by the dependency relationship on the variables. One difference is that the Promela translation uses nondeterministic choice to implement the branches in a table, which is impossible in C. This is not a significant difference (i.e., it does not result in a possibly different semantics), since the conditions on the branches of a table are required to be disjoint [21], a requirement that is verified by the consistency checker in the SCR toolset.

6 Conclusion and Future Work

This paper described our experiments in developing code generators using APTS. Two different strategies to generate C code from SCR requirements were implemented. One strategy transforms a parse tree in the specification language into a parse tree in the target language, while the other accumulates the generated code in a relation associated with nodes in the specification language parse tree. Both APTS implementations generate the exact same code and perform significant analysis before generating code. Though APTS is currently too slow for use in a production-quality system, implementing the improvements that Paige suggested should produce a system that uses a relational database and rewrite rules to generate code at an acceptable speed.

The results of these case studies inspired our recent work on generating optimized code from SCR specifications [43]. In the future, we plan to explore further techniques for optimizing code generated from SCR specifications, such as finite differencing. We also plan to investigate techniques, such as [40], for generating provably correct code from SCR specifications.

Acknowledgments

We are very grateful for Bob Paige's work on the APTS system and the assistance he provided us in using APTS. Bob provided us with a first version of the SCR grammar for APTS and also answered many of our questions about APTS. Much of the initial work on the SCR abstract grammar was done by Ramesh Bharadwaj. Discussions with Myla Archer were very useful in our development of the grammar. The algorithms for constructing invariants from SCR specifications and the idea of using invariants to identify parts of the specification that would lead to redundant or dead code are due to Ralph Jeffords. We thank our colleagues Myla Archer and Ralph Jeffords for useful comments on drafts of this book chapter and Annie Liu and Scott Stoller for helpful discussions. We also acknowledge the contributions of Tom Rothamel who implemented the optimized code generator described in Section 4.

References

1. A. V. Aho, M. Ganapathi, and S. W. K. Tjiang: Code generation using tree matching and dynamic programming. *ACM Transactions on Programming Languages and Systems*, 11(4):491–516, Oct. 1989.
2. R. Alur, F. Ivančić, J. Kim, I. Lee, and L. Sokolsky: Generating embedded software from heirarchical hybrid models. *SIGPLAN Not.*, 38:171–182, 2003.
3. M. Archer: TAME: Using PVS strategies for special-purpose theorem proving. *Annals of Mathematics and Artificial Intelligence*, 29(1–4), Feb. 2001.
4. R. Bharadwaj and C. Heitmeyer: Model checking complete requirements specifications using abstraction. *Automated Software Engineering*, 6(1), Jan. 1999.
5. J. Cai and R. Paige: Program derivation by fixed point computation. *Science of Computer Programming*, 11:3, 197–261, 1989.
6. J. Cai and R. Paige: Towards increased productivity of algorithm implementation. *Proceedings ACM SIGSOFT 1993, Software Engineering Notes*, 18(5):71–78, Dec. 1993.
7. P.-J. Courtois and D. L. Parnas: Documentation for safety critical software. In *Proc. 15th Int'l Conf. on Softw. Eng. (ICSE '93)*, 315–323, Baltimore, MD, 1993.
8. J. B. Dabney and T. L. Harman: *Mastering Simulink*. Prentice-Hall, 2004.
9. S. Easterbrook, R. Lutz, R. Covington, Y. Ampo, and D. Hamilton: Experiences using lightweight formal methods for requirements modeling. *IEEE Transactions on Software Engineering*, 24(1), Jan. 1998.
10. S. R. Faulk, L. Finneran, J. Kirby, Jr., S. Shah, and J. Sutton: Experience applying the CoRE method to the Lockheed C-130J. In *Proc. 9th Annual Conf. on Computer Assurance (COMPASS '94)*, Gaithersburg, MD, June 1994.
11. C. W. Fraser, D. R. Hanson, and T. A. Proebsting: Engineering a simple, efficient code generator generator. *ACM Letters on Programming Languages and Systems*, 1(3):213–226, Sept. 1992.
12. N. Halbwachs, F. Lagnier, and C. Ratel: Programming and verifying real-time systems by means of the synchronous data-flow language LUSTRE. *IEEE Trans. Softw. Eng.*, 18(9):785–793, Sept. 1992.

13. N. Halbwachs, P. Raymond, and C. Ratel: Generating efficient code from data-flow programs. In *Third Intern. Symposium on Programming Language Implementation and Logic Programming*, Passau (Germany), Aug. 1991.
14. D. Harel: Statecharts: A visual formulation for complex systems. *Sci. Comput. Program.*, 8(3):231–274, 1987.
15. D. Harel, H. Lachover, A. Naamad, A. Pnueli, M. Politi, R. Sherman, A. Shtull-Trauring, and M. B. Trakhtenbrot: Statemate: A working environment for the development of complex reactive systems. *IEEE Trans. Softw. Eng.*, SE-16(4), Apr. 1990.
16. M. P. E. Heimdahl and D. J. Keenan: Generating code from hierarchical state-based requirements. In *Proc. IEEE International Symposium on Requirements Engineering*, Jan. 1997.
17. C. Heitmeyer: Software cost reduction. In J. J. Marciniak, editor, *Encyclopedia of Software Engineering*. John Wiley & Sons, Inc., New York, NY, second edition, 2002.
18. C. Heitmeyer, M. Archer, R. Bharadwaj, and R. Jeffords: Tools for constructing requirements specifications: The SCR toolset at the age of ten. *Computer Systems Science and Engineering*, 20(1):19–35, Jan. 2005.
19. C. Heitmeyer, J. Kirby, B. Labaw, M. Archer, and R. Bharadwaj: Using abstraction and model checking to detect safety violations in requirements specifications. *IEEE Trans. on Softw. Eng.*, 24(11), Nov. 1998.
20. C. Heitmeyer, J. Kirby, Jr., B. Labaw, and R. Bharadwaj: SCR*: A toolset for specifying and analyzing software requirements. In *Proc. Computer-Aided Verification, 10th Annual Conf. (CAV '98)*, Vancouver, Canada, 1998.
21. C. L. Heitmeyer, R. D. Jeffords, and B. G. Labaw: Automated consistency checking of requirements specifications. *ACM Transactions on Software Engineering and Methodology*, 5(3):231–261, Apr.–June 1996.
22. K. Heninger, D. L. Parnas, J. E. Shore, and J. W. Kallander: Software requirements for the A-7E aircraft. Technical Report 3876, NRL, Washington, DC, 1978.
23. S. D. Hester, D. L. Parnas, and D. F. Utter: Using documentation as a software design medium. *Bell System Tech. J.*, 60(8):1941–1977, Oct. 1981.
24. R. Jeffords and C. Heitmeyer: Automatic generation of state invariants from requirements specifications. In *Proc. Sixth ACM SIGSOFT Symp. on Foundations of Software Engineering*, Nov. 1998.
25. R. Jeffords and E. I. Leonard: Using invariants to optimize formal specifications before code synthesis. In *Proc. 2nd ACM/IEEE Conference on Formal Methods and Models for Co-Design (MEMOCODE '04), San Diego, CA*, June 2004.
26. R. D. Jeffords: Personal communication. Oct. 2001.
27. R. D. Jeffords: Personal communication. Jan. 2002.
28. R. D. Jeffords and C. L. Heitmeyer: An algorithm for strengthening state invariants generated from requirements specifications. In *Proc. of the Fifth IEEE International Symposium on Requirements Engineering*, Aug. 2001.
29. J. Kirby, Jr., M. Archer, and C. Heitmeyer: SCR: A practical approach to building a high assurance COMSEC system. In *Proceedings of the 15th Annual Computer Security Applications Conference (ACSAC '99)*. IEEE Computer Society Press, Dec. 1999.
30. D. E. Knuth: Semantics of context-free languages. *Mathematical Systems Theory*, 2(2):127–145, 1968.
31. E. I. Leonard and C. L. Heitmeyer: Program synthesis from formal requirements specifications using APTS. *Higher-Order and Symbolic Computation*, 16(1-2):63–92, 2003.
32. M. R. Lowry and J. V. V. Baalen: Meta-amphion: Synthesis of efficient domain-specific program synthesis systems. *Automated Software Engineering*, 4:199–241, 1997.
33. S. Meyers and S. White: Software requirements methodology and tool study for A6-E technology transfer. Technical report, Grumman Aerospace Corp., Bethpage, NY, July 1983.

34. S. Miller: Specifying the mode logic of a flight guidance system in CoRE and SCR. In *Proc. 2nd ACM Workshop on Formal Methods in Software Practice* (*FMSP '98*), 1998.
35. R. Paige: Programming with invariants. *IEEE Software*, 3(1):56–69, Jan. 1986.
36. R. Paige: Symbolic finite differencing — part 1. In N. Jones, editor, *Proc. ESOP 90, LNCS 432*. Springer, 1990.
37. R. Paige: APTS external specification manual (rough draft). Unpublished manuscript, available at http://www.cs.nyu.edu/jessie/, 1993.
38. R. Paige: Viewing a program transformation system at work. In *Proc. Joint 6th Int'l Conf. on Programming Language Implementation and Logic Programming* (*PLICLP*) *and 4th Int'l Conf. on Algebraic and Logic Programming* (*ALP*). LNCS 844, Springer, Sept. 1994.
39. D. L. Parnas, G. Asmis, and J. Madey: Assessment of safety-critical software in nuclear power plants. *Nuclear Safety*, 32(2), Apr.–June 1991.
40. A. Pnueli, M. Siegel, and E. Singerman: Translation validation. In *International Conference on Tools and Algorithms for the Construction and Analysis of Systems* (*TACAS 1998*), LNCS 1384, Springer, 151–166, 1998.
41. Reasoning Systems. *Refine User's Guide Version* 3.0, May 1990.
42. T. W. Reps and T. Teitelbaum: *The Synthesizer Generator: A System for Constructing Language-Based Editors.* Springer, New York, NY, 1989.
43. T. Rothamel, C. L. Heitmeyer, E. I. Leonard, and Y. A. Liu: Generating optimized code from SCR specifications. In *Proceedings of the 2006 ACM Sigplan/Sigbed Conference on Languages, Compilers, and Tools for Embedded Systems* (*LCTES '06*), *Ottawa, Ontario, Canada, June 14-16, 2006*, 135–144. ACM, 2006.
44. D. R. Smith: Kids: A semiautomatic program development system. *IEEE Transactions on Software Engineering*, 16:1024–1043, Sept. 1990.
45. K. Snyder: The setl2 programming language. Technical Report 490, Courant Institute/ New York University, NY, 1990.
46. J. A. Tauber, N. A. Lynch, and M. J. Tsai: Compiling IOA without global synchronization. In *Proc. of the Third IEEE Int'l Symp. on Network Computing and Applications* (*NCA '04*), *Washington, DC, 2004*, 2004.
47. Z. Wan, W. Taha, and P. Hudak: Event-driven frp. In *Proc. Fourth International Symposium on Practical Aspects of Declarative Languages* (*PADL 02*), Jan. 2002.
48. M. W. Whalen: High-integrity code generation for state-based formalisms. In *Proc. of the 22nd Int'l Conf. on Software Eng.* (*ICSE '00*), *New York, NY, USA*, 2000.
49. V. L. Winter, D. Kapur, and R. S. Berg: Refinement-based derivation of train controllers. In V. L. Winter and S. Bhattacharya, editors, *High Integrity Software*, chapter 9, 197–240. Kluwer Academic Publishers, Norwell, MA, 2001.

Universal Regular Path Queries

Oege de Moor[1], David Lacey[2], and Eric Van Wyk[3]

[1] Computing Laboratory, Oxford University, Oxford, England. oege@comlab.ox.ac.uk
[2] Department of Computer Science, University of Warwick, Coventry, England. david.lacey@dcs.warwick.ac.uk
[3] Department of Computer Science and Engineering, University of Minnesota, Minneapolis, Minnesota, USA. evw@cs.umn.edu

Summary. Given are a directed edge-labelled graph G with a distinguished node n_0, and a regular expression P which may contain variables. We wish to compute all substitutions ϕ (of symbols for variables), together with all nodes n such that all paths $n_0 \rightarrow n$ are in $\phi(P)$. We derive an algorithm for this problem using relational algebra, and show how it may be implemented in Prolog. The motivation for the problem derives from a declarative framework for specifying compiler optimisations.

Keywords: relational algebra, program analysis, regular expressions, query languages, program transformation.

1 Bob Paige and IFIP WG 2.1

Bob Paige was a long-standing member of IFIP Working Group 2.1 on Algorithmic Languages and Calculi. In recent years, the main aim of this group has been to investigate the derivation of algorithms from specifications by program transformation. Already in the mid-eighties, Bob was way ahead of the pack: instead of applying transformational techniques to well-worn examples, he was applying his theories of program transformation to new problems, and discovering new algorithms [8, 39, 43]. The secret of his success lay partly in his insistence on the study of general algorithm design strategies (in particular finite differencing [38, 42] and data structure selection [11]) rather than the study of tiny derivational steps that some of the working group had focused on.

His success in the systematic discovery of new algorithms was in itself remarkable, but perhaps even more impressive was the fact that he succeeded in automating his derivations in the APTS system [12, 40]. This provided the ultimate proof that he had succeeded in identifying deep principles in algorithm design: an automated derivation leaves no room for cheating. The mechanism for applying transformations in the APTS system was that of rewrite rules, and a fast pattern matching algorithm (invented, naturally, by Bob himself [13]) provided the basic engine. The rules could have side conditions expressed as queries on a "database" of facts about the program under consideration. The facts in the database could be any result of program analyses.

A major difficulty, which we repeatedly discussed with Bob, was to express the queries in a declarative metalanguage, and to maintain the database incrementally as the rules are applied [41]. These remain major problems in the field of automated program transformation, and the present paper is a small contribution towards solving them. We follow Bob's example in our attempt to derive the relevant algorithm itself in a transformational fashion.

Bob's influence on working group 2.1 has been immense, and he provided much inspiration for its members. He will be much missed. His work, however, lives on in the current research of the group, and this paper is but a small example of that.

O. Danvy et al. (eds.), Automatic Program Development – A Tribute to Robert Paige, 114–129.

1.1 Specifying Compiler Optimisations

Several of the phases of a compiler can be generated from declarative specifications: for instance, there are commonly used tools for syntax analysis (lex and yacc), for semantic analysis (attribute grammar systems such as FNC-2 and the SG [25, 47]) and also for instruction selection (IBURG [21]). There is however no such widely accepted tool for the declarative specification of optimising transformations, although there have been many proposals, e.g. [3, 6, 9, 17, 20, 27, 31, 52, 54, 55].

In a traditional compiler, the optimising transformations are typically performed as rewrites on the flow graph [1, 2, 37]. The difficulty lies in the specification of the necessary side conditions. For example, consider *constant propagation*. In essence, it is simply the rewrite rule

$$x := y \;\Rightarrow\; x := c$$

where y is a program variable, and c a constant. The rule is applicable only if on all execution paths to the assignment $x := y$, the last modification to y was an assignment of the form $y := c$. How can one conveniently express this condition?

Let us assume that edges in the flow graph are labelled with atomic propositions about their target statement. For example, each edge to a node that is of the form $y := E$ (or otherwise modifies y) would be labelled with the proposition $def(y)$. We can think of a path in the flow graph as a sequence of edges, or alternatively as a sequence of edge-labelling propositions. The side condition of constant propagation then becomes the requirement that all paths from program entry to $x := y$ are in the regular language

$$P = (\_)* \; ; \; y := c \; ; \; (\neg def(y))* \; ; \; x := y$$

Here (_) denotes a wildcard, (;) is sequential composition and (_)* is the usual closure operation. The symbols y, c and x are pattern variables: we seek to compute substitutions ϕ that instantiate these variables, coupled with nodes n such that all paths to n are in the regular language $\phi(P)$. In Figure 1, an example is shown that has two solutions, namely $(\{y \to q, c \to 0, x \to s\}, n_1)$ and $(\{y \to q, c \to 0, x \to t\}, n_2)$. Note the difference between *pattern* variables (x, y, c) which appear in the regular expression and *program* variables (p, q, r) which appear in the flow graph. Also notice that one of the paths in the solution, from n_0 to n_2, cannot occur in actual program runs.

```
// entry n0
if p > 3
then  q := 0;
      r := 1
else  r := 2;
      q := 0;
s := q;    // node n1
if q > 3
then  t := q    // node n2
else  t := r
```

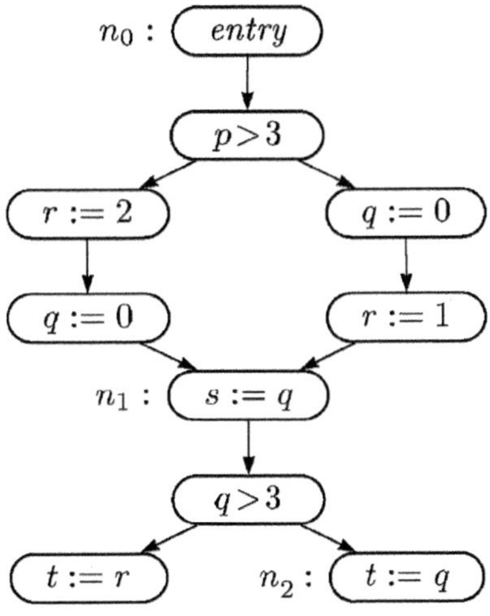

Fig. 1. An example flow graph.

As another example consider common subexpression elimination. Here the rewrite is equally simple, we replace an assignment to a complex expression with an assignment to a variable

$$x := y + z \Rightarrow x := w$$

The side condition must say that every path to the assignment $x := y + z$ passes through the assignment $w := y + z$ and nothing must interfere between these two assignments. The pattern that all paths from the entry to the $x := y + z$ node must match is

$$P_{cse} = (\_)*; w := y + z; (\neg def(w) \wedge \neg def(y) \wedge \neg def(z))*; x := y + z$$

Furthermore, the variable w should not be equal to y or z. That additional requirement could have been encoded in the above formula, but we prefer to treat it separately for expository reasons.

It is now apparent that our initial description of the problem was somewhat over-simplified, because in general we are interested in many different facts about edges in the flow graph, not just the atomic propositions such as $def(x)$. We therefore need to consider edges to be labelled by *composite* propositions that are true of the target node. Composite propositions are built from atomic propositions, and the usual logical connectives $(\neg, \wedge, \vee)$. A path in the flow graph thus corresponds to a sequence $[p_0, p_1, \ldots, p_{n-1}]$, where each p_i is a composite proposition. Similarly, the alphabet in the regular expression is that of composite propositions. We seek to compute all (substitution, node) pairs (ϕ, n) that satisfy the following condition. For every path to n in the flow graph (say $[p_0, p_1, \ldots, p_{m-1}]$) there exists a word $[q_0, q_1, \ldots, q_{m-1}]$ in the language of the pattern $\phi(P)$ such that $p_i \Rightarrow q_i$ for each $0 \leq i < m$.

To illustrate, the program in Figure 2 is annotated with composite propositions about the use and definition of different program variables. Consider the pattern P_{cse} that we introduced above. One solution is (ϕ, n_5), where $\phi = \{x \rightarrow t, w \rightarrow p, y \rightarrow 5, z \rightarrow q\}$. For example, one of the paths from the entry to n_5 is through n_1,n_2 and n_4. This path is labelled by the propositions

$$[\, p := 5 * q \wedge def(p) \wedge use(q), \quad r > 4 \wedge use(r), \quad t := 3 \wedge def(t), \quad t := 5 * q \wedge def(t) \wedge use(q)\,]$$

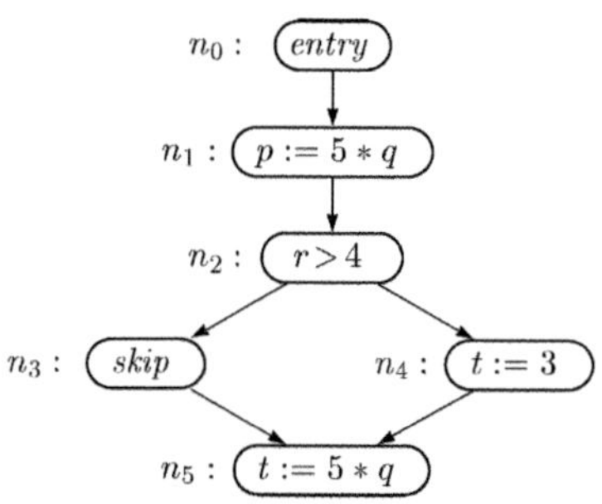

Edge	Proposition
$n_0 \rightarrow n_1$	$p := 5 * q \wedge def(p) \wedge use(q)$
$n_1 \rightarrow n_2$	$r > 4 \wedge use(r)$
$n_2 \rightarrow n_3$	*skip*
$n_2 \rightarrow n_4$	$t := 3 \wedge def(t)$
$n_3 \rightarrow n_5$	$t := 5 * q \wedge def(t) \wedge use(q)$
$n_4 \rightarrow n_5$	$t := 5 * q \wedge def(t) \wedge use(q)$

Fig. 2. A program annotated with composite propositions.

Each element of this path implies an associated element in a path in $\phi(P_{cse})$, namely:

$$[\, p := 5 * q, \quad \neg def(p) \wedge \neg def(5) \wedge \neg def(q), \quad \neg def(p) \wedge \neg def(5) \wedge \neg def(q), \quad t := 5 * q\,]$$

In this paper, we shall initially ignore the propositional structure of the alphabet, and solve a (seemingly) simpler problem first, as a stepping stone towards the above application. We aim to develop an algorithm for solving *universal regular path queries* of the following form. Given a regular expression P that contains a number of variables, an edge-labelled directed graph G and a distinguished node n_0 of G, it is required to compute all (substitution, node) pairs (ϕ, n) so that all paths $n_0 \rightarrow n$ are in the regular language $\phi(P)$. Naturally we are only interested in those pairs where n is actually reachable, so that there exists at least one path $n_0 \rightarrow n$ in $\phi(P)$.

The structure of this paper is as follows. First we derive an algorithm for the case that P does not contain variables. Our purpose in presenting this derivation is to promote the use of universal algebra in reasoning about problems in automata theory; using a number of well-understood concepts from universal algebra, the derivation is a calculation of merely six steps. Next, we encode that algorithm as a Prolog program.

This paper is an exploratory step towards a tool for programming optimising transformations in a declarative style, and we conclude with a discussion of the further work required. We also briefly discuss some intriguing connections with other fields, in particular that of query languages for semistructured data.

2 Specification and Derivation

2.1 Specification

Relations. We write $R : X \leftarrow Y$ to indicate that R is a subset of $X \times Y$. This slightly unusual notation (with arrows pointing backwards from source Y to target X) makes it a little easier to read formulae involving composition, defined below. The predicate xRy is shorthand for $(x, y) \in R$. Two relations $R : X \leftarrow Y$ and $S : Y \leftarrow Z$ can be composed to form $R \cdot S$:

$$x(R \cdot S)z \equiv \exists y \in Y : xRy \wedge ySz$$

A relation $R : X \leftarrow Y$ is said to be a *function* if it relates each $y \in Y$ to exactly one $x \in X$. A particular example of a function is the identity relation $id_X : X \leftarrow X$, which maps each element of X to itself.

Automata and folds. To formulate the specification of our problem as a relation, we shall first need to cast the familiar notion of an automaton in relational calculus. Functional programmers know that automata are very similar to the *fold-left* function $(\![init, step]\!)$, which takes a constant *init*, a transition function *step*. When applied to a list, it sums the elements from left to right using *step*, starting with the constant *init*:

$$(\![init, step]\!)[a_0, a_1, \ldots, a_{n-1}] = (\ldots((init \text{ `}step\text{' } a_0) \text{ `}step\text{' } a_1) \ldots \text{ `}step\text{' } a_{n-1})$$

Indeed, fold-left exactly operates like a deterministic state machine, with initial state *init* and transition relation *step*. In a pioneering paper on algebra and automata theory, Eilenberg and Wright [19] have shown that fold-left can be generalised to take relational arguments. This has the obvious intuitive interpretation, where each application of *step* makes a non-deterministic choice among the possible transitions. All the familiar identities of functional programming generalise to the relational setting. We shall see several examples of such identities shortly.

For convenience, we shall think of *init* and *step* as relations with types:

$$init : S \leftarrow 1 \qquad \text{and} \qquad step : S \leftarrow S \times A$$

Here 1 denotes a set that has only one element, which we denote as $\bullet$. If *init* is a function, it picks out exactly one element in S. In general, *init* corresponds to a subset of S, the set of all initial states of a nondeterministic state machine. That is, $s(init)\bullet$ if s is an initial state. In what follows, we shall refer to a pair $(init, step)$ (that has the above signature for some sets S and A) as a *machine*.

In our problem, both the flow graph and the pattern can be modelled as machines. To wit, the flow graph is a machine

$$G = (G_0 : N \leftarrow 1, \quad G_1 : N \leftarrow N \times A)$$

Here G_0 is the distinguished start node of the flow graph, and G_1 specifies the edges. Thus $(\![G]\!) : N \leftarrow A^*$. The pattern is a machine

$$P = (P_0 : S \leftarrow 1, \quad S_1 : S \leftarrow S \times A)$$

corresponding to the regular expression, and thus we have $(\![P]\!) : S \leftarrow A^*$. In addition to this machine for the pattern, we also need a specification of its final states. This we chose to model as a relation: $F : 1 \leftarrow S$. Note that (as in the case of initial states) we can identify such a relation F with a subset of S. That is, $\bullet(F)s$ if s is a final state. The advantage of defining the final states in this way is the following concise definition of the language of the pattern:

$$F \cdot (\![P]\!) : 1 \leftarrow A^*$$

In words, a string x (a list with elements drawn from A) is in the language of the pattern if $(\![P]\!)$ relates some final state to x. Below we shall sometimes write $L(F, P)$ for the subset of A^* defined in this way.

To complete the specification of our problem, we need an operator that encodes universal quantification in the relational calculus. Given two relations $R : X \leftarrow Z$ and $S : Y \leftarrow Z$ that share the same source type, the *division* of $R/S : X \leftarrow Y$ is defined by

$$x(R/S)y \equiv \forall z : ySz \Rightarrow xRz$$

That is, R/S is the largest relation T such that $T \cdot S \subseteq R$. Expressed as an equivalence, that means $T \subseteq R/S \equiv T \cdot S \subseteq R$, for all $T : X \leftarrow Y$.

Readers familiar with relational semantics of imperative programs will recognise the weakest prespecification [22, 23] in this formula.

Specification. Here is the problem that we wish to solve: compute each node n of the flow graph such that $\forall xs \in A^* : n(\![G]\!)xs \Rightarrow xs \in L(F, P)$.

We could also have formulated that requirement thus:

$$\forall xs \in A^* : n(\![G]\!)xs \Rightarrow \bullet(F \cdot (\![P]\!))xs$$

Hardened veterans of the relational calculus will spot that this formula can be expressed more concisely with division, thus obtaining

$$\bullet\ (F \cdot (\![P]\!))/(\![G]\!)\, n$$

Now we have arrived at the official specification from which we wish to derive an algorithm:

$$(F \cdot (\![P]\!))/(\![G]\!) : 1 \leftarrow N$$

The conciseness of this expression may appear somewhat forbidding. As we shall see, however, it allows us to give a very compact presentation of the algorithm that solves our problem.

2.2 Derivation

From infinite to finite universal quantification. It is worth noting that our starting point is non-executable. To see why, consider the universal quantification

$$\forall xs \in A^* : n(\![G]\!)xs \Rightarrow xs \in L(F, P)$$

Here we quantify over an infinite range, namely all strings with elements drawn from A. It stands to reason, therefore, that our first step towards an algorithm is to try and reduce that infinite quantification to a finite one. In terms of the official specification

$$(F \cdot (\![P]\!))/(\![G]\!)$$

we aim to achieve that by shunting $(\![P]\!)$ from the numerator to the denominator.

This is the purpose of the so-called *shunting* law:

$$(R \cdot f)/S = R/(S \cdot f^\circ) \tag{1}$$

Here f is required to be a function, and f° stands for the converse of f (the relation f with all pairs flipped round). In the left hand side the quantification is over the source type of f, whereas one the right hand side, the quantification is over the source type of R. Unfortunately the shunting law is not applicable here, because the relation $(\![P]\!)$ that we wish to shunt is not necessarily a function.

There is hope, however, because every relation of $R : X \leftarrow Y$ can be represented as a function $\Lambda R : \mathsf{P}X \leftarrow Y$ that maps Y to the powerset of $X(\mathsf{P}X)$:

$$\Lambda R y = \{\, x \mid xRy \,\}$$

We call ΛR the *power transpose* of R. The function (Λ) is a bijection, and the original relation can be retrieved by composing with the membership relation $mem : X \leftarrow \mathsf{P}X$:

$$mem \cdot \Lambda R = R\,. \tag{2}$$

Let us now return to our original goal, namely to reduce the universal quantification in the specification from finite to infinite. We calculate:

$$\begin{array}{ll} & (F \cdot (\!|P|\!))/(\!|G|\!) \\ = & \{\text{Equation (2)}\} \quad (F \cdot mem \cdot \Lambda(\!|P|\!))/(\!|G|\!) \\ = & \{\text{Equation (1)}\} \quad (F \cdot mem)/((\!|G|\!) \cdot (\Lambda(\!|P|\!))^\circ) \end{array}$$

This has achieved the desired reduction, because instead of universally quantifying over all strings, we are now quantifying over all sets of states. As the set of states is finite, so is the collection of all its subsets. It may be helpful to spell out the details of this reduction from an infinite to a finite quantification. Consider the types in the division $(F \cdot mem)/((\!|G|\!) \cdot (\Lambda(\!|P|\!))^\circ)$. The right hand side operand has type $1 \leftarrow \mathsf{P}S$ and the left hand operator has type $N \leftarrow \mathsf{P}S$. The division thus quantifies over the elements of $\mathsf{P}S$, which is a finite set. By contrast, in the original specification we quantified over the source type of $(\!|P|\!)$, which is the infinite set of words A^*.

Eliminating converse. Unfortunately it appears that we have created a new obstacle to executability in the denominator, however. The composition $(\!|G|\!) \cdot (\Lambda(\!|P|\!))^\circ$ now involves an existential quantification over all strings since composition $(\cdot)$ quantifies over the target type of $\Lambda(\!|P|\!)^\circ$ (the source of $(\!|G|\!)$) which is A^*. To get rid of this new infinity, we first aim to massage the converse operator away. For that, we shall need some auxiliary facts about splits, and the range of a relation. Some readers will recognise that at this point we are heading for the construction of a product automaton — the precise sense in which that is true will become apparent shortly.

Given two relations $R : X \leftarrow Y$ and $S : Z \leftarrow Y$, we can form a new relation $\langle X, Y \rangle : (X \times Z) \leftarrow Y$ such that

$$(x, z)\langle R, S\rangle y = xRy \wedge zSy\,.$$

This is called the *split* of R and S. As an example, $\langle (\!|P|\!), \Lambda(\!|G|\!)\rangle$ is a relation of type

$$(N \times \mathsf{P}S) \leftarrow A^*$$

The *range* (denoted by $ran(...)$) of a relation $T : U \leftarrow V$ is a subset of the identity relation on U, defined by

$$u(ran(R))u' = u = u' \wedge \exists v : uRv$$

Writing $outl : X \leftarrow X \times Z$ and $outr : Z \leftarrow X \times Z$ for the obvious projection functions, we have

$$R \cdot S^\circ = outl \cdot ran\langle R, S\rangle \cdot outr^\circ \tag{3}$$

Both sides of this equation are merely ways of formulating the familiar predicate

$$x(R \cdot S^\circ)z \quad = \quad \exists y : xRy \wedge zSy \quad = \quad x(outl \cdot ran\langle R, S\rangle \cdot outr^\circ)z$$

Applying Equation (3) to our programming problem, we obtain

$$(F \cdot mem)/((\!|G|\!) \cdot (\Lambda(\!|P|\!))^\circ) \;=\; (F \cdot mem)/(outl \cdot ran\langle (\!|G|\!), \Lambda(\!|P|\!)\rangle \cdot outr^\circ)$$

At first it might appear that little has been gained here. After all, the range operator itself is not executable. However, we have now set the scene for applying some well-known identities from functional programming to the fold-left operators. In what follows, we shall exclusively concentrate on obtaining an executable expression for the relation

$$ran\langle (\!|G|\!), \Lambda(\!|P|\!)\rangle$$

which we can regard as a subset of $(N \times \mathsf{P}S)$.

Simplifying the fold-lefts, and range. First, consider $\Lambda(\!|P|\!)$. This is the function corresponding to a non-deterministic machine. It is well-known that this function can itself be expressed in terms of a deterministic machine. That is, there exists a function P' such that

$$\Lambda(\!|P|\!) = (\!|P'|\!) \tag{4}$$

In fact, the general proof of this equation was one of the main achievements of the paper by Eilenberg and Wright [19] that we mentioned earlier. We refrain from spelling out the detailed definition of P'.

Using the above equation in our programming problem, we obtain a subexpression of the form $\langle (\!|G|\!), (\!|P'|\!) \rangle$. As every functional programmer knows, this is an inefficient program, because it makes two independent traversals of its input list. That is, given $xs \in A^*$ we traverse the xs list twice; once to compute $(\!|G|\!)xs$ and once to compute $(\!|P'|\!)xs$. Using the *tupling* transformation, the same result can be achieved in a single pass over xs. Formally, there exists a composite machine $G \otimes P'$ such that

$$\langle (\!|G|\!), (\!|P'|\!) \rangle = (\!|G \otimes P'|\!) \tag{5}$$

This identity is affectionately known as the *banana-split* law [35]. It was not invented to reason about automata, but rather to capture the tupling transformation [15,45] in a calculational style. Again we leave it to the interested reader to work out the detailed definition of $G \otimes P'$. In formal language theory, it is known as the *product machine* of G and P' [24, page 59]. An example of a graph G, a pattern P, deterministic pattern P' and the cross product $G \otimes P'$ is shown in Figure 3. The final states of the pattern and cross product have thickly drawn edges.

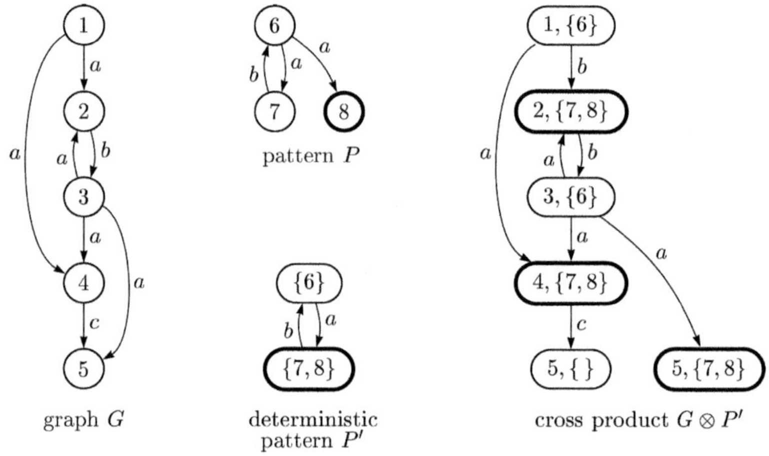

Fig. 3. A graph, pattern, deterministic pattern, and cross product.

We have achieved our task if we can give an executable expression for $ran(\!|G \otimes P'|\!)$. Before continuing however, we need a closure operator on relations. The *closure* of a relation $R : X \leftarrow X$ is the smallest reflexive and transitive relations that contains R. We write R^* for the closure of R: $R^* = id \cup R \cup (R \cdot R) \cup (R \cdot R \cdot R) \cup \ldots$

In describing $ran(\!|G \otimes P'|\!)$ in terms of machines, we are asking for the set of reachable states of the machine $G \otimes P'$. Given that reading, it should not come as a surprise that

$$ran(\!|M|\!) = ran((M_1 \cdot outl^\circ)^* \cdot M_0) \tag{6}$$

In words, to find the reachable states of a machine M, proceed as follows. Start with the initial states M_0. Then find all states reachable via zero or more transitions in the step relation M_1. Naturally this reachability problem can be implemented through depth-first search. Summarising the results of this section, we derived that

$$ran\langle (\!|G|\!), \Lambda(\!|P|\!) \rangle = ran(((G \otimes P')_1 \cdot outl^\circ)^* \cdot (G \otimes P')_0)$$

This completes our derivation.

2.3 Summary and Complexity Analysis

The above exposition was leisurely, aimed at readers who are unfamiliar with relational calculus. We now repeat the same calculation as an expert would have written it in his notebook, and we analyse the result. First, the problem is reduced to computing the range of a relation:

$$
\begin{array}{ll}
 & (F \cdot (\![P]\!))/(\![G]\!) \\
= & \{\text{cancelling } \Lambda \text{ (reverse)}\} \quad (F \cdot mem \cdot \Lambda(\![P]\!))/(\![G]\!) \\
= & \{\text{shunting}\} \quad (F \cdot mem)/((\![G]\!) \cdot (\Lambda(\![P]\!))^{\circ}) \\
= & \{\text{split and range}\} \quad (F \cdot mem)/(outl \cdot ran\langle(\![G]\!), \Lambda(\![P]\!)\rangle \cdot outr^{\circ})
\end{array}
$$

Next, we elaborate the range expression:

$$
\begin{array}{ll}
 & ran\langle(\![G]\!), \Lambda(\![P]\!)\rangle \\
= & \{\text{Eilenberg-Wright}\} \quad ran\langle(\![G]\!), (\![P']\!)\rangle \\
= & \{\text{banana split}\} \quad ran(\![G \otimes P']\!) \\
= & \{\text{range of fold-left}\} \quad ran(((G \otimes P')_1 \cdot outl^{\circ})^* \cdot (G \otimes P')_0)
\end{array}
$$

In words, we have derived an algorithm that proceeds in four steps:

1. Let P' be the deterministic equivalent of P.
2. Take product machine $G \otimes P'$.
3. Compute the reachable states of $G \otimes P'$.
4. Return the set $\{n \mid \forall s : (n, s) \text{ reachable} : s \text{ final in } P'\}$.

What is the time complexity of this algorithm? There are a number of characteristics of our application that simplify the analysis. First, the pattern is very small compared to the flow graph, so we can regard its size as a constant. Furthermore, in a typical flow graph all nodes have a bounded out-degree, so the number of edges is linear in the number of nodes. It thus stands to reason that we measure the complexity in terms of the number of nodes in G. The crucial step is the third, where we compute the reachable states of the product machine. The size of that machine is linear in the size of G. Furthermore, the reachable states can be computed by depth-first search, again linear in G. We conclude that the overall complexity of the algorithm is linear. It is also worth noting that when using this algorithm in an optimising compiler, the patterns for recognising optimisation opportunities are fixed. Its only input is the program which is converted to the flowgraph G. Thus, the computational expense imposed by the patterns can be considered constant.

3 Implementing Universal Regular Path Queries

Let us now return to the original problem from the introduction, which may have seemed slightly more general than the algorithm that we have just derived.

First of all, our machines operate on predicates, not symbols that are drawn from a finite set. That is only a seeming generalisation, for we did not exploit the finiteness of the alphabet anywhere in our proofs.

The notion of acceptance of sequences of propositions is also a special case of our earlier definitions. A transition $a \xrightarrow{p} b$ in the pattern is possible on input q precisely when $q \Rightarrow p$. This is a slightly more complicated way of mapping labelled edges to transitions, but there is nothing special about the resulting transition relation.

All that remains, therefore, is to cater for the presence of free variables, and finding substitutions for those variables. This we can do by regarding the pattern with variables as mere shorthand for a family of ground patterns. By writing the algorithm we have derived above in a logic programming language, with the pattern as a predicate that takes the variables as explicit arguments, we get a program that exhibits exactly the desired behaviour.

However, as we shall see, the program employs logical negation, which is logically sound only if all variables have been instantiated to ground terms.

Astute readers will have noticed a subtle discrepancy between our informal discussion of the problem in the introduction, and its formalisation in relational calculus. In the introduction, we stipulated that there must exist a path $v \to w$ in the language of the pattern, as well as requiring that all paths $v \to w$ are in that language. The difference is only important, of course, if w is not reachable from v. Here we shall take advantage of the insistence on the existence of a suitable path (which is dictated by our application to program transformation), and first run an existential path query to instantiate the variables appropriately. We can then run the ground universal query to check that the instantiation is indeed a valid answer.

Below we shall show how a particular query can be compiled to a Prolog program. Such compilation happens when the transformations and the associated path queries are known; the queries are then run when we also have a flow graph to transform. The particular implementation of Prolog that we have chosen is called XSB [46,48]. It is particularly suitable for experiments in program transformation and analysis [16].

3.1 Common Subexpression Elimination

Common subexpression elimination is applicable at node N if all paths from program entry to N can be split into four parts:

- First there is a part that we do not care about, consisting of zero or more edges.
- Next, we encounter an edge whose target is an assignment $W := A$, where A is a non-trivial expression and the set of variables used in A is Vs. Also, the variable W should not occur in Vs.
- Then we have zero or more edges to nodes that do not define W, nor any of the variables in Vs.
- Finally, we have an edge target at N, where the statement is of the form $X := A$.

If this condition is satisfied, and there exists at least one path of the appropriate form, the statement at node N can be replaced by the assignment $X := W$.

We could write the above condition as the following regular path query.

```
{}*;
{tgt'(assign(W,A)),  not(triv(A)),  uses(A,Vs),  not(elem(V,Ws))} ;
{not(def'(W)), not(somedefs'(Vs))}* ;
{tgt'(assign(X,A))}
```

Here each of the goals in curly brackets matches a single edge. Each predicate that is marked with an apostrophe takes the edge as an implicit argument. Below we show how to map this query into Prolog, and that mapping makes the implicit edge arguments explicit. To illustrate the use of the query, an example program is shown in Figure 4(a), and the results of running the query in Prolog are shown in Figure 4(b). Note that it is a property of this query that it can succeed only by instantiating all the free variables to ground terms. We shall rely on that property in what follows. The query will depend on its free variables (`W,A,Vs,X`). We can represent these variables as free variables in XSB. We can also put them in a term wrapper (`subst`) to give us a representation of substitution `Phi` which can then be passed between predicates

```
Phi = subst(W,A,Vs,X)
```

To define the flow graph itself in Prolog, we declare the nodes and edges with clauses such as these:

```
node(0) .
node(1) .
edge(0,e0,1) .
```

```
0:  entry
1:  w := a + g(b,c)
2:  i := g(a,b)
3:  if i < 10 then goto 4
              else goto 6
4:  x := a + g(b,c)
5:  i := i + x; goto 3
6:  a := g(a,b)
7:  w := a + g(b,d)
8:  i := g(a,b)
9:  if i < 10 then goto 10
              else goto 12
10: x := a + g(b,d)
11: i := i + x; goto 9
12: a := g(a,b)
13: exit
```

(a)

```
| ?- univpaths(subst(W,A,Vs,X),0,N).

W = w
A = plus(var(a),g(var(b),var(c)))
Vs = [a,b,c]
X = x
N = 4;

W = w
A = plus(var(a),g(var(b),var(d)))
Vs = [a,b,d]
X = x
N = 10;

no
```

(b)

Fig. 4. Example program (a) and an example run of query (b).

The middle argument of the `edge` predicate is an identifier for the relevant edge, so `edge(N_0,E,M_0)` and `edge(N_1,E,M_1)` implies that `N_0 = N_1` and `M_0 = M_1`.

The statements at each node are specified by the relevant clauses of the `stmt` predicate:

```
stmt(assign(i,g(var(a),var(b))),2) .
stmt(if(less(var(i),const(10))),3) .
```

We can then check for the statement at the target of an edge as follows:

```
tgt(S,E) :- edge(_N,E,M), stmt(S,M) .
```

Note how the edge `E` has been made explicit. Similar definitions can be made for all the constituents of the above regular path query, and a summary can be found in Figure 5. Using these definitions, we can define the predicate `goal`, with a separate case for each of the constituents of our query:

```
:- table goal/3 .
goal(0,subst(_W,_A,_Vs,_X),_E) .
goal(1,subst(W,A,Vs,_X),E)  :- tgt(assign(W,A),E), not(triv(A)),
                               not(elem(W,Vs)), uses(A,Vs) .
goal(2,subst(W,_A,Vs,_X),E) :- not(def(W,E)), not(somedefs(Vs,E)) .
goal(3,subst(_W,A,_Vs,X),E) :- tgt(assign(X,A),E) .
```

The tabling directive for `goal` is important for efficiency, as the same goal may be evaluated many times at a particular edge.

```
elem(A,[A|_As]) .
elem(A,[_B|As]) :- elem(A,As) .
src(S,E) :- edge(N,E,_M),stmt(S,N) .
tgt(S,E) :- edge(_N,E,M),stmt(S,M) .
def(V,E) :- tgt(assign(V,_X),E) .
somedefs(Vs,E) :- elem(V,Vs),def(V,E) .
triv(const(_C)) .
triv(var(_V)) .
append([],X,X) .
append([A|X],Y,[A|Z]) :- append(X,Y,Z) .
uses(const(_C),[]) .
uses(var(V),[V]) .
uses(plus(P,Q),Z) :- uses(P,X), uses(Q,Y), append(X,Y,Z) .
uses(g(P,Q),Z)    :- uses(P,X), uses(Q,Y), append(X,Y,Z) .
```

Fig. 5. Prolog preliminaries.

With the definition of `goal` in hand, we can construct the nondeterministic automaton that corresponds to the pattern. We name the states of the pattern `{a,b,c}`, with `a` the initial state, and `c` the final state. One can now define the possible transitions thus:

```
npattern(Phi,a,a,E) :- goal(0,Phi,E) .
npattern(Phi,a,b,E) :- goal(1,Phi,E) .
npattern(Phi,b,b,E) :- goal(2,Phi,E) .
npattern(Phi,b,c,E) :- goal(3,Phi,E) .
```

3.2 Solving Existential Path Queries

To solve an existential path query, we first construct the product of the flow graph and the non-deterministic pattern:

```
nproduct(Phi,N1,A1,N2,A2) :- edge(N1,E,N2), npattern(Phi,A1,A2,E) .
```

This definition says that there exists a transition from `(N1,A1)` to `(N2,A2)` in the product if there exists corresponding transitions from `N1` to `N2` and from `A1` to `A2`.

We now need to compute what states are reachable in the product. Using *tabling*, we can write this simply as the definition of reflexive transitive closure in Prolog:

```
:- table nreach/5 .
nreach(_Phi,N1,P1,N1,P1) .
nreach(Phi,N1,P1,N2,P2) :- nproduct(Phi,N1,P1,Na,Pa),
                           nreach(Phi,Na,Pa,N2,P2) .
```

There exists a path from `N1` to `N2` in the flow graph in the language of the pattern if there exists a path from `(N1,a)` to `(N2,c)` in the product. Accordingly, we define:

```
somepaths(Phi,N1,N2) :- nreach(Phi,N1,a,N2,c) .
```

Interested readers may find it an amusing exercise to formally derive this program in relational calculus: the derivation for existential queries is much simpler than that for universal ones.

3.3 Solving Universal Regular Path Queries

Our next task is to write a program for solving universal regular path queries. For that, we need a deterministic version of the pattern automaton. The construction of that automaton is a little tricky, so we defer its discussion till later. For now it suffices to know that it is given by three predicates, namely `state`, `final` and `pattern`. The first of these corresponds to `node` and it is true of all the states in the pattern automaton. The second predicate singles out those states that are final. The last predicate `pattern(Phi,P1,P2,E)` checks whether for a particular substitution `Phi`, the transition from state `P1` to `P2` in the pattern automaton is implied by the edge `E`.

Recall that in our algorithm, we need to compute the product automaton of the flow graph and the pattern. The definition is analogous to that in the nondeterministic case:

```
product(Phi,N1,P1,N2,P2) :- edge(N1,E,N2),
                            pattern(Phi,P1,P2,E) .
```

That is, we can make the transition from the product state `(N1,P1)` to `(N2,P2)` if there are relevant edges in the flow graph, and in the pattern.

The next step in the algorithm is to compute the reachable states in the product automaton. Again we use tabling:

```
:- table reach/5 .
reach(_Phi,N1,P1,N1,P1) .
reach(Phi,N1,P1,N2,P2) :- product(Phi,N1,P1,Na,Pa), reach(Phi,Na,Pa,N2,P2) .
```

Finally, we need to ensure that all paths from `N1` to `N2` are in the language of our pattern, that is

```
allpaths(Phi,N1,N2) :- bagof(P, reach(Phi,N1,p1,N2,P), Ps),
                        all(final)(Ps) .
```

The bagof primitive collects all P that can be reached (with the given instantiations of the variables) in Ps. It then only remains to check that all states in Ps are final. The higher-order predicate all is defined by

```
:- hilog all .
all(_P)([]).
all(P)([X|Xs]) :- P(X), all(P)(Xs) .
```

In the introduction of this section, we already indicated that the determinization of the pattern requires negation, and that this imposes the requirement that all variables in the substitution Phi are ground in the predicate allpaths(Phi,N1,N2). To ensure that this requirement is indeed satisfied, we define

```
univpaths(Phi,N1,N2) :- somepaths(Phi,N1,N2), ground(Phi),
                         allpaths(Phi,N1,N2) .
```

The program given here is close to that we derived in Section 2.2. Although we have not worked out the formal details, we believe the transition from relational algebra to a Prolog program could be mechanised. Obviously the difficult point in such a mechanisation would be the treatment of negation in association with non-ground queries.

3.4 Deterministic Pattern Automaton

It now remains to define the transition relation of the deterministic pattern automaton, which we have called pattern(Phi,P1,P2,E). We shall split its definition into two parts:

```
pattern(Phi,P1,P2,E) :- goals(P1,I,Phi,E), pat(I,P1,P2) .
```

The role of the two parts on the right hand side is as follows:

- The second part is a predicate pat(I,P1,P2). Here I is a 4-bit number that we shall interpret as a bit vector: the least significant bit indicates whether goal(0,...) was proved at edge E, the second bit indicates whether goal(1,...) was proved, and so on. Based on the bitvector I, pat(I,P1,P2) gives the transition relation between the states p0..p7.
- The first part of pattern is goals(P1,I,Phi,E). It sets the bit vector I according to the provability of the four goals.

As an example, let us consider the state p2, which corresponds to the singleton set {b}. We can make transitions according to goal2 or goal3 from this state:

```
goals(p2,I,Phi,E) :- try(2,I2,Phi,E), try(3,I3,Phi,E), I is 4*I2 + 8*I3 .
```

The predicate try(K,B,Phi,E) attempts to prove goal number K, setting bit B to 0 or 1 accordingly:

```
try(K,1,Phi,E) :- goal(K,Phi,E) .
try(K,0,Phi,E) :- tnot(goal(K,Phi,E)) .
```

Here tnot is the special version of negation necessary for the correct handling of tabled predicates. Like its ordinary counterpart in logic programming, in general we require that it is applied only to ground arguments. This is the reason that we needed to resort to existential queries to do the variable binding before calling the algorithm we derived earlier.

It is interesting to note that if we restrict our attention to variables whose values are drawn from a finite domain, one could employ a constraint logic programming language to implement the algorithm for universal queries. The logical negation operator tnot could then be replaced by negation for constraints. We are currently investigating such an implementation, and weighing its advantages against the restrictions it places on the expressiveness of path queries.

4 Discussion

This paper (originally published as [18]) is part of a larger effort to construct a toolkit for easy experimentation with compiler optimisations. In previous work, we have specified the side conditions of such transformations in a variant of temporal logic [30], inspired by the work of Steffen and his coworkers on specifying data flow analyses through temporal logic [50]. The formulae in temporal logic are verified using a model checker, which also finds instantiations of free variables. We noticed that many of our examples do not use the full power of temporal logic, and this motivated the exploration of universal regular path queries. In hindsight it is a very obvious thing to do, given the close connection between path problems, regular algebra and program analysis [5, 51, 52]. It remains to be seen how the algorithm presented here compares in practice with our use of a model checker.

Verbarere, Ettinger, and de Moor [53] have developed these ideas further as part of JunGL, a scripting language that allows programmers to write scripts that implement refactorings. Refactorings are source-to-source program transformations that use side conditions that capture dataflow properties. These dataflow properties use syntax inspired by [34] but are based on the path queries as described here. Sittampalam, de Moor, and Larsen [49] have also developed a incremental algorithm for solving universal regular path queries. When they are used to specify non-trivial program analyses as side conditions to program transformations it is important to solve these queries incremenatally as it can be prohibitively expensive to re-run such analyses after each transformation.

Liu and her colleagues [32, 33] have built upon the work presented here by specifying a complete set of algorithms and data structures to more efficiently solve both universal and existential path queries. This has resulted in improvements in time complexity. Other contributions of this work include a precise complexity analysis of these algorithms.

Two of us (Van Wyk and Lacey) have been working on a method of proving the correctness of program transformations whose side conditions are stated in temporal logic, in collaboration with Neil Jones and Carl Cristian Frederiksen [28, 29]. A draft of their paper prompted the exploration of a simplified formalism for the side conditions. It appears that proof methods for the temporal formalism carry over without much modification to regular expressions.

There are multiple ways in which the present algorithm could be improved. In particular, it will pay to start out with a minimal deterministic automaton for the pattern [1]. It is still possible for the composite automaton to become nondeterministic during the processing of substitutions, but at least the nondeterminism will be restricted to those places where it is actually necessary.

We did not set out to apply our previous work on relational algebra to this problem, but the proof turned out to be so pretty that we decided to include it in this paper. It would be interesting to see a further exploration of automata and language theory in this style (see also [4, 7, 44]). The application of program transformation to algorithms involving automata was pioneered by Bob Paige [8, 14, 26].

Very recently we became aware that *existential* regular path queries are a well studied subject in the database community. In particular, the query language UnQL provides the possibility of querying hierarchically structured data through the use of regular expressions that contain variables [10]. The queries are existential in that it suffices to find *some* path that is in the regular language, rather than requiring that *all* paths (between the relevant vertices) are in the regular language. It would be interesting to investigate whether the techniques used to speed up existential regular path queries [36] can be adapted for universal queries as well. This connection with database research also follows a lead of Bob Paige: his own language for expressing side conditions of transformations in APTS [40] was inspired by DataLog.

Acknowledgements

Three anonymous referees suggested many improvements to the presentation of the original version of this paper which appeared in Higher Order and Symbolic Computation [18]. Oege de Moor would like to thank Annie Liu for many constructive criticisms on the work presented here; she especially helped to improve the presentation of the derivation. She also made us aware of the existence of XSB as an appropriate tool for this type of experiment. Shin-Cheng Mu and Richard Bird provided valuable comments on an early draft. We would like to thank Microsoft Research for its generous support of this research programme, as part of the Intentional Programming project.

References

1. A. V. Aho, R. Sethi, and J. D. Ullman: *Compilers: Principles, Techniques and Tools.* Addison-Wesley, 1985.
2. A. W. Appel: *Modern Compiler Implementation in ML.* Cambridge University Press, 1998.
3. U. Assmann: How to uniformly specify program analysis and transformation with graph rewrite systems. In P. Fritzson, editor, *Compiler Construction 1996*, Lecture Notes in Computer Science 1060. Springer, 1996.
4. R. Backhouse: Fusion on languages. In 10*th European Symposium on Programming, ESOP 2001*, Lecture Notes in Computer Science 2028, 107–121. Springer, 2001.
5. R. C. Backhouse and B. A. Carré: Regular algebra applied to path-finding problems. *Journal of the Institute of Mathematics and its Applications*, 15:161–186, 1975.
6. A. J. C. Bik, P. J. Brinkhaus, P. M. W. Knijnenburg, and H. A. G. Wijshoff: Transformation mechanisms in MT1. Technical report, Leiden Institute of Advanced Computer Science, 1998.
7. R. S. Bird and O. De Moor: Relational program derivation and context-free language recognition. In A. W. Roscoe, editor, *A Classical Mind: Essays dedicated to C.A.R. Hoare*, 17–35. Prentice-Hall International, 1994.
8. B. Bloom and R. Paige: Transformational design and implementation of a new efficient solution to the ready simulation problem. *Science of Computer Programming*, 24(3):189–220, 1995.
9. J. M. Boyle, K. W. Dritz, M. M. Muralidharan, and R. Taylor: Deriving sequential and parallel programs from pure LISP specifications by program transformation. In L. G. L. T. Meertens, editor, *Proceedings of the IFIP TC2/WG* 2.1 *Working Conference on Program Specification and Transformation*, 1–19. North-Holland, 1987.
10. P. Buneman, M. Fernandez, and D. Suciu: UnQL: A query language and algebra for semistructured data based on structural recursion. *VLDB Journal*, 9(1):76–110, 2000.
11. J. Cai, P. Facon, F. Henglein, R. Paige, and E. Schonberg: Type analysis and data structure selection. In B. Möller, editor, *Constructing Programs from Specifications*, 126–164. North-Holland, 1991.
12. J. Cai and R. Paige: Towards increased productivity of algorithm implementation. *ACM Software Engineering Notes*, 18(5):71–78, 1993.
13. J. Cai, R. Paige, and R. Tarjan: More efficient bottom-up multi-pattern matching in trees. *Theoretical Computer Science*, 106(1):21–60, 1992.
14. C. Chang and R. Paige: From regular expressions to DFAs using compressed NFAs. *Theoretical Computer Science*, 178(1-2):1–36, 1997.
15. W. N. Chin: Fusion and tupling transformations: Synergies and conflicts (Invited paper). In *Fuji International Workhsop on Functional and Logic Programming*, 176–195. World Scientific, 1995.
16. M. Codish, B. Demoen, and K. Sagonas: Xsb as the natural habitat for general purpose program analysis. Technical report, KU Leuven, 1996.

17. J. R. Cordy, I. H. Carmichael, and R. Halliday: The TXL programming language, version 8. Legasys Corporation, April 1995.
18. O. de Moor, D. Lacey, and E. Van Wyk: Universal regular path queries. *Higher Order and Symbolic Computation*, 16(1–2):15–35, Special issue dedicated to Robert Paige. March – June 2003.
19. S. Eilenberg and J. B. Wright: Automata in general algebras. *Information and Control*, 11(4):452–470, 1967.
20. R. E. Faith, L. S. Nyland, and J. F. Prins: KHEPERA: A system for rapid implementation of domain-specific languages. In *Proceedings USENIX Conference on Domain-Specific Languages*, 243–255, 1997.
21. D. Hanson, C. W. Fraser, and T. A. Proebsting: Engineering a simple, efficient code generator generator. *ACM Letters on Programming Languages and Systems*, 1(3):213–226, 1992.
22. C. A. R. Hoare and J. He: The weakest prespecification, I. *Fundamenta Informaticae*, 9(1):51–84, 1986.
23. C. A. R. Hoare and J. He: The weakest prespecification, II. *Fundamenta Informaticae*, 9(2):217–251, 1986.
24. J. E. Hopcroft and J. D. Ullman: *Introduction to Automata Theory, Languages, and Computation.* Addison-Wesley, 1979.
25. M. Jourdan, D. Parigot, Julié, O. Durin, and C. Le Bellec: Design, implementation and evaluation of the FNC-2 attribute grammar system. In *Conference on Programming Languages Design and Implementation.* Published as *ACM Sigplan Notices*, 25(6), 209–222, 1990.
26. J. P. Keller and R. Paige: Program derivation with verified transformations — a case study. *Communications on Pure and Applied Mathematics*, 48(9–10), 1996.
27. M. Klein, J. Knoop, D. Koschützski, and B. Steffen: DFA & OPT-METAFrame: a toolkit for program analysis and optimization. In *Tools and Algorithms for the Construction and Analysis of Systems (TACAS '96)*, Lecture Notes in Computer Science 1055, 418–421. Springer, 1996.
28. D. Lacey, N. D. Jones, E. Van Wyk, and C. C. Frederiksen: Proving correctness of compiler optimizations by temporal logic. In *Proc. 29th ACM Symposium on Principles of Programming Languages*, 283–294. ACM Press, 2002.
29. D. Lacey, N. D. Jones, E. Van Wyk, and C. C. Frederiksen: Compiler optimization correctness by temporal logic. *Higher Order and Symbolic Computation*, 17(3):173–206, September 2003.
30. D. Lacey and O. de Moor. Imperative program transformation by rewriting: In R. Wilhelm, editor, *Proceedings of the 10th International Conference on Compiler Construction*, Lecture Notes in Computer Science 2027, 52–68. Springer, 2001.
31. P. Lipps, U. Mönke, and R. Wilhelm: OPTRAN — a language/system for the specification of program transformations: system overview and experiences. In *Proceedings 2nd Workshop on Compiler Compilers and High Speed Compilation*, Lecture Notes in Computer Science 371, 52–65, 1988.
32. Y. A. Liu, T. Rothamel, F. Yu, S. D. Stoller, and N. Hu: Parametric regular path queries. In *PLDI '04: Proceedings of the ACM Sigplan 2004 Conference on Programming language design and implementation*, 219–230, New York, ACM Press, 2004.
33. Y. A. Liu and S. D. Stoller: From datalog rules to efficient programs with time and space guarantees. In *PPDP '03: Proceedings 5th ACM Sigplan International Conference on Principles and practice of declaritive programming*, 172–183, New York, ACM Press, 2003.
34. Y. A. Liu and S. D. Stoller: Querying complex graphs. In P. V. Hentenryck, editor, *8th International Symposium on Practical Aspects of Declarative Languages (PADL)*, 16–30. ACM Press, 2006.

35. E. Meijer, M. Fokkinga, and R. Paterson: Functional programming with bananas, lenses, envelopes and barbed wire. In J. Hughes, editor, *Proceedings of the 1991 ACM Conference on Functional Programming Languages and Computer Architecture*, Lecture Notes in Computer Science 523, 124–144. Springer, 1991.
36. T. Milo and D. Suciu: Index structures for path expressions. In *International Conference on Database Theory '99*, Lecture Notes in Computer Science 1540, 277–295. Springer, 1999.
37. S. Muchnick: *Advanced Compiler Design and Implementation.* Morgan Kaufmann, 1997.
38. R. Paige: Programming with invariants. *IEEE Software*, 3(1):56–69, 1986.
39. R. Paige: Real-time simulation of a set machine on a RAM. In N. Janicki and W. Koczkodaj, editors, *Computing and Information*, volume 2, 69–73. Canadian Scholars' Press, 1989.
40. R. Paige: Viewing a program transformation system at work. In M. Hermenegildo and J. Penjam, editors, *Joint 6th International Conference on Programming Language Implementation and Logic Programming (PLILP) and 4th International conference on Algebraic and Logic Programming (ALP)*, Lecture Notes in Computer Science 844, 5–24. Springer, 1991.
41. R. Paige: Future directions in program transformations. *Computing Surveys*, 28A(4), 1996.
42. R. Paige and S. Koenig: Finite differencing of computable expressions. *ACM Transactions on Programming Languages and Systems*, 4(3):401–454, 1982.
43. R. Paige, R. Tarjan, and R. Bonic: A linear time solution to the single function coarsest partition problem. *Theoretical computer science*, 40(1):67–84, 1985.
44. H. A. Partsch: Transformational program development in a particular problem domain. *Science of Computer Programming*, 7(2):99–241, 1986.
45. A. Pettorossi: Methodologies for transformations and memoing in applicative languages. Ph.D. thesis CST-29-84, University of Edinburgh, Scotland, 1984.
46. I. V. Ramakrishnan, P. Rao, K. F. Sagonas, T. Swift, and D. S. Warren: Efficient tabling mechanisms for logic programs. In *International Conference on Logic Programming*, 697–711, 1995.
47. T. W. Reps and T. Teitelbaum: *The Synthesizer Generator: A system for constructing language-based editors*. Texts and Monographs in Computer Science. Springer, 1989.
48. K. Saganos: The XSB system v2.4: Programmer's Manual. 2001.
49. G. Sittampalam, O. de Moor, and K. F. Larsen: Incremental execution of transformation specifications. In *POPL '04: Proceedings of the 31st ACM Sigplan-Sigact Symposium on Principles of programming languages*, 26–38. ACM Press, 2004.
50. B. Steffen: Data flow analysis as model checking. In *Proceedings of Theoretical Aspects of Computer Science (TACS '91)*, Lecture Notes in Computer Science 526, 346–364. Springer, 1991.
51. R. E. Tarjan: Fast algorithms for solving path problems. *Journal of the Association for Computing Machinery*, 28(3):594–614, 1981.
52. S. W. K. Tjiang and J. L. Hennessy: Sharlit — a tool for building optimizers. In *ACM Sigplan Conference on Programming Language Design and Implementation*, 1992.
53. M. Verbaere, R. Ettinger, and O. de Moor: JunGL: a scripting language for refactoring. In D. Rombach and M. L. Soffa, editors, *28th International Conference on Software Engineering (ICSE)*. ACM Press, 2006.
54. E. Visser, Z. Benaissa, and A. Tolmach: Building program optimizers with rewriting strategies. In *International Conference on Functional Programming '98*, ACM Sigplan, 13–26. ACM Press, 1998.
55. D. Whitfield and M. L. Soffa: An approach for exploring code-improving transformations. *ACM Transactions on Programming Languages and Systems*, 19(6):1053–1084, 1997.

Derivation of Efficient Logic Programs by Specialization and Reduction of Nondeterminism

Alberto Pettorossi[1], Maurizio Proietti[2], and Sophie Renault[3]

[1] DISP, University of Roma Tor Vergata, Roma, Italy. pettorossi@info.uniroma2.it
[2] IASI-CNR, Roma, Italy. proietti@iasi.rm.cnr.it
[3] European Patent Office, Rijswijk, The Netherlands. srenault@epo.org

Summary. Program specialization is a program transformation methodology which improves program efficiency by exploiting the information about the input data which are available at compile time. We show that current techniques for program specialization based on partial evaluation do not perform well on nondeterministic logic programs. We then consider a set of transformation rules which extend the ones used for partial evaluation, and we propose a strategy for guiding the application of these extended rules so to derive very efficient specialized programs. The efficiency improvements which sometimes are exponential, are due to the reduction of nondeterminism and to the fact that the computations which are performed by the initial programs in different branches of the computation trees, are performed by the specialized programs within single branches. In order to reduce nondeterminism we also make use of mode information for guiding the unfolding process. To exemplify our technique, we show that we can automatically derive very efficient matching programs and parsers for regular languages. The derivations we have performed could not have been done by previously known partial evaluation techniques.

Keywords: automatic program derivation, logic programming, program specialization, program transformation, transformation rules and strategies.

1 Introduction

The goal of *program specialization* [21] is the adaptation of a generic program to a specific context of use. *Partial evaluation* [7,21] is a well-established technique for program specialization which from a program and its *static input* (that is, the portion of the input which is known at compile time), allows us to derive a new, more efficient program in which the portion of the output which depends on the static input, has already been computed. Partial evaluation has been applied in several areas of computer science, and it has been applied also to logic programs [13,26,29], where it is also called *partial deduction*. In this paper we follow a *rule-based* approach to the specialization of logic programs [4,36,37,41]. In particular, we consider definite logic programs [28] and we propose new program specialization techniques based on unfold/fold transformation rules [6,46]. In our approach, the process of program specialization can be viewed as the construction of a sequence, say $P_0, \ldots, P_n$, of programs, where P_0 is the program to be specialized, P_n is the derived, specialized program, and every program of the sequence is obtained from the previous one by applying a transformation rule.

As shown in [36,41], partial deduction can be viewed as a particular rule-based program transformation technique using the definition, unfolding, and folding rules [46] with the following two restrictions: (i) each new predicate introduced by the definition rule is defined by precisely one non-recursive clause whose body consists of precisely one atom (in this sense, according to the terminology of [16], partial deduction is said to be *monogenetic*), and (ii) the folding rule uses only clauses introduced by the definition rule. In what follows

O. Danvy et al. (eds.), Automatic Program Development – A Tribute to Robert Paige, 130–177.

the definition and folding rules which comply with restrictions (i) and (ii), are called *atomic definition* and *atomic folding*, respectively.

In Section 3 we will see that the use of these restricted transformation rules makes it easier to automate the partial deduction process, but it may limit the program improvements which can be achieved during program specialization. In particular, when we perform partial deduction of nondeterministic programs using atomic definition, unfolding, and atomic folding, it is impossible to combine information present in different branches of the computation trees, and as a consequence, it is often the case that we cannot reduce the nondeterminism of the programs.

This weakness of partial deduction is demonstrated in Section 3.3 where we revisit the familiar problem of looking for occurrences of a pattern in a string. It has been shown in [11, 13, 15] that by partial deduction of a string matching program, we may derive a deterministic finite automaton (DFA, for short), similarly to what is done by the Knuth–Morris–Pratt algorithm [22]. However, in [11,13,15] the string matching program to which partial deduction is applied, is *deterministic*. We show that by applying partial deduction to a *nondeterministic* version of the matching program, one cannot derive a specialized program which is deterministic, and thus, one cannot get a program which corresponds to a DFA.

Conjunctive partial deduction [8] is a program specialization technique which extends partial deduction by allowing the specialization of logic programs w.r.t. conjunctions of atoms, instead of a single atom. Conjunctive partial deduction can be realized by the definition, unfolding, and folding rules where each new predicate introduced by the definition rule is defined by precisely one nonrecursive clause whose body is a conjunction of atoms (in this sense conjunctive partial deduction is said to be *polygenetic*).

Conjunctive partial deduction may sometimes reduce nondeterminism. In particular, it may transform generate-and-test programs into programs where the generation phase and the test phase are interleaved. However, as shown in Section 3.3, conjunctive partial deduction is not capable to derive from the nondeterministic version of the matching program a new program which corresponds to a DFA.

In our paper, we propose a specialization technique which enhances both partial deduction and conjunctive partial deduction by making use of more powerful transformation rules. In particular, in Section 4 we consider a version of the definition introduction rule so that a new predicate may be introduced by means of several non-recursive clauses whose bodies consist of conjunctions of atoms, and we allow folding steps which use these predicate definitions consisting of several clauses. We also consider the following extra rules: *head generalization*, *case split*, *equation elimination*, and *disequation replacement*. These rules may introduce, replace, and eliminate equations and negated equations between terms.

Similarly to [14, 40, 46], our extended set of program transformation rules preserves the least Herbrand model semantics. For the logic language with equations and negated equations considered in this paper, we adopt the usual Prolog operational semantics with the left-to-right selection rule, where equations are evaluated by using unification. Unfortunately, the unrestricted use of the extended set of transformation rules may not preserve the Prolog operational semantics. To overcome this problem, we consider: (i) the class of *safe programs* and (ii) suitably restricted transformation rules, called *safe transformation rules*. Through some examples we show that the class of safe programs and the safe transformation rules are general enough to allow significant program specializations.

Our notions of safe programs and transformation rules, and also the notion of determinism are based on the *modes* which are associated with predicate calls [32, 49]. We describe these notions in Section 5, where we also prove that the application of safe transformation rules preserve the operational semantics of safe programs.

Then, in Section 6, we introduce a strategy, called *Determinization*, for applying our safe transformation rules in an automatic way, so to specialize programs and reduce their nondeterminism. The new features of our strategy w.r.t. other specialization techniques are: (i) the use of mode information for unfolding and producing deterministic programs, (ii) the use of the case split rule for deriving *mutually exclusive* clauses (e.g. from the clause $H \leftarrow Body$ we may derive the two clauses: $(H \leftarrow Body)\{X/t\}$ and $H \leftarrow X \neq t, Body$), and (iii) the use of the enhanced definition and folding rules for replacing many clauses by one clause only, thereby reducing nondeterminism.

Finally, in Section 7, we show by means of some examples which refer to parsing and matching problems, that our strategy is more powerful than both partial deduction and conjunctive partial deduction. In particular, given a nondeterministic version of the matching program, by using our strategy one can derive a specialized program which corresponds to a DFA.

2 Logic Programs with Equations and Disequations between Terms

In this section we introduce an extension of definite logic programs with equations and negated equations between terms. Negated equations will also be called *disequations*. The introduction of equations and disequations during program specialization allows us to derive mutually exclusive clauses. The declarative semantics we consider, is a straightforward extension of the usual least Herbrand model of definite logic programs. The operational semantics essentially is SLD-resolution as implemented by most Prolog systems: atoms are selected from left to right, and equations are evaluated by using unification. This operational semantics is sound w.r.t. the declarative semantics (see Theorem 2 below). However, since non-ground disequations can be selected, a goal evaluated according to our operational semantics can fail, even if it is true according to the declarative semantics. In this sense, the operational semantics is not complete w.r.t. the declarative semantics.

For the notions of *substitution*, *composition* of substitutions, *identity* substitution, *domain* of a substitution, *restriction* of a substitution, *instance*, *most general unifier* (abbreviated as *mgu*), *ground expression*, *ground substitution*, *renaming substitution*, *variant*, and for other notions not defined here, we refer to [28].

2.1 Syntax

The syntax of our language is defined starting from the following infinite and pairwise disjoint sets:
(i) *variables*: $X, Y, Z, X_1, X_2, \ldots,$
(ii) *function symbols* (with arity): $f, f_1, f_2, \ldots,$ and
(iii) *predicate symbols* (with arity): *true*, $=$, $\neq$, $p, p_1, p_2, \ldots$ The predicate symbols *true*, $=$, and $\neq$ are said to be *basic*, and the other predicate symbols are said to be *non-basic*. Predicate symbols will also be called *predicates*, for short.

Now we introduce the following sets: (iv) *Terms*: $t, t_1, t_2, \ldots,$ (v) *Basic atoms*: $B, B_1, B_2, \ldots,$ (vi) *Non-basic atoms*: $A, A_1, A_2, \ldots,$ and (vii) *Goals*: $G, G_1, G_2, \ldots$ Their syntax is as follows:

Terms :	$t ::= X \mid f(t_1, \ldots, t_n)$
Basic Atoms :	$B ::= true \mid t_1 = t_2 \mid t_1 \neq t_2$
Non-basic Atoms :	$A ::= p(t_1, \ldots, t_m)$
Goals :	$G ::= B \mid A \mid G_1, G_2$

Basic and nonbasic atoms are collectively called *atoms*. Goals made out of basic atoms only are said to be *basic goals*. Goals with at least one non-basic atom are said to be *nonbasic goals*. The binary operator "," denotes *conjunction* and it is assumed to be associative with neutral element *true*. Thus, a goal G is the same as goal $(true, G)$, and it is also the same as goal $(G, true)$.

Clauses: $C, C_1, C_2, \ldots$ have the following syntax:

$C ::= A \leftarrow G$

Given a clause C of the form: $A \leftarrow G$, the non-basic atom A is called the *head* of C and it is denoted by $hd(C)$, and the goal G is called the *body* of C and it is denoted by $bd(C)$. A clause $A \leftarrow G$ where G is a basic goal, is called a *unit clause*. We write a unit clause of the form: $A \leftarrow true$ also as: $A \leftarrow$.
We say that C *is a clause for* a predicate p iff C is a clause of the form $p(\ldots) \leftarrow G$.

Programs: $P, P_1, P_2, \ldots$ are *sets* of clauses.

In what follows we will feel free to use different meta-variables to denote our syntactic expressions, and in particular, we will also denote non-basic atoms by $H, H_1, \ldots$, and goals by $K, K_1, Body, Body_1, \ldots$

Given a program P, we consider the relation δ_P over pairs of predicates such that $\delta_P(p,q)$ holds iff there exists in P a clause for p whose body contains an occurrence of q. Let δ_P^+ be the transitive closure of δ_P. We say that p *depends on* q *in* P iff $\delta_P^+(p,q)$ holds. We say that a predicate p depends on a clause C in a program P iff *either* C is a clause for p *or* C is a clause for a predicate q and p depends on q in P.

Terms, atoms, goals, clauses, and programs are collectively called *expressions*, ranged over by $e, e_1, e_2, \ldots$ By $vars(e)$ we denote the set of variables occurring in an expression e. We say that X is a *local* variable of a goal G in a clause $C : H \leftarrow G_1, G, G_2$ iff $X \in vars(G) - vars(H, G_1, G_2)$.

The application of a renaming substitution to an expression is also called a *renaming of variables*. A renaming of variables can be applied to a clause whenever needed, because it preserves the least Herbrand model semantics which we define below. Given a clause C, a *renamed apart* clause C' is any clause obtained from C by a renaming of variables, so that each variable of C' is a fresh new variable. (For a formal definition of this concept, see the definition of *standardized apart* clause in [1,28])

For any two unifiable terms t_1 and t_2, there exists at least one mgu ϑ which is *relevant* (that is, each variable occurring in ϑ also occurs in $vars(t_1) \cup vars(t_2)$) and *idempotent* (that is, $\vartheta\vartheta = \vartheta$) [1]. Without loss of generality, we assume that all mgu's considered in this paper are relevant and idempotent.

2.2 Declarative Semantics

In this section we extend the definition of least Herbrand model of definite logic programs [28] to logic programs with equations and disequations between terms. We follow the approach usually taken when defining the least $\mathcal{D}$-model of a CLP program (see, for instance, [20]). According to this approach, we consider a class of Herbrand models, called $\mathcal{H}$*-models*, where the predicates *true*, $=$, and $\neq$ have a fixed interpretation. In particular, the predicate $=$ is interpreted as the identity relation over the Herbrand universe and the predicate $\neq$ is interpreted as the complement of the identity relation. Then we define the least Herbrand model of a logic program with equations and disequations between terms as the least $\mathcal{H}$-model of the program.

The *Herbrand base* $\mathcal{HB}$ is the set of all ground *nonbasic* atoms. An $\mathcal{H}$-interpretation is a subset of $\mathcal{HB}$. Given an $\mathcal{H}$-interpretation I and a ground goal, or ground clause, or program φ, the relation $I \models \varphi$, read as φ *is true in* I, is inductively defined as follows (as usual, by $I \not\models \varphi$ we indicate that $I \models \varphi$ does not hold):

(i) $I \models true$
(ii) for every ground term t, $I \models t{=}t$
(iii) for every pair of distinct ground terms t_1 and t_2, $I \models t_1 \neq t_2$
(iv) for every nonbasic ground atom A, $I \models A$ iff $A \in I$
(v) for every pair of ground goals G_1 and G_2, $I \models G_1, G_2$ iff $I \models G_1$ and $I \models G_2$
(vi) for every ground clause C, $I \models C$ iff either $I \models hd(C)$ or $I \not\models bd(C)$
(vii) for every program P, $I \models P$ iff for every ground instance C of a clause in P, $I \models C$.

As a consequence of the above definition, a ground basic goal is true in an $\mathcal{H}$-interpretation iff it is true in all $\mathcal{H}$-interpretations. We say that a ground basic goal *holds* iff it is true in all $\mathcal{H}$-interpretations.

An $\mathcal{H}$-interpretation I is said to be an *$\mathcal{H}$-model* of a program P iff $I \models P$. Since the model intersection property holds for $\mathcal{H}$-models, similarly to [20, 28], we can prove the following important result.

Theorem 1 *For any program P there exists an $\mathcal{H}$-model of P which is the least (w.r.t. set inclusion) $\mathcal{H}$-model.*

The *least Herbrand model* of a program P is defined as the least $\mathcal{H}$-model of P and is denoted by $M(P)$.

2.3 Operational Semantics

We define the operational semantics of our programs by introducing, for each program P, a relation $G_1 \overset{\vartheta}{\longmapsto}_P G_2$, where G_1 and G_2 are goals and ϑ is a substitution, defined as follows:

(1) $(t_1{=}t_2,\ G) \overset{\vartheta}{\longmapsto}_P\ G\vartheta$ iff t_1 and t_2 are unifiable via an mgu ϑ

(2) $(t_1 \neq t_2,\ G) \overset{\varepsilon}{\longmapsto}_P\ G$ iff t_1 and t_2 are not unifiable and ε is the identity substitution

(3) $(A, G) \overset{\vartheta}{\longmapsto}_P\ (bd(C), G)\vartheta$ iff (i) A is a nonbasic atom,
(ii) C is a renamed apart clause in P, and
(iii) A and $hd(C)$ are unifiable via an mgu ϑ.

A sequence $G_0 \overset{\vartheta_1}{\longmapsto}_P \cdots \overset{\vartheta_n}{\longmapsto}_P G_n$, with $n \geq 0$, is called a *derivation* using P. If G_n is *true* then the derivation is said to be *successful*. If there exists a successful derivation $G_0 \overset{\vartheta_1}{\longmapsto}_P \cdots \overset{\vartheta_n}{\longmapsto}_P$ *true* and ϑ is the substitution obtained by restricting the composition $\vartheta_1 \ldots \vartheta_n$ to the variables of G_0, then we say that the goal G_0 *succeeds* in P with *answer substitution* ϑ.

When denoting derivations, we will feel free to omit their associated substitutions. In particular, given two goals G_1 and G_2, we write $G_1 \longmapsto_P G_2$ iff there exists a substitution ϑ such that $G_1 \overset{\vartheta}{\longmapsto}_P G_2$. We say that G_2 is *derived in one step* from G_1 (*using* P) iff $G_1 \longmapsto_P G_2$ holds. In particular, if G_2 is derived in one step from G_1 according to Point (3) of the operational semantics by using a clause C, then we say that G_2 is derived in one step from G_1 *using* C. The relation $\longmapsto_P^*$ is the reflexive and transitive closure of $\longmapsto_P$. Given two goals G_1 and G_2 such that $G_1 \longmapsto_P^* G_2$ holds, we say that G_2 is *derived* from G_1 (using P). We will feel free to omit the reference to program P when it is understood from the context.

The operational semantics presented above can be viewed as an abstraction of the usual Prolog semantics, because: (i) given a goal G_1, in order to derive a goal G_2 such that $G_1 \longmapsto_P G_2$, we consider the leftmost atom in G_1, (ii) the predicate $=$ is interpreted as unifiability of terms, and (iii) the predicate $\neq$ is interpreted as non-unifiability of terms. Similarly to [28], we have the following relationship between the declarative and the operational semantics.

Theorem 2 *For any program P and ground goal G, if G succeeds in P then $M(P) \models G$.*

The converse of Theorem 2 does not hold. Indeed, consider the program P consisting of the clause $p(1) \leftarrow X \neq 0$ only. We have that $M(P) \models p(1)$ because there exists a value for X, namely 1, which is syntactically different from 0. However, $p(1)$ does not succeed in P, because X and 0 are unifiable terms.

2.4 Deterministic Programs

Various notions of determinism have been proposed for logic programs in the literature (see, for instance, [10,18,31,43]). They capture various properties such as: "the program succeeds at most once", or "the program succeeds exactly once", or "the program will never backtrack to find alternative solutions".

Let us now present the definition of deterministic program used in this paper. This definition is based on the operational semantics described in Section 2.3.

We first need the following notation. Given a program P, a clause $C \in P$, and two goals (A_0, G_0) and (A_n, G_n), where A_0 is a nonbasic atom, we write $(A_0, G_0) \Rightarrow_C (A_n, G_n)$ iff there exists a derivation $(A_0, G_0) \longmapsto_P \ldots \longmapsto_P (A_n, G_n)$, such that: (i) $n>0$, (ii) (A_1, G_1) is derived in one step from (A_0, G_0) using C, (iii) for $i = 1, \ldots, n-1$, A_i is a basic atom, and (iv) either A_n is a nonbasic atom or (A_n, G_n) is the basic atom *true*. We write $G_0 \Rightarrow_P^* G_n$ iff there exist clauses $C_1, \ldots, C_n$ in P such that $G_0 \Rightarrow_{C_1} \cdots \Rightarrow_{C_n} G_n$.

Definition 1 (Determinism) *A program P is* deterministic *for a nonbasic atom A iff for each goal G such that $A \Rightarrow_P^* G$, there exists* at most one *clause C such that $G \Rightarrow_C G'$ for some goal G'.*

We say that a program P is *nondeterministic* for a nonbasic atom A iff it is not the case that P is deterministic for A, that is, there exists a goal G derivable from A, and there exist at least two goals G_1 and G_2, and two distinct clauses C_1 and C_2 in P, such that $G \Rightarrow_{C_1} G_1$ and $G \Rightarrow_{C_2} G_2$.

According to Definition 1, the following program is deterministic for any atom of the form $non\_zero(Xs,\ Ys)$ where Xs is a ground list.

1. $non\_zero([\,],[\,]) \leftarrow$
2. $non\_zero([0|Xs],\ Ys) \leftarrow non\_zero(Xs,\ Ys)$
3. $non\_zero([X|Xs],[X|\,Ys]) \leftarrow X \neq 0, non\_zero(Xs,\ Ys)$

Notice that the above definition of a deterministic program for a nonbasic atom A allows some search during the construction of a derivation starting from A. Indeed, there may be a goal G derived from A such that from G we can derive in one step two or more new goals using distinct clauses. However, if the program is deterministic for A, after evaluating the basic atoms occurring at leftmost positions in these new goals, at most one derivation can be continued and at most one successful derivation can be constructed. For instance, from the goal $non\_zero([0,0,1],\ Ys)$ we can derive in one step two distinct goals: (i) $non\_zero([0,1],\ Ys)$ (using clause 2), and (ii) $0 \neq 0, non\_zero([0,1],\ Ys')$ (using clause 3). However, there exists only one clause C (that is, clause 2) such that $non\_zero([0,0,1],\ Ys) \Rightarrow_C G'$ for some goal G' (that is, $non\_zero([0,1],\ Ys')$).

3 Partial Deduction via Unfold/Fold Transformations

In this section we recall the rule-based approach to partial deduction. We also point out some limitations of partial deduction [36, 41] and conjunctive partial deduction [8]. These limitations motivate the introduction of the new, enhanced rules and strategies for program specialization presented in Sections 4, 5, and 6.

3.1 Transformation Rules and Strategies for Partial Deduction

In the rule-based approach, partial deduction can be viewed as the construction of a sequence $P_0, \ldots, P_n$ of programs, called a *transformation sequence*, where P_0 is the initial program to be specialized, P_n is the final, specialized program, and for $k = 0, \ldots, n-1$, program P_{k+1} is *derived* from program P_k by by applying one of the following *transformation rules* PD1–PD4.

Rule PD1 (Atomic Definition Introduction). We introduce a clause D, called *atomic definition clause*, of the form

$newp(X_1, \ldots, X_h) \leftarrow A$

where (i) *newp* is a nonbasic predicate symbol not occurring in $P_0, \ldots, P_k$, (ii) A is a nonbasic atom whose predicate occurs in program P_0, and (iii) $\{X_1, \ldots, X_h\} = vars(A)$. Program P_{k+1} is the program $P_k \cup \{D\}$.

We denote by $Defs_k$ the set of atomic definition clauses which have been introduced by the definition introduction rule during the construction of the transformation sequence $P_0, \ldots, P_k$. Thus, in particular, we have that $Defs_0 = \emptyset$.

Rule PD2 (Definition Elimination). Let p be a predicate symbol. By *definition elimination w.r.t.* p we derive the program $P_{k+1} = \{C \in P_k \mid p \text{ depends on } C\}$.

Rule PD3 (Unfolding). Let C be a renamed apart clause of P_k of the form: $H \leftarrow G_1, A, G_2$, where A is a nonbasic atom. Let $C_1, \ldots, C_m$, with $m \geq 0$, be the clauses of P_k such that, for $i = 1, \ldots, m$, A is unifiable with the head of C_i via the mgu ϑ_i. By *unfolding* C *w.r.t.* A, for $i = 1, \ldots, m$, we derive the clause $D_i : (H \leftarrow G_1, bd(C_i), G_2)\vartheta_i$. Program P_{k+1} is the program $(P_k - \{C\}) \cup \{D_1, \ldots, D_m\}$.

Rule PD4 (Atomic Folding). Let C be a renamed apart clause of P_k of the form: $H \leftarrow G_1, A\vartheta, G_2$, where: (i) A is a nonbasic atom, and (ii) ϑ is a substitution, and let D be an atomic definition clause in $Defs_k$ of the form: $N \leftarrow A$. By *folding* C *w.r.t.* $A\vartheta$ *using* D we derive the nonbasic atom $N\vartheta$ and we derive the clause $E : H \leftarrow G_1, N\vartheta, G_2$. Program P_{k+1} is the program $(P_k - \{C\}) \cup \{E\}$.

The partial deduction of a program P may be realized by applying the atomic definition introduction, definition elimination, unfolding, and atomic folding rules, according to the so called *partial deduction strategy* which we will describe below. Our partial deduction strategy uses two subsidiary strategies: (1) an *Unfold* strategy, which derives new sets of clauses by repeatedly applying the unfolding rule, and (2) a *Define-Fold* strategy, which introduces new atomic definition clauses and it folds the clauses derived by the *Unfold* strategy. These subsidiary strategies use an *unfolding selection function* and a *generalization function*, which we now define. Let us first introduce the following notation: (i) *NBAtoms* is the set of all nonbasic atoms, (ii) *Clauses* is the set of all clauses, (iii) *Clauses** is the set of all finite sequences of clauses, (iv) $\mathcal{P}(Clauses)$ is the powerset of *Clauses*, (v) a sequence of clauses is denoted by $C_1, \ldots, C_n$, and (vi) the empty sequence of clauses is denoted by ().

An *unfoldingselection function* is a total function $Select: Clauses^* \times Clauses \rightarrow NBAtoms \cup \{halt\}$, where *halt* is a symbol not occurring in *NBAtoms*. We assume that, for $C_1, \ldots, C_n \in Clauses^*$ and $C \in Clauses$, $Select((C_1, \ldots, C_n), C)$ is a nonbasic atom in the body of C.

When applying the *Unfold* strategy the *Select* function is used as follows. During the unfolding process starting from a set *Cls* of clauses, we consider a clause, say C, to be unfolded, and the sequence of its *ancestor clauses*, that is, the sequence $C_1, \ldots, C_n$ of clauses such that: (i) $C_1 \in Cls$, (ii) for $k = 1, \ldots, n{-}1$, C_{k+1} is derived by unfolding C_k, and (iii) C is derived by unfolding C_n. Now, (i) if $Select((C_1, \ldots, C_n), C) = A$, where A is a nonbasic atom in the body of C, then C is unfolded w.r.t. A, and (ii) if $Select((C_1, \ldots, C_n), C) = halt$ then C is not unfolded.

A *generalization function* is a function $Gen : \mathcal{P}(Clauses) \times NBAtoms \rightarrow Clauses$ which is defined for any set *Defs* of atomic definition clauses and for any nonbasic atom A. $Gen(Defs, A)$ is either a clause in *Defs* or a clause of the form $g(X_1, \ldots, X_h) \leftarrow GenA$, where: (i) $\{X_1, \ldots, X_h\} = vars(GenA)$, (ii) A is an instance of *GenA*, and (iii) g is a new predicate, that is, it occurs neither in P nor in *Defs*.

When applying the *Define-Fold* strategy the generalization function *Gen* is used as follows: when we want to fold a clause C w.r.t. a nonbasic atom A in its body, we consider the set *Defs* of all atomic definition clauses introduced so far and we apply the folding rule using $Gen(Defs, A)$. This application of the folding rule is indeed possible because, by construction, A is an instance of the body of $Gen(Defs, A)$.

Partial Deduction Strategy

Input: A program P and a non-basic atom $p(t_1, \ldots, t_h)$ w.r.t. which we want to specialize P.
Output: A program P_{pd} and a non-basic atom $p_{pd}(X_1, \ldots, X_r)$, such that: (i) $\{X_1, \ldots, X_r\} = vars(p(t_1, \ldots, t_h))$, and (ii) for every ground substitution $\vartheta = \{X_1/u_1, \ldots, X_r/u_r\}$,

$M(P) \models p(t_1, \ldots, t_h)\vartheta$ iff $M(P_{pd}) \models p_{pd}(X_1, \ldots, X_r)\vartheta$.

Initialize: Let S be the clause $p_{pd}(X_1, \ldots, X_r) \leftarrow p(t_1, \ldots, t_h)$. Let $Ancestors(S)$ be the empty sequence of clauses.
$TransfP := P$; $Defs := \{S\}$; $Cls := \{S\}$;
while $Cls \neq \emptyset$ **do**
(1) *Unfold*:
while there exists a clause $C \in Cls$ with $Select(Ancestors(C), C) \neq halt$ **do**
Let $Unf(C) = \{E \mid E$ is derived by unfolding C w.r.t. $Select(Ancestors(C), C)\}$.
$Cls := (Cls - \{C\}) \cup Unf(C)$;
for each $E \in Unf(C)$ let $Ancestors(E)$ be the sequence $Ancestors(C)$ followed by C
end-while;
(2) *Define-Fold*:
$NewDefs := \emptyset$;
while there exists a clause $C \in Cls$ and there exists a non-basic atom $A \in bd(C)$ which has not been derived by folding **do**
Let G be the atomic definition clause $Gen(Defs, A)$ and F be the clause derived by folding C w.r.t. A using G.
$Cls := (Cls - \{C\}) \cup \{F\}$;
if $G \notin Defs$ **then** ($Defs := Defs \cup \{G\}$; $NewDefs := NewDefs \cup \{G\}$)
end-while;
$TransfP := TransfP \cup Cls$; $Cls := NewDefs$
end-while;
We derive the final program P_{pd} by applying the definition elimination rule and keeping only the clauses of $TransfP$ on which p_{pd} depends.

A given unfolding selection function *Select* is said to be *progressive* iff for the empty sequence () of clauses and for any clause C whose body contains at least one nonbasic atom, we have that $Select((), C) \neq halt$.

We have the following correctness result which is a straightforward corollary of Theorem 5 of Section 4.2.

Theorem 3 (Correctness of Partial Deduction w.r.t. the Declarative Semantics) *Let Select be a progressive unfolding selection function. Given a program P and a nonbasic atom $p(t_1, \ldots, t_h)$, if the partial deduction strategy using Select terminates with output program P_{pd} and output atom $p_{pd}(X_1, \ldots, X_r)$, then for every ground substitution $\vartheta = \{X_1/u_1, \ldots, X_r/u_r\}$, $M(P) \models p(t_1, \ldots, t_h)\vartheta$ iff $M(P_{pd}) \models p_{pd}(X_1, \ldots, X_r)\vartheta$.*

We say that an unfolding selection function *Select* is *halting* iff for any infinite sequence $C_1, C_2, \ldots$ of clauses, there exists $n \geq 0$ such that $Select((C_1, C_2, \ldots, C_n), C_{n+1}) = halt$.

Given an infinite sequence $A_1, A_2, \ldots$ of nonbasic atoms, its *image* under the generalization function *Gen*, is the sequence of sets of clauses defined as follows:

$G_1 = \{newp(X_1, \ldots, X_n) \leftarrow A_1\}$, where $\{X_1, \ldots, X_n\} = vars(A_1)$
$G_{i+1} = G_i \cup \{Gen(G_i, A_{i+1})\}$ for $i \geq 1$.

We say that *Gen* is *stabilizing* iff for any infinite sequence $A_1, A_2, \ldots$ of nonbasic atoms whose image under *Gen* is $G_1, G_2, \ldots$, there exists $n > 0$ such that $G_k = G_n$ for all $k \geq n$.

We have the following theorem whose proof is similar to the one in [25].

Theorem 4 (Termination of Partial Deduction) *Let Select be a halting unfolding selection function and Gen be a stabilizing generalization function. Then for any input program P and non-basic atom* $p(t_1, \ldots, t_h)$*, the partial deduction strategy using Select and Gen terminates.*

The following example shows that the unfolding rule (and thus, the partial deduction strategy) is not correct w.r.t. the operational semantics.

Example 1 Let us consider the following program P_1:

1. $p \leftarrow X \neq a,\ q(X)$
2. $q(b) \leftarrow$

By unfolding clause 1 w.r.t. q(X) we derive the following program P_2:

3. $p \leftarrow b \neq a$
2. $q(b) \leftarrow$

We have that the goal p does not succeed in P_1, while it succeeds in P_2.

We will address this correctness issue in detail in Section 5, where we will present a set of transformation rules which are correct w.r.t. the operational semantics for the class of *safe* programs (see Theorem 6).

3.2 An Example of Partial Deduction: String Matching

In this section we illustrate the partial deduction strategy by means of a well-known program specialization example which consists in specializing a general string matching program w.r.t. a given pattern (see [11,13,44] for a similar example). Given a program for searching a pattern in a string, and a fixed ground pattern p, we want to derive a new, specialized program for searching the pattern p in a given string. Now we present a general program, called *Match*, for searching a pattern P in a string S in $\{a, b\}^*$. Strings in $\{a, b\}^*$ are denoted by lists of a's and b's. This program is deterministic for atoms of the form $match(P, S)$, where P and S are ground lists.

Program *Match* (initial, deterministic)

1. $match(P, S) \leftarrow match1(P, S, P, S)$
2. $match1([\,], S, Y, Z) \leftarrow$
3. $match1([C|P], [C|S], Y, Z) \leftarrow match1(P, S, Y, Z)$
4. $match1([a|P], [b|S], Y, [C|Z]) \leftarrow match1(Y, Z, Y, Z)$
5. $match1([b|P], [a|S], Y, [C|Z]) \leftarrow match1(Y, Z, Y, Z)$

Let us assume that we want to specialize this program *Match* w.r.t. the goal $match([a, a, b], S)$, that is, we want to derive a program which tells us whether or not the pattern $[a, a, b]$ occurs in the string S.

We apply our partial deduction strategy using the following unfolding selection function *DetU* and generalization function *Variant*.

(1) The function $DetU : Clauses^* \times Clauses \rightarrow NBAtoms \cup \{halt\}$ is defined as follows:

(i) $DetU((), C) = A$ if A is the leftmost nonbasic atom in the body of clause C,
(ii) $DetU((C_1, C_2, \ldots, C_n), C) = A$ if $n \geq 1$ and A is the leftmost nonbasic atom in the body of C such that A is unifiable with at most one clause head in the program to be partially evaluated, and
(iii) $DetU((C_1, C_2, \ldots, C_n), C) = halt$ if there exists no nonbasic atom in the body of C which is unifiable with at most one clause head in the program to be partially evaluated.

(2) The function $Variant : \mathcal{P}(Clauses) \times NBAtoms \rightarrow Clauses$ is defined as follows:

(i) $Variant(Defs, A)$ is a clause C such that $bd(C)$ is a variant of A, if in *Defs* there exists any such clause C, and
(ii) $Variant(Defs, A)$ is the clause $newp(X_1, \ldots, X_h) \leftarrow A$, where *newp* is a new predicate symbol and $\{X_1, \ldots, X_h\} = vars(A)$, otherwise.

The function *DetU* corresponds to the *determinate unfolding rule* considered in [13]. We have that *DetU* is not halting and *Variant* is not stabilizing. Nevertheless, in our example, as the reader may verify, the partial deduction strategy using *DetU* and *Variant* terminates and generates the following specialized program:

Program $Match_{pd}$ (specialized by partial deduction, deterministic)

6. $match_{pd}(S) \leftarrow new1(S)$
7. $new1([a|S]) \leftarrow new2(S)$
8. $new1([b|S]) \leftarrow new1(S)$
9. $new2([a|S]) \leftarrow new3(S)$
10. $new2([b|S]) \leftarrow new1(S)$
11. $new3([b|S]) \leftarrow$
12. $new3([a|S]) \leftarrow new3(S)$

The program $Match_{pd}$ is deterministic for atoms of the form $match_{pd}(S)$, where S is a ground list, and it corresponds to a DFA in the sense that: (i) each predicate corresponds to a state, (ii) each clause, except for clause 6 and 11, corresponds to a transition from the state corresponding to the predicate of the head to the state corresponding to the predicate of the body, (iii) each transition is labeled by the symbol (either a or b) occurring in the head of the corresponding clause, (iv) by clause 6 we have that *new*1 is the initial state for goals of the form $match_{pd}(w)$, where w is any ground list representing a word in $\{a, b\}^*$, and (v) clause 11 corresponds to a transition, labeled by b, to an unnamed final state where any remaining portion of the input word is accepted.

Thus, via partial deduction we can derive a DFA from a deterministic string matching program. The derived program corresponds to the Knuth-Morris-Pratt string matching algorithm [22].

3.3 Some Limitations of Partial Deduction

The fact that the partial deduction strategy derives a DFA is a consequence of the fact that the initial string matching program *Match* is rather sophisticated and, indeed, the correctness proof of the program *Match* is not straightforward. Actually, the partial deduction strategy does *not* derive a DFA if we consider, instead of the program *Match*, the following naive initial program for string matching:

Program *Naive_Match* (initial, nondeterministic)

1. $naive\_match(P, S) \leftarrow append(X, R, S),\ append(L, P, X)$
2. $append([\,], Y, Y) \leftarrow$
3. $append([A|X], Y, [A|Z]) \leftarrow append(X, Y, Z)$

This program is nondeterministic for atoms of the form $naive\_match(P, S)$, where P and S are ground lists. The correctness of this naive program is straightforward because for a given pattern P and a string S, *Naive_Match* tests whether or not P occurs in S by looking in a nondeterministic way for two strings L and R such that S is the concatenation of L, P, and R in this order.

The reader may verify that the partial deduction strategy does not derive a DFA when starting from the program *Naive_Match*. Indeed, if we specialize *Naive_Match* w.r.t. the goal $naive\_match([a, a, b], S)$ by applying the partial deduction strategy using the unfolding selection function *DetU* and the generalization function *Variant*, then we derive the following program $Naive\_Match_{pd}$ which does *not* correspond to a DFA and it is nondeterministic:

Program $Naive\_Match_{pd}$ (specialized by partial deduction, nondeterministic)

4. $naive\_match_{pd}(S) \leftarrow new1(X, R, S),\ new2(L, X)$
5. $new1([\ ], Y, Y) \leftarrow$
6. $new1([A|X], Y, [A|Z]) \leftarrow new1(X, Y, Z)$
7. $new2([\ ], [a, a, b]) \leftarrow$
8. $new2([A|X], [A|Z]) \leftarrow new2(X, Z)$

Indeed, this $Naive\_Match_{pd}$ program looks in a nondeterministic way for two strings L and R such that S is the concatenation of L, $[a, a, b]$, and R. If the pattern $[a, a, b]$ is not found within the string S at a given position, then the search for $[a, a, b]$ is restarted after a shift of one character to the right of that position.

From the program *Naive_Match* we can derive a specialized program which is much more efficient than $Naive\_Match_{pd}$ by applying *conjunctive* partial deduction, instead of partial deduction. Conjunctive partial deduction, viewed as a sequence of applications of transformation rules, enhances partial deduction because: (i) one may introduce a definition clause whose body is a conjunction of atoms, instead of one atom only (see Rule PD1), and (ii) one may fold a clause w.r.t. a conjunction of atoms in its body, instead of one atom only (see Rule PD4). By applying conjunctive partial deduction one may avoid intermediate data structures, such as the list X constructed by using clause 1 of program *Naive_Match*. Indeed, by using the ECCE system for conjunctive partial deduction [24], from the *Naive_Match* program we derive the following specialized program:

Program $Naive\_Match_{cpd}$ (specialized by conjunctive partial deduction, nondeterministic)

9. $naive\_match_{cpd}([X, Y, Z|S]) \leftarrow new1(X, Y, Z, S)$
10. $new1(a, a, b, S) \leftarrow$
11. $new1(X, Y, Z, [C|S]) \leftarrow new1(Y, Z, C, S)$

This $Naive\_Match_{cpd}$ program searches for the pattern $[a, a, b]$ in the input string by looking at the first three elements of that string. If they are a, a, and b, in this order, then the search succeeds, otherwise the search for the pattern continues in the tail of the string. Although this $Naive\_Match_{cpd}$ program is much more efficient than the initial *Naive_Match* program, it does not correspond to a DFA because, when searching for the pattern $[a, a, b]$, it looks at a prefix of length 3 of the input string, instead of one symbol only.

The failure of partial deduction and conjunctive partial deduction to derive a DFA when starting from the *Naive_Match* program, is due to some limitations which can be overcome by using the enhanced transformation rules we will present in the next section. By applying these enhanced rules we can define a new predicate by introducing *several* clauses whose bodies are non-atomic goals, while by applying the rules for partial deduction or conjunctive partial deduction, a new predicate can be defined by introducing *one* clause only. By folding using definition clauses of the enhanced form, we can derive specialized programs where non-determinism is reduced and intermediate data structures are avoided. Among our enhanced rules we also have the so called *case split rule* which, given a clause, produces two mutually exclusive instances of that clause by introducing negated equations. The application of this rule allows subsequent folding steps which reduce nondeterminism.

By applying the enhanced transformation rules according to the *Determinization Strategy* we will present in Section 6, one can automatically specialize the nondeterministic program *Naive_Match* w.r.t. the goal $naive\_match([a, a, b], S)$ thereby deriving the following deterministic program (this derivation is not presented here and it is similar to the one presented in Section 7.1):

Program *Naive_Match*$_s$	(specialized by Determinization, deterministic)
12. $naive\_match_s(S) \leftarrow new1(S)$	
13. $new1([a\|S]) \leftarrow new2(S)$	
14. $new1([C\|S]) \leftarrow C \neq a, new1(S)$	
15. $new2([a\|S]) \leftarrow new3(S)$	
16. $new2([C\|S]) \leftarrow C \neq a, new1(S)$	
17. $new3([b\|S]) \leftarrow new4(S)$	
18. $new3([a\|S]) \leftarrow new3(S)$	
19. $new3([C\|S]) \leftarrow C \neq b, C \neq a, new1(S)$	
20. $new4(S) \leftarrow$	

The program *Naive_Match*$_s$ corresponds in a straightforward way to a DFA. Moreover, since the clauses of *Naive_Match*$_s$ are pairwise mutually exclusive, the disequations in their bodies can be dropped in favor of *cuts* (or equivalently, *if-then-else* constructs) as follows:

Program *Naive_Match*$_{cut}$	(specialized, with cuts)
21. $naive\_match_s(S) \leftarrow new1(S)$	
22. $new1([a\|S]) \leftarrow !,\ new2(S)$	
23. $new1([C\|S]) \leftarrow new1(S)$	
24. $new2([a\|S]) \leftarrow !,\ new3(S)$	
25. $new2([C\|S]) \leftarrow new1(S)$	
26. $new3([b\|S]) \leftarrow !,\ new4(S)$	
27. $new3([a\|S]) \leftarrow !,\ new3(S)$	
28. $new3([C\|S]) \leftarrow new1(S)$	
29. $new4(S) \leftarrow$	

Computer experiments confirm that the final *Naive_Match*$_{cut}$ program is indeed more efficient than the *Naive_Match*, *Naive_Match*$_{pd}$, and *Naive_Match*$_{cpd}$ programs. In Section 7 we will present more experimental results which demonstrate that the specialized programs derived by our technique are more efficient than those derived by partial deduction or conjunctive partial deduction.

4 Transformation Rules for Logic Programs with Equations and Disequations between Terms

In this section we present the program transformation rules which we use for program specialization. These rules extend the unfold/fold rules considered in [14, 40, 46] to logic programs with atoms which denote equations and disequations between terms. The transformation rules we present in this section enhance in several respects the rules PD1-PD4 for partial deduction which we have considered in Section 3. In particular, we consider a definition introduction rule (see Rule 1) which allows the introduction of new predicates defined by *several* clauses whose bodies are *nonatomic* goals, while by Rule PD1 a new predicate can be defined by introducing *one* clause whose body is an *atomic* goal. We also consider a folding rule (see Rule 4) by which we can fold several clauses at a time, while by Rule PD4 we can fold one clause only. In addition, we consider the subsumption rule and the following transformation rules for introducing and eliminating equations and disequations: (i) head generalization, (ii) case split, (iii) equation elimination, and (iv) disequation replacement. Our rules preserve the least Herbrand model as indicated in Theorem 5 below.

4.1 Transformation Rules

Similarly to Section 3, the process of program transformation is viewed as a transformation sequence constructed by applying some transformation rules. However, as already mentioned, in this section we consider an enhanced set of transformation rules. A transformation sequence $P_0, \ldots, P_n$ is constructed from a given initial program P_0 by applications of the transformation rules 1–9 given below, as follows. For $k = 0, \ldots, n-1$, program P_{k+1} is derived from program P_k by: (i) selecting a (possibly empty) subset γ_1 of clauses of P_k, (ii) deriving a set γ_2 of clauses by applying a transformation rule to γ_1, and (iii) replacing γ_1 by γ_2 in P_k.

Notice that Rules 2 and 3 are in fact equal to Rules PD2 and PD3, respectively. However, we rewrite them below for the reader's convenience.

Rule 1 (Definition Introduction) We introduce m (≥ 1) new clauses, called *definition clauses*, of the form:

$$\begin{cases} D_1.\ \mathit{newp}(X_1, \ldots, X_h) \leftarrow \mathit{Body}_1 \\ \ldots \\ D_m.\ \mathit{newp}(X_1, \ldots, X_h) \leftarrow \mathit{Body}_m \end{cases}$$

where: (i) *newp* is a nonbasic predicate symbol not occurring in $P_0, \ldots, P_k$, (ii) the variables $X_1, \ldots, X_h$ are all distinct and for all $i \in \{1, \ldots, h\}$ there exists $j \in \{1, \ldots, m\}$ such that X_i occurs in the goal Body_j, (iii) for all $j \in \{1, \ldots, m\}$, every nonbasic predicate occurring in Body_j also occurs in P_0, and (iv) for all $j \in \{1, \ldots, m\}$, there exists at least one nonbasic atom in Body_j.
Program P_{k+1} is the program $P_k \cup \{D_1, \ldots, D_m\}$.

As in Section 3, we denote by Defs_k the set of definition clauses introduced by the definition introduction rule during the construction of the transformation sequence $P_0, \ldots, P_k$. In particular, we have that $\mathit{Defs}_0 = \emptyset$.

Rule 2 (Definition Elimination) Let p be a predicate symbol. By *definition elimination w.r.t. p* we derive the program $P_{k+1} = \{C \in P_k \mid p \text{ depends on } C\}$.

Rule 3 (Unfolding) Let C be a renamed apart clause of P_k of the form: $H \leftarrow G_1, A, G_2$, where A is a nonbasic atom. Let $C_1, \ldots, C_m$, with $m \geq 0$, be the clauses of P_k such that, for $i = 1, \ldots, m$, A is unifiable with the head of C_i via the mgu ϑ_i. By *unfolding C w.r.t. A*, for $i = 1, \ldots, m$, we derive the clause $D_i : (H \leftarrow G_1, bd(C_i), G_2)\vartheta_i$.
Program P_{k+1} is the program $(P_k - \{C\}) \cup \{D_1, \ldots, D_m\}$.

Notice that an application of the unfolding rule to clause C amounts to the deletion of C iff $m=0$. Sometimes in the literature this particular instance of the unfolding rule is treated as an extra rule.

Rule 4 (Folding) Let

$$\begin{cases} C_1.\ H \leftarrow G_1, \mathit{Body}_1\vartheta, G_2 \\ \ldots \\ C_m.\ H \leftarrow G_1, \mathit{Body}_m\vartheta, G_2 \end{cases}$$

be renamed clauses of P_k, for a suitable substitution ϑ, and let

$$\begin{cases} D_1.\ \mathit{newp}(X_1, \ldots, X_h) \leftarrow \mathit{Body}_1 \\ \ldots \\ D_m.\ \mathit{newp}(X_1, \ldots, X_h) \leftarrow \mathit{Body}_m \end{cases}$$

be all clauses in Defs_k which have *newp* as head predicate. Suppose that for $i = 1, \ldots, m$, the following condition holds: for every variable X occurring in the goal Body_i and not in $\{X_1, \ldots, X_h\}$, we have that: (i) $X\vartheta$ is a variable which does not occur in (H, G_1, G_2), and (ii) $X\vartheta$ does not occur in $Y\vartheta$, for any variable Y occurring in Body_i and different from X. By *folding $C_1, \ldots, C_m$ using $D_1, \ldots, D_m$* we derive the single clause E: $H \leftarrow G_1, \mathit{newp}(X_1, \ldots, X_h)\vartheta, G_2$.
Program P_{k+1} is the program $(P_k - \{C_1, \ldots, C_m\}) \cup \{E\}$.

For instance, the clauses C_1: $p(X) \leftarrow q(t(X),Y), r(Y)$ and C_2: $p(X) \leftarrow s(X), r(Y)$ can be folded (by considering the substitution $\vartheta = \{U/X, V/Y\}$) using the two definition clauses D_1: $a(U,V) \leftarrow q(t(U),V)$ and D_2: $a(U,V) \leftarrow s(U)$, and we replace C_1 and C_2 by the clause E: $p(X) \leftarrow a(X,Y), r(Y)$.

Rule 5 (Subsumption) (i) Given a substitution ϑ, we say that a clause $H \leftarrow G_1$ *subsumes* a clause $(H \leftarrow G_1, G_2)\vartheta$.
Program P_{k+1} is derived from program P_k by deleting a clause which is subsumed by another clause in P_k.

Rule 6 (Head Generalization) Let C be a clause of the form: $H\{X/t\} \leftarrow Body$ in P_k, where $\{X/t\}$ is a substitution such that X occurs in H and X does not occur in C. By *head generalization*, we derive the clause $GenC$: $H \leftarrow X\!=\!t, Body$.
Program P_{k+1} is the program $(P_k - \{C\}) \cup \{GenC\}$.

Rule 6 is a particular case of the rule of *generalization + equality introduction* considered, for instance, in [38].

Rule 7 (Case Split) Let C be a clause in P_k of the form: $H \leftarrow Body$. By *case split of C w.r.t.* the binding X/t, where X does not occur in t, we derive the following two clauses:

C_1. $(H \leftarrow Body)\{X/t\}$
C_2. $H \leftarrow X \neq t, Body$.

Program P_{k+1} is the program $(P_k - \{C\}) \cup \{C_1, C_2\}$.

In this Rule 7 we do not assume that X occurs in C. However, in the Determinization Strategy of Section 6, we will always apply the case split rule to a clause $C : H \leftarrow Body$ w.r.t. a binding X/t where X occurs in H. This use of the case split rule will be sufficient to derive mutually exclusive clauses. Indeed, according to our operational semantics, if $G \longmapsto_{P_{k+1}} G_1$ using clause C_1 and X occurs in H, then no G_2 exists such that $G \longmapsto_{P_{k+1}} G_2$ using clause C_2. The same holds by interchanging C_1 and C_2. We will return to this property in Definitions 8 (Semideterminism) and 12 (Mutual Exclusion) below.

Rule 8 (Equation Elimination) Let C_1 be a clause in P_k of the form:

C_1. $H \leftarrow G_1,\ t_1\!=\!t_2,\ G_2$

If t_1 and t_2 are unifiable via the most general unifier ϑ, then by equation elimination we derive the following clause:

C_2. $(H \leftarrow G_1, G_2)\vartheta$

Program P_{k+1} is the program $(P_k - \{C_1\}) \cup \{C_2\}$.
If t_1 and t_2 are not unifiable then by equation elimination we derive program P_{k+1} which is $P_k - \{C_1\}$.

Rule 9 (Disequation Replacement) Let C be a clause in program P_k. Program P_{k+1} is derived from P_k by either removing C or replacing C as we now indicate:

9.1 if C is of the form: $H \leftarrow G_1, t_1 \neq t_2, G_2$ and t_1 and t_2 are not unifiable, then C is replaced by $H \leftarrow G_1, G_2$
9.2 if C is of the form: $H \leftarrow G_1,\ f(t_1,\ldots,t_m) \neq f(u_1,\ldots,u_m),\ G_2$, then C is replaced by the following m (≥ 0) clauses: $H \leftarrow G_1, t_1 \neq u_1, G_2,\ \ldots,\ H \leftarrow G_1, t_m \neq u_m, G_2$
9.3 if C is of the form: $H \leftarrow G_1, X \neq X, G_2$, then C is removed from P_k
9.4 if C is of the form: $H \leftarrow G_1, t \neq X, G_2$, then C is replaced by $H \leftarrow G_1, X \neq t, G_2$
9.5 if C is of the form: $H \leftarrow G_1, X \neq t_1, G_2, X \neq t_2, G_3$ and there exists a substitution ρ which is a bijective mapping from the set of the local variables of $X \neq t_1$ in C onto the set of the local variables of $X \neq t_2$ in C such that $t_1\rho = t_2$, then C is replaced by $H \leftarrow G_1, X \neq t_1, G_2, G_3$.

In particular, by Rule 9.5, if a disequation occurs twice in the body of a clause, then we can remove the rightmost occurrence.

4.2 Correctness of the Transformation Rules w.r.t. the Declarative Semantics

In this section we show that, under suitable hypotheses, our transformation rules preserve the declarative semantics presented in Section 2.2. In that sense we also say that our transformation rules are *correct* w.r.t. the given declarative semantics. The following correctness theorem extends similar results holding for logic programs [14, 40, 46] to the case of logic programs with equations and disequations.

Theorem 5 (Correctness of the Rules w.r.t. the Declarative Semantics) *Let P_0, ..., P_n be a transformation sequence constructed by using the transformation rules* 1–9 *and let* p *be a nonbasic predicate in* P_n. *Let us assume that:*

1. if *the folding rule is applied for the derivation of a clause* C *in program* P_{k+1} *from clauses* $C_1, \ldots, C_m$ *in program* P_k *using clauses* $D_1, \ldots, D_m$ *in* $Defs_k$, *with* $0 \leq k < n$, then *for every* $i \in \{1, \ldots, m\}$ *there exists* $j \in \{1, \ldots, n-1\}$ *such that* D_i *occurs in* P_j *and* P_{j+1} *is derived from* P_j *by unfolding* D_i;
2. *during the transformation sequence* $P_0, \ldots, P_n$ *the definition elimination rule* either *is never applied* or *it is applied w.r.t. predicate* p *once only, in the last step, that is, when deriving* P_n *from* P_{n-1}.

Then, for every ground atom A *with predicate* p, *we have that* $M(P_0 \cup Defs_n) \models A$ *iff* $M(P_n) \models A$.

Proof: It is a simple extension of a similar result presented in [14] for the case where we use the unfolding, folding, and *generalization + equality introduction* rules. The proof technique used in [14] can be adapted to prove also the correctness of our extended set of rules. □

In Example 1 of Section 3 we have shown that the unfolding rule may not preserve the operational semantics. The following examples show that also other transformation rules may not preserve the operational semantics.

Example 2 Let us consider the following program P_1:

1. $p(X) \leftarrow q(X),\ X \neq a$
2. $q(X) \leftarrow$
3. $q(X) \leftarrow X = b$

By Rule 5 we may delete clause 3 which is subsumed by clause 2 and we derive a new program P_2. Now, we have that $p(X)$ succeeds in P_1, while it does not succeed in P_2.

Example 3 Let us consider the following program P_3:

1. $p(X) \leftarrow$

By the case split rule we may replace clause 1 by the two clauses:

2. $p(a) \leftarrow$
3. $p(X) \leftarrow X \neq a$

and we derive a new program P_4. The goal $p(X), X = b$ succeeds in P_3, while it does not succeed in P_4.

Example 4 Let us consider the following program P_5:

1. $p \leftarrow X \neq a,\ X = b$

By Rule 8 we may replace clause 1 by:

2. $p \leftarrow b \neq a$

and we derive a new program P_6. The goal p does not succeed in P_5, while it succeeds in P_6.

Finally, let us consider the following two operations on the body of a clause: (i) removal of a duplicate atom, and (ii) reordering of atoms. The following examples show that these two operations, which preserve the declarative semantics, may not preserve the operational semantics. Notice, however, that the removal of a duplicate atom and the reordering of atoms cannot be accomplished by the transformation rules listed in Section 4, except for the special case considered at Point 9.5 of the disequation replacement rule.

Example 5 Let us consider the program P_7:

1. $p \leftarrow q(X,Y),\ q(X,Y),\ X \neq Y$
2. $q(X,b) \leftarrow$
3. $q(a,Y) \leftarrow$

and the program P_8 obtained from P_7 by replacing clause 1 by the following clause:

4. $p \leftarrow q(X,Y),\ X \neq Y$

The goal p succeeds in P_7, while it does not succeed in P_8. Indeed, (i) for program P_7 we have that: $p \longmapsto_{P_7} q(X,Y), q(X,Y), X \neq Y \longmapsto_{P_7} q(X,b), X \neq b \longmapsto_{P_7} a \neq b \longmapsto_{P_7} true$, and (ii) for program P_8 we have that: *either* $p \longmapsto_{P_8} X \neq b$ *or* $p \longmapsto_{P_8} a \neq Y$. In Case (ii), since X and Y are unifiable with b and a, respectively, we have that $p \longmapsto^*_{P_8} true$ does not hold.

Example 6 Let us consider the program P_9:

1. $p \leftarrow q(X),\ r(X)$
2. $q(a) \leftarrow$
3. $r(X) \leftarrow X \neq b$

and the program P_{10} obtained from P_9 by replacing clause 1 by the following clause:

4. $p \leftarrow r(X),\ q(X)$

The goal p succeeds in P_9, while it does not succeed in P_{10}.

In the next section we will introduce a class of programs and a class of goals for which our transformation rules preserve both the declarative semantics and the operational semantics. In order to do so, we associate a *mode* with every predicate. A mode of a predicate specifies the *input* arguments of that predicate, and we assume that whenever the predicate is called, its input arguments are bound to ground terms. We will see that, if some suitable conditions are satisfied, compliance to modes guarantees the preservation of the operational semantics. This fact is illustrated by the above Examples 2 and 3, and indeed, in each of them, if we restrict ourselves to calls of the predicate p with ground arguments, then the initial program and the derived program have the same operational semantics.

Notice, however, that the incorrectness of the transformation of Example 4 does not depend on the modes. Thus, in order to ensure correctness w.r.t. the operational semantics we have to rule out clauses such as clause 1 of program P_5. Indeed, as we will see in the next section, the clauses we will consider satisfy the following condition: each variable which occurs in a disequation *either* occurs in an input argument of the head predicate *or* it is a local variable of the disequation.

5 Program Transformations based on Modes

Modes provide information about the directionality of predicates, by specifying whether an argument should be used as input or output (see, for instance, [32, 49]). Mode information is very useful for specifying and verifying logic programs [2, 10] and it is used in existing compilers, such as Ciao and Mercury, to generate very efficient code [19, 45]. Mode information has also been used in the context of program transformation to provide sufficient conditions which ensure that reorderings of atoms in the body of a clause preserve program termination [5].

In this paper we use mode information for: (i) specifying classes of programs and goals w.r.t. which the transformation rules we have presented in Section 4.1 preserve the operational semantics (see Section 2.3), and (ii) designing our strategy for specializing programs and reducing nondeterminism.

5.1 Modes

A *mode for a nonbasic predicate* p of arity h (≥ 0) is an expression of the form $p(m_1, \ldots, m_h)$, where for $i = 1, \ldots, h$, m_i is either $+$ (denoting *any ground* term) or ? (denoting *any* term). In particular, if $h=0$, then p has a unique mode which is p itself. Given an atom $p(t_1, \ldots, t_h)$ and a mode $p(m_1, \ldots, m_h)$,
(1) for $i = 1, \ldots, h$, the term t_i is said to be an *input argument* of p iff m_i is $+$, and
(2) a variable of $p(t_1, \ldots, t_h)$ with an occurrence in an input argument of p, is said to be an *input variable* of $p(t_1, \ldots, t_h)$.

A *mode for a program* P is a set of modes for nonbasic predicates containing exactly one mode for every distinct, nonbasic predicate p occurring in P.

Notice that a mode for a program P may or may not contain modes for nonbasic predicates which do *not* occur in P. Thus, if M is a mode for a program P_1 and, by applying a transformation rule, from P_1 we derive a new program P_2 where all occurrences of a predicate have been eliminated, then M is a mode also for P_2. The following rules may eliminate occurrences of predicates: definition elimination, unfolding, folding, subsumption, disequation replacement (case 9.5). Clearly, if from P_1 we derive P_2 by applying the definition introduction rule, then in order to obtain a mode for P_2 we should add to M a mode for the newly introduced predicate (unless it is already in M).

Example 7 Given the program P:

$p(0, 1) \leftarrow$
$p(0, Y) \leftarrow q(Y)$

the set $M_1 = \{p(+, ?), q(?)\}$ is a mode for P. $M_2 = \{p(+, ?), q(+), r(+)\}$ is a different mode for P.

Definition 2 Let M be a mode for a program P and p a nonbasic predicate. We say that an atom $p(t_1, \ldots, t_h)$ *satisfies* the mode M iff (1) a mode for p belongs to M and (2) for $i = 1, \ldots, h$, if the argument t_i is an input argument of p according to M, then t_i is a ground term. In particular, when $h=0$, we have that p *satisfies* M iff $p \in M$.
The program P *satisfies* the mode M iff for each non-basic atom A_0 which satisfies M, and for each non-basic atom A and goal G such that $A_0 \longmapsto_P^* (A, G)$, we have that A satisfies M.

With reference to Example 7 above, program P satisfies mode M_1, but it does *not* satisfy mode M_2.

In general, the property that a program satisfies a mode is undecidable. Two approaches are usually followed for verifying this property: (i) the first one uses *abstract interpretation* methods (see, for instance, [9, 32]) which always terminate, but may return a *don't know* answer, and (ii) the second one checks suitable syntactic properties of the program at hand, such as *well-modedness* [2], which imply that the mode is satisfied.

Our technique is independent of any specific method used for verifying that a program satisfies a mode. However, as the reader may verify, all programs presented in the examples of Section 7 are well-moded and, thus, they satisfy the given modes.

5.2 Correctness of the Transformation Rules w.r.t. the Operational Semantics

Now we introduce a class of programs, called *safe* programs, and we prove that if the transformation rules are applied to a safe program and suitable restrictions hold, then the given program and the derived program are equivalent w.r.t. the operational semantics.

Definition 3 (Safe Programs) Let M be a mode for a program P. We say that a clause C in P is *safe* w.r.t. M iff for each disequation $t_1 \neq t_2$ in the body of C, we have that: for each variable X occurring in $t_1 \neq t_2$ *either* X is an input variable of $hd(C)$ *or* X is a local variable of $t_1 \neq t_2$ in C. Program P is safe w.r.t. M iff all its clauses are safe w.r.t. M.

For instance, let us consider the mode $M = \{p(+), q(?)\}$. Clause $p(X) \leftarrow X \neq f(Y)$ is safe w.r.t. M and clause $p(X) \leftarrow X \neq f(Y),\ q(Y)$ is not safe w.r.t. M because Y occurs both in $f(Y)$ and in $q(Y)$.

When mentioning the safety property w.r.t. a given mode M, we feel free to omit the reference to M, if it is irrelevant or understood from the context.

In order to get our desired correctness result (see Theorem 6 below), we need to restrict the use of our transformation rules as indicated in Definitions 4–7 below. In particular, these restrictions ensure that, by applying the transformation rules, program safety and mode satisfaction are preserved (see Propositions 3 and 4 in [35, Appendix A]).

Definition 4 (Safe Unfolding) Let P_k be a program and M be a mode for P_k. Let us consider an application of the unfolding rule (see Rule 3 in Section 4.1) whereby from the following clause of P_k:

$H \leftarrow G_1, A, G_2$

we derive the clauses:

$$\begin{cases} D_1.\ (H \leftarrow G_1, bd(C_1), G_2)\vartheta_1 \\ \dots \\ D_m.\ (H \leftarrow G_1, bd(C_m), G_2)\vartheta_m \end{cases}$$

where $C_1, \dots, C_m$ are the clauses in P_k such that, for $i \in \{1, \dots, m\}$, A is unifiable with the head of C_i via the mgu ϑ_i.

We say that this application of the unfolding rule is *safe* w.r.t. mode M iff for all $i = 1, \dots, m$, for all disequations d in $bd(C_i)$, and for all variables X occurring in $d\vartheta_i$, we have that either X is an input variable of $H\vartheta_i$ or X is a local variable of d in C_i.

To see that unrestricted applications of the unfolding rule may not preserve safety, let us consider the following program:

1. $p \leftarrow q(X),\ r(X)$
2. $q(1) \leftarrow$
3. $r(X) \leftarrow X \neq 0$

and the mode $M = \{p,\ q(?),\ r(+)\}$ for it. By unfolding clause 1 w.r.t. the atom $r(X)$ we derive the clause:

4. $p \leftarrow q(X),\ X \neq 0$

This clause is not safe w.r.t. M because X does not occur in its head.

Definition 5 (Safe Folding) Let us consider a program P_k and a mode M for P_k. Let us also consider an application of the folding rule (see Rule 4 in Section 4.1) whereby from the following clauses in P_k:

$$\begin{cases} C_1.\ H \leftarrow\ G_1, (A_1, K_1)\vartheta, G_2 \\ \dots \\ C_m.\ H \leftarrow\ G_1, (A_m, K_m)\vartheta, G_2 \end{cases}$$

and the following definition clauses in $Defs_k$:

$$\begin{cases} D_1.\ newp(X_1, \dots, X_h) \leftarrow A_1, K_1 \\ \dots \\ D_m.\ newp(X_1, \dots, X_h) \leftarrow A_m, K_m \end{cases}$$

we derive the new clause:

$H \leftarrow G_1, newp(X_1, \dots, X_h)\vartheta, G_2$

We say that this application of the folding rule is *safe* w.r.t. mode M iff the following Property Σ holds:

(*Property* Σ) Each input variable of $newp(X_1, \ldots, X_h)\vartheta$ is also an input variable of at least one of the non-basic atoms occurring in $(H, G_1, A_1\vartheta, \ldots, A_m\vartheta)$.

Unrestricted applications of the folding rule may not preserve modes. Indeed, let us consider the following initial program:

1. $p \leftarrow q(X)$
2. $q(1) \leftarrow$

Suppose that first we introduce the definition clause:

3. $new(X) \leftarrow q(X)$

and then we apply the clause split rule, thereby deriving:

4. $new(0) \leftarrow q(0)$
5. $new(X) \leftarrow X \neq 0,\ q(X)$

The program made out of clauses 1, 2, 4, and 5 satisfies the mode $M = \{p,\ q(?),\ new(+)\}$. By folding clause 1 using clause 3 we derive:

6. $p \leftarrow new(X)$

This application of the folding rule is not safe and the program we have derived, consisting of clauses 2, 4, 5, and 6, does not satisfy M.

Definition 6 (Safe Head Generalization) Let us consider a program P_k and a mode M for P_k. We say that an application of the head generalization rule (see Rule 6 in Section 4.1) to a clause of P_k is *safe* iff X is not an input variable w.r.t. M.

The restrictions considered in Definition 6 are needed to preserve safety. For instance, the clause $p(t(X)) \leftarrow X \neq 0$ is safe w.r.t. the mode $M = \{p(+)\}$, while $p(Y) \leftarrow Y = t(X), X \neq 0$ is not.

Definition 7 (Safe Case Split) Let us consider a program P_k and a mode M for P_k. Let us consider also an application of the case split rule (see Rule 7 in Section 4.1) whereby from a clause C in P_k of the form: $H \leftarrow Body$ we derive the following two clauses:

C_1. $(H \leftarrow Body)\{X/t\}$
C_2. $H \leftarrow X \neq t, Body$.

We say that this application of the case split rule is *safe* w.r.t. mode M iff X is an input variable of H, X does not occur in t, and for all variables $Y \in vars(t)$, *either* Y is an input variable of H *or* Y does not occur in C.

When applying the safe case split rule, X occurs in H and thus, given a goal G, it is not the case that for some goals G_1 and G_2, we have both $G \longmapsto G_1$ using clause C_1 and $G \longmapsto G_2$ using clause C_2. In Definition 12 below, we will formalize this property by saying that the clauses C_1 and C_2 are *mutually exclusive*.

Similarly to the unfolding and head generalization rules, the unrestricted use of the case split rule may not preserve safety. For instance, from the clause $p(X) \leftarrow$ which is safe w.r.t. the mode $M = \{p(?)\}$, we may derive the two clauses $p(0) \leftarrow$ and $p(X) \leftarrow X \neq 0$, and this last clause is not safe w.r.t. M.

We have shown in Section 4.1 (see Example 6), that the reordering of atoms in the body of a clause may not preserve the operational semantics. Now we prove that a particular reordering of atoms, called *disequation promotion*, which consists in moving to the left the disequations occurring in the body of a safe clause, preserves the operational semantics. Disequation promotion (not included, for reason of simplicity, among the transformation rules) allows us to rewrite the body of a safe clause so that every disequation occurs to the left of every atom different from a disequation thereby deriving the *normal form* of that clause (see Section 6). The use of normal forms will simplify the proof of Theorem 6 below and the presentation of the Determinization Strategy in Section 6.

Proposition 1 (Correctness of Disequation Promotion) *Let M be a mode for a program P_1. Let us assume that P_1 is safe w.r.t. M and P_1 satisfies M. Let C_1: $H \leftarrow G_1,\ G_2,\ t_1 \neq t_2,\ G_3$ be a clause in P_1. Let P_2 be the program derived from P_1 by replacing clause C_1 by clause C_2: $H \leftarrow G_1,\ t_1 \neq t_2,\ G_2,\ G_3$. Then: (i) P_2 is safe w.r.t. M, (ii) P_2 satisfies M, and (iii) for each non-basic atom A which satisfies mode M, A succeeds in P_1* iff *A succeeds in P_2.*

Proof: Point (i) follows from the fact that safety does not depend on the position of the disequation in a clause. Moreover, the evaluation of goal G_2 in program P_1 according to our operational semantics, does not bind any variable in $t_1 \neq t_2$, and thus, we get Point (ii). Point (iii) is a consequence of Points (i) and (ii) and the fact that the evaluation of $t_1 \neq t_2$ does not bind any variable in the goals G_2 and G_3. □

The above proposition does not hold if we interchange clause C_1 and C_2. Consider, in fact, the following clause which is safe w.r.t. mode $M = \{p(+), q(+)\}$:

C_3. $p(X) \leftarrow X \neq Y,\ q(Z)$

This clause satisfies M because for all derivations starting from a ground instance $p(t)$ of $p(X)$ the atom $t \neq Y$ does not succeed. In contrast, if we use the clause C_4: $p(X) \leftarrow q(Z), X \neq Y$, we have that in the derivation starting from $p(t)$, the variable Z is not bound to a ground term and thus, clause C_4 does not satisfy the mode M which has the element $q(+)$.

In Theorem 6 below we will show that if we apply our transformation rules and their safe versions in a restricted way, then a program P which satisfies a mode M and is safe w.r.t. M, is transformed into a new program, say Q, which satisfies M and is safe w.r.t. M. Moreover, the programs P and Q have the same operational semantics.

Theorem 6 (Correctness of the Rules w.r.t. the Operational Semantics) *Let $P_0, \ldots, P_n$ be a transformation sequence constructed by using the transformation rules 1–9 and let p be a non-basic predicate in P_n. Let M be a mode for $P_0 \cup Defs_n$ such that: (i) $P_0 \cup Defs_n$ is safe w.r.t. M, (ii) $P_0 \cup Defs_n$ satisfies M, and (iii) the applications of the unfolding, folding, head generalization, and case split rules during the construction of $P_0, \ldots, P_n$ are all safe w.r.t. M. Suppose also that Conditions 5 and 5 of Theorem 5 hold. Then: (i) P_n is safe w.r.t. M, (ii) P_n satisfies M, and (iii) for each atom A which has predicate p and satisfies mode M, A succeeds in $P_0 \cup Defs_n$ iff A succeeds in P_n.*

Proof: See [35, Appendix A]. □

5.3 Semideterministic Programs

In this section we introduce the concept of *semideterminism* which characterizes the class of programs which can be obtained by using the Determinization Strategy of Section 6. (The reader should not confuse the notion of semideterminism presented here with the one considered in [18].)

We have already noticed that if a program P is deterministic for an atom A according to Definition 1, then there is at most one successful derivation starting from A, and A succeeds in P with at most one answer substitution. Thus, if an atom succeeds in a program with more than one answer substitution, and none of these substitutions is more general than another, then there is no chance to transform that program into a new program which is deterministic for that atom.

For instance, let us consider the following generalization of the problem of Sections 3.2 and 3.3: Given a pattern P and a string S we want to compute the *position*, say N, of an occurrence of P in S, that is, we want to find two strings L and R such that: (i) S is the concatenation of L, P, and R, and (ii) the length of L is N. The following program *Match_Pos* computes N for any given P and S:

Program *Match_Pos* (initial, nondeterministic)

1. $match\_pos(P,S,N) \leftarrow append(Y,R,S),\ append(L,P,Y),\ length(L,N)$
2. $length([\,],0) \leftarrow$
3. $length([H|T],s(N)) \leftarrow length(T,N)$
4. $append([\,],Y,Y) \leftarrow$
5. $append([A|X],Y,[A|Z]) \leftarrow append(X,Y,Z)$

The *Match_Pos* program is nondeterministic for atoms of the form $match\_pos(P,S,N)$ where P and S are ground lists, and it computes one answer substitution for each occurrence of P in S.

Suppose that we want to specialize *Match_Pos* w.r.t. the atom $match\_pos([a,a,b],S,N)$. Thus, we want to derive a new, specialized program $Match\_Pos_s$ and a new binary predicate $match\_pos_s$. This new program should be able to compute multiple answer substitutions for a goal. For instance, for the atom $match\_pos_s([a,a,b,a,a,b],N)$ the program $Match\_Pos_s$ should compute the two substitutions $\{N/0\}$ and $\{N/s(s(s(0)))\}$ and, thus, $Match\_Pos_s$ cannot be deterministic for the atom $match\_pos_s([a,a,b,a,a,b],N)$.

Now, in order to deal with programs which may return multiple answer substitutions, we introduce the notion of semideterminism, which is weaker than that of determinism. Informally, we may say that a semideterministic program has the minimum amount of nondeterminism which is needed to compute multiple answer substitutions. In Section 6 we will prove that the Determinization Strategy, if it terminates, derives a semideterministic program.

Definition 8 (Semideterminism) A program P is *semideterministic* for a nonbasic atom A iff for each goal G such that $A \Rightarrow_P^* G$, there exists *at most one* clause C such that $G \Rightarrow_C G'$ for some goal G' different from *true*.
Given a mode M for a program P, we say that P is *semideterministic* w.r.t. M iff P is semideterministic for each non-basic atom which satisfies M.

In Section 7.1 below we will show that by applying the Determinization Strategy, from $Match\_Pos_s$ we derive the following specialized program $Match\_Pos_s$ which is semideterministic for atoms of the form $match\_pos_s(S,N)$, where S is a ground list.

Program $Match\_Pos_s$ (specialized, semideterministic)

9. $match\_pos_s(S,N) \leftarrow new1(S,N)$
20. $new1([a|S],M) \leftarrow new2(S,M)$
21. $new1([C|S],s(N)) \leftarrow C{\neq}a,\ new1(S,N)$
32. $new2([a|S],M) \leftarrow new3(S,M)$
33. $new2([C|S],s(s(N))) \leftarrow C{\neq}a,\ new1(S,N)$
46. $new3([a|S],s(M)) \leftarrow new3(R,S)$
47. $new3([b|S],M) \leftarrow new4(R,S)$
48. $new3([C|S],s(s(s(N)))) \leftarrow C{\neq}a,\ C{\neq}b,\ new1(S,N)$
49. $new4(S,0) \leftarrow$
55. $new4([a|S],s(s(s(M)))) \leftarrow new2(S,M)$
56. $new4([C|S],s(s(s(s(N))))) \leftarrow C{\neq}a,\ new1(S,N)$

Now we give a simple sufficient condition which ensures semideterminism. It is based on the concept of *mutually exclusive* clauses which we introduce below. We need some preliminary definitions.

Definition 9 (Satisfiability of Disequations w.r.t. a Set of Variables) Given a set V of variables, we say that a conjunction D of disequations, is *satisfiable w.r.t.* V iff there exists a ground substitution σ with domain V, such that every ground instance of $D\sigma$ holds (see Section 2.2). In particular, D is satisfiable w.r.t. $\emptyset$ iff every ground instance of D holds.

The satisfiability of a conjunction D of disequations w.r.t. a given set V of variables, can be checked by using the following algorithm defined by structural induction:

(1) *true*, i.e., the empty conjunction of disequations, is satisfiable w.r.t. V,

(2) (D_1, D_2) is satisfiable w.r.t. V iff both D_1 and D_2 are satisfiable w.r.t. V,

(3) $X \neq t$ is satisfiable w.r.t. V iff X occurs in V and t is either a nonvariable term or a variable occurring in V distinct from X,

(4) $t \neq X$ is satisfiable w.r.t. V iff $X \neq t$ is satisfiable w.r.t. V,

(5) $f(\ldots) \neq g(\ldots)$, where f and g are distinct function symbols, is satisfiable w.r.t. V, and

(6) $f(t_1, \ldots, t_m) \neq f(u_1, \ldots, u_m)$ is satisfiable w.r.t. V iff at least one disequation among $t_1 \neq u_1, \ldots, t_m \neq u_m$ is satisfiable w.r.t. V.

The correctness of this algorithm relies on the fact that the set of function symbols is infinite (see Section 2.1).

Definition 10 (Linearity) A program P is said to be *linear* iff every clause of P has at most one nonbasic atom in its body.

Definition 11 (Guard of a Clause) The *guard* of a clause C, denoted $grd(C)$, is $bd(C)$ if all atoms in $bd(C)$ are disequations, otherwise $grd(C)$ is the (possibly empty) conjunction of the disequations occurring in $bd(C)$ to the left of the leftmost atom which is not a disequation.

Definition 12 (Mutually Exclusive Clauses) Let us consider a mode M for the following two, renamed apart clauses:

C_1. $p(t_1, u_1) \leftarrow G_1$
C_2. $p(t_2, u_2) \leftarrow G_2$

where: (i) p is a predicate of arity k (≥ 0) whose first h arguments, with $0 \leq h \leq k$, are input arguments according to M, (ii) t_1 and t_2 are h-tuples of terms denoting the input arguments of p, and (iii) u_1 and u_2 are $(k-h)$-tuples of terms.
We say that C_1 and C_2 are *mutually exclusive* w.r.t. mode M iff either (i) t_1 is not unifiable with t_2 or (ii) t_1 and t_2 are unifiable via an mgu ϑ and $(grd(C_1), grd(C_2))\vartheta$ is not satisfiable w.r.t. $vars(t_1, t_2)$.
If $h = 0$ we stipulate that the empty tuples t_1 and t_2 are unifiable via an mgu which is the identity substitution.

The following proposition is useful for proving that a program is semideterministic.

Proposition 2 (Sufficient Condition for Semideterminism) *If (i) P is a linear program, (ii) P is safe w.r.t. a given mode M, (iii) P satisfies M, and (iv) the nonunit clauses of P are pairwise mutually exclusive w.r.t. M, then P is semideterministic w.r.t. M.*

Proof: See [35, Appendix B]. □

In Section 6, we will present a strategy for deriving specialized programs which satisfies the hypotheses (i)–(iv) of the above Proposition 2, and thus, these derived programs are semideterministic.

The following examples show that in Proposition 2 no hypothesis on program P can be discarded.

Example 8 Consider the following program P and the mode $M = \{p, q\}$ for P:

1. $p \leftarrow q,\ q$
2. $q \leftarrow$
3. $q \leftarrow q$

P is not linear, but P is safe w.r.t. M and P satisfies M. The nonunit clauses of P which are the clauses 1 and 3, are pairwise mutually exclusive. However, P is not semideterministic w.r.t. M, because $p \longmapsto_P^* (q, q)$, and there exist two nonbasic goals, namely q and (q, q), such that $(q, q) \Rightarrow_P q$ and $(q, q) \Rightarrow_P (q, q)$.

Example 9 Consider the following program Q and the mode $M = \{p(?), q_1, q_2\}$ for Q:

1. $p(X) \leftarrow X \neq 0,\ q_1$
2. $p(1) \leftarrow q_2$

Q is linear and it satisfies M, but Q is not safe w.r.t. M because X is not an input variable of p. Clauses 1 and 2 are mutually exclusive w.r.t. M, because the set of input variables in $p(X)$ is empty and $X \neq 0$ is not satisfiable w.r.t. $\emptyset$. However, Q is not semideterministic w.r.t. M, because $p(1) \longmapsto_Q^* p(1)$, and there exist two non-basic goals, namely q_1 and q_2, such that $p(1) \Rightarrow_Q q_1$ and $p(1) \Rightarrow_Q q_2$.

Example 10 Consider the following program R and the mode $M = \{p, r(+), r_1, r_2\}$ for R:

1. $p \leftarrow r(X)$
2. $r(1) \leftarrow r_1$
3. $r(2) \leftarrow r_2$

R is linear and safe w.r.t. M, but R does not satisfy M, because $p \longmapsto_R r(X)$ and X is not a ground term. Clauses 1, 2, and 3 are pairwise mutually exclusive. However, R is not semideterministic w.r.t. M, because $p \longmapsto_R^* r(X)$ and there exist two nonbasic goals, namely r_1 and r_2, such that $r(X) \Rightarrow_R r_1$ and $r(X) \Rightarrow_R r_2$.

Example 11 Consider the following program S and the mode $M = \{p, r_1, r_2\}$ for S:

1. $p \leftarrow r_1$
2. $p \leftarrow r_2$

S is linear and safe w.r.t. M, and S satisfies M. Clauses 1 and 2 are not pairwise mutually exclusive. S is not semideterministic w.r.t. M, because $p \longmapsto_S^* p$, and there exist two nonbasic goals, namely r_1 and r_2, such that $p \Rightarrow_S r_1$ and $p \Rightarrow_S r_2$.

We conclude this section by observing that when a program consists of mutually exclusive clauses and, thus, it is semideterministic, it may be executed very efficiently on standard Prolog systems by inserting cuts in a suitable way. We will return to this point in Section 8 when we discuss the speedups obtained by our specialization technique.

6 A Transformation Strategy for Specializing Programs and Reducing Nondeterminism

In this section we present a strategy, called *Determinization*, for guiding the application of the transformation rules presented in Section 4.1. Our strategy pursues the following objectives. (1) The specialization of a program w.r.t. a particular goal. This is similar to what partial deduction does. (2) The elimination of multiple or intermediate data structures. This is similar to what the strategies for eliminating *unnecessary variables* [38] and conjunctive partial deduction do. (3) The reduction of nondeterminism. This is accomplished by deriving programs whose nonunit clauses are mutually exclusive w.r.t. a given mode, that is, by Proposition 2, semideterministic programs.

The Determinization Strategy is based upon three subsidiary strategies: (i) the *Unfold-Simplify* subsidiary strategy, which uses the safe unfolding, equation elimination, disequation replacement, and subsumption rules, (ii) the *Partition* subsidiary strategy, which uses the safe case split, equation elimination, disequation replacement, subsumption, and safe head generalization rules, and (iii) the *Define-Fold* subsidiary strategy which uses the definition introduction and safe folding rules. For reasons of clarity, during the presentation of the Determinization Strategy we use high-level descriptions of the subsidiary strategies. These descriptions are used to establish the correctness of Determinization (see Theorem 7). Full details of the subsidiary strategies will be given in Sections 6.2, 6.3, and 6.4, respectively.

6.1 The Determinization Strategy

Given an initial program P, a mode M for P, and an atom $p(t_1,\ldots,t_h)$ w.r.t. which we want to specialize P, we introduce by the definition introduction rule, the clause

S: $p_s(X_1, \ldots, X_r) \leftarrow p(t_1,\ldots,t_h)$

where $X_1,\ldots,X_r$ are the distinct variables occurring in $p(t_1,\ldots,t_h)$.
We also define a mode $p_s(m_1, \ldots, m_r)$ for the predicate p_s by stipulating that, for any $j = 1,\ldots,r$, m_j is $+$ iff X_j is an input variable of $p(t_1,\ldots,t_h)$ according to the mode M. We assume that the program P is safe w.r.t. M. Thus, also program $P \cup \{S\}$ is safe w.r.t. $M \cup \{p_s(m_1, \ldots, m_r)\}$. We also assume that P satisfies mode M and thus, program $P \cup \{S\}$ satisfies mode $M \cup \{p_s(m_1,\ldots,m_r)\}$.

Our Determinization Strategy is presented below as an iterative procedure that, at each iteration, manipulates the following three sets of clauses: (1) *TransfP*, which is the set of clauses from which we will construct the specialized program, (2) *Defs*, which is the set of clauses introduced by the definition introduction rule, and (3) *Cls*, which is the set of clauses to be transformed during the current iteration. Initially, *Cls* consists of the single clause S: $p_s(X_1,\ldots,X_r) \leftarrow p(t_1,\ldots,t_h)$ which is constructed as we have indicated above.

The Determinization Strategy starts off each iteration by applying the Unfold-Simplify subsidiary strategy to the set *Cls*, thereby deriving a new set of clauses called *UnfoldedCls*. The Unfold-Simplify strategy first unfolds the clauses in *Cls*, and then it simplifies the derived set of clauses by applying the equation elimination, disequation replacement, and subsumption rules.

Then the set *UnfoldedCls* is divided into two sets: (i) *UnitCls*, which is the set of unit clauses, and (ii) *NonunitCls*, which is the set of non-unit clauses. The Determinization Strategy proceeds by applying the Partition subsidiary strategy to *NonunitCls*, thereby deriving a new set of clauses called *PartitionedCls*. The Partition strategy consists of suitable applications of the case split, equation elimination, disequation replacement, and head generalization rules such that the set *PartitionedCls* has the following property: it can be partitioned into sets of clauses, called *packets*, such that two clauses taken from different packets are mutually exclusive (w.r.t. a suitable mode).

The Determinization Strategy continues by applying the Define-Fold subsidiary strategy to the clauses in *PartitionedCls*, thereby deriving a new, semideterministic set of clauses called *FoldedCls*. The Define-Fold subsidiary strategy introduces a (possibly empty) set *NewDefs* of definition clauses such that each packet can be folded into a single clause by using a set of definition clauses in *Defs* $\cup$ *NewDefs*. We have that clauses derived by folding different packets are mutually exclusive and, thus, *UnitCls* $\cup$ *FoldedCls* is semideterministic.

At the end of each iteration, *UnitCls* $\cup$ *FoldedCls* is added to *TransfP*, *NewDefs* is added to *Defs*, and the value of the set *Cls* is updated to *NewDefs*.

The Determinization Strategy terminates when *Cls* $= \emptyset$, that is, no new predicate is introduced during the current iteration.

Determinization Strategy

Input: A program P, an atom $p(t_1,\ldots,t_h)$ w.r.t. which we want to specialize P, and a mode M for P such that P is safe w.r.t. M and P satisfies M.

Output: A specialized program P_s, and an atom $p_s(X_1,\ldots,X_r)$, with $\{X_1,\ldots,X_r\} = vars(p(t_1,\ldots,t_h))$ such that: (i) for every ground substitution $\vartheta = \{X_1/u_1,\ldots,X_r/u_r\}$, $M(P) \models p(t_1,\ldots,t_h)\vartheta$ iff $M(P_s) \models p_s(X_1,\ldots,X_r)\vartheta$, and (ii) for every substitution $\sigma = \{X_1/v_1,\ldots,X_r/v_r\}$ such that the atom $p(t_1,\ldots,t_h)\sigma$ satisfies mode M, we have that: (ii.1) $p(t_1,\ldots,t_h)\sigma$ succeeds in P iff $p_s(X_1,\ldots,X_r)\sigma$ succeeds in P_s, and (ii.2) P_s is semideterministic for $p_s(X_1,\ldots,X_r)\sigma$.

Initialize: Let S be the clause $p_s(X_1,\ldots,X_r) \leftarrow p(t_1,\ldots,t_h)$.
$\mathit{TransfP} := P$; $\mathit{Defs} := \{S\}$; $\mathit{Cls} := \{S\}$; $M_s := M \cup \{p_s(m_1,\ldots,m_r)\}$, where for any $j = 1,\ldots,r$, $m_j = +$ iff X_j is an input variable of $p(t_1,\ldots,t_h)$ according to the mode M;

while $\mathit{Cls} \neq \emptyset$ **do**

(1) *Unfold-Simplify*:
We apply the safe unfolding, equation elimination, disequation replacement, and subsumption rules according to the Unfold-Simplify Strategy given in Section 6.2 below, and from *Cls* we derive a new set of clauses *UnfoldedCls*.

(2) *Partition*:
Let *UnitCls* be the unit clauses occurring in *UnfoldedCls*, and *NonunitCls* be the set of non-unit clauses in *UnfoldedCls*.
We apply the safe case split, equation elimination, disequation replacement, and safe head generalization rules according to the Partition Strategy given in Section 6.3 below, and from *NonunitCls* we derive a set *PartitionedCls* of clauses which is the union of disjoint subsets of clauses. Each subset is called a *packet*. The packets of *PartitionedCls* enjoy the following properties:
(2a) each packet is a set of clauses of the form (modulo renaming of variables):

$$\begin{cases} H \leftarrow \mathit{Diseqs}, G_1 \\ \quad \ldots \\ H \leftarrow \mathit{Diseqs}, G_m \end{cases}$$

where *Diseqs* is a conjunction of disequations and for $k = 1,\ldots,m$, no disequation occurs in G_k, and
(2b) for any two clauses C_1 and C_2, if the packet of C_1 is different from the packet of C_2, then C_1 and C_2 are mutually exclusive w.r.t. mode M_s.

(3) *Define-Fold*:
We apply the definition introduction and the safe folding rules according to the Define-Fold subsidiary strategy given in Section 6.4 below. According to that strategy, we introduce a (possibly empty) set *NewDefs* of new definition clauses and a set M_{new} of modes such that:
(3a) in M_{new} there exists exactly one mode for each distinct head predicate in *NewDefs*, and
(3b) from each packet in *PartitionedCls* we derive a single clause of the form:

$$H \leftarrow \mathit{Diseqs}, \mathit{newp}(\ldots)$$

by an application of the folding rule, which is safe w.r.t. M_{new}, using the clauses in $\mathit{Defs} \cup \mathit{NewDefs}$.
Let *FoldedCls* be the set of clauses derived by folding the packets in *PartitionedCls*.

(4) $\mathit{TransfP} := \mathit{TransfP} \cup \mathit{UnitCls} \cup \mathit{FoldedCls}$; $\mathit{Defs} := \mathit{Defs} \cup \mathit{NewDefs}$; $\mathit{Cls} := \mathit{NewDefs}$; $M_s := M_s \cup M_{new}$

end-while

We derive the specialized program P_s by applying the definition elimination rule and keeping only the clauses of *TransfP* on which p_s depends.

The Determinization Strategy may fail to terminate for two reasons: (i) the Unfold-Simplify subsidiary strategy may not terminate, because it may perform infinitely many unfolding steps, and (ii) the condition $Cls \neq \emptyset$ for exiting the while-do loop may always be false, because at each iteration the Define-Fold subsidiary strategy may introduce new definition clauses. We will discuss these issues in more detail in Section 9.

Now we show that, if the Determinization Strategy terminates, then the least Herbrand model and the operational semantics are preserved. Moreover, the derived specialized program P_s is semideterministic for $p_s(X_1,\ldots,X_r)\sigma$ as indicated by the following theorem.

Theorem 7 (Correctness of the Determinization Strategy) *Let us consider a program P, a nonbasic atom $p(t_1,\ldots,t_h)$, and a mode M for P such that:* (1) *P is safe w.r.t. M and* (2) *P satisfies M. If the Determinization Strategy terminates with output program P_s and output atom $p_s(X_1,\ldots,X_r)$ where $\{X_1,\ldots,X_r\} = vars(p(t_1,\ldots,t_h))$, then*

(i) for every ground substitution $\vartheta = \{X_1/u_1,\ldots,X_r/u_r\}$,
$M(P) \models p(t_1,\ldots,t_h)\vartheta$ *iff* $M(P_s) \models p_s(X_1,\ldots,X_r)\vartheta$ *and*

(ii) for every substitution $\sigma = \{X_1/v_1,\ldots,X_r/v_r\}$ such that the atom $p(t_1,\ldots,t_h)\sigma$ satisfies mode M,

(ii.1) $p(t_1,\ldots,t_h)\sigma$ succeeds in P iff $p_s(X_1,\ldots,X_r)\sigma$ succeeds in P_s, and

(ii.2) P_s is semideterministic for $p_s(X_1,\ldots,X_r)\sigma$.

Proof: Let *Defs* and P_s be the set of definition clauses and the specialized program obtained at the end of the Determinization Strategy.

(i) Since $p_s(X_1,\ldots,X_r) \leftarrow p(t_1,\ldots,t_h)$ is the only clause for p_s in $P \cup Defs$ and $\{X_1,\ldots,X_r\} = vars(p(t_1,\ldots,t_h))$, for every ground substitution $\vartheta = \{X_1/u_1,\ldots,X_r/u_r\}$ we have that $M(P) \models p(t_1,\ldots,t_h)\vartheta$ iff $M(P \cup Defs) \models p_s(X_1,\ldots,X_r)\vartheta$. By the correctness of the transformation rules w.r.t. the least Herbrand model (see Theorem 5), we have that $M(P \cup Defs) \models p_s(X_1,\ldots,X_r)\vartheta$ iff $M(P_s) \models p_s(X_1,\ldots,X_r)\vartheta$.

Point (ii.1) follows from Theorem 6 because during the Determinization Strategy, each application of the unfolding, folding, head generalization, and case split rule is safe.

(ii.2) We first observe that, by construction, for every substitution σ, the atom $p(t_1,\ldots,t_h)\sigma$ satisfies mode M iff $p_s(X_1,\ldots,X_r)\sigma$ satisfies mode M_s, where M_s is the mode obtained from M at the end of the Determinization Strategy. Thus, Point (ii.2) can be shown by proving that P_s is semideterministic w.r.t. M_s. In order to prove this fact, it is enough to prove that $TransfP_w - P$ is semideterministic w.r.t. M_s, where $TransfP_w$ is the set of clauses which is the value of the variable *TransfP* at the end of the while-do statement of the Determinization Strategy. Indeed, P_s is equal to $TransfP_w - P$ because, by construction, p_s does not depend on any clause of P, and thus, by the final application of the definition elimination rule, all clauses of P are removed from $TransfP_w$.

By Proposition 2, it is enough to prove that: (a) $TransfP_w - P$ is linear, (b) $TransfP_w - P$ is safe w.r.t. M_s, (c) $TransfP_w - P$ satisfies M_s, and (d) the nonunit clauses of $TransfP_w - P$ are pairwise mutually exclusive w.r.t. M_s.

Property (a) holds because according to the Determinization Strategy, after every application of the safe folding rule we get a clause of the form: $H \leftarrow Diseqs, newp(\ldots)$, where a single nonbasic atom occurs in the body. All other clauses in $TransfP_w - P$ are unit clauses.

Properties (b) and (c) follow from Theorem 6 recalling that the application of the unfolding, folding, head generalization, and case split rules are all safe.

Property (d) can be proved by showing that, during the execution of the Determinization Strategy, the following Property (I) holds:

(I): all the non-unit clauses of $TransfP - P$ are pairwise mutually exclusive w.r.t. M_s. Indeed, initially $TransfP - P$ is empty and thus, Property (I) holds. Furthermore, Property (I) is an invariant of the while-do loop. Indeed, at the end of each execution of the body of the while-do (see Point (4) of the strategy), the nonunit clauses which are added to the current value of *TransfP* are the elements of the set *FoldedCls* and those nonunit clauses are derived by applying the Partition and Define-Fold subsidiary strategies at Points (3) and (4), respectively. By construction, the clauses in *FoldedCls* are pairwise mutually exclusive w.r.t. M_{new}, and their head predicates do not occur in *TransfP*. Thus, the clauses of *TransfP* $\cup$ *UnitCls* $\cup$ *FoldedCls* are pairwise mutually exclusive w.r.t. $M_s \cup M_{new}$. As a consequence, after the two assignments (see Point (4) of the strategy) *TransfP* := *TransfP* $\cup$ *UnitCls* $\cup$ *FoldedCls* and $M_s := M_s \cup M_{new}$, we have that Property (I) holds. □

Now we describe the three subsidiary strategies for realizing the *Unfold-Simplify*, *Partition*, and *Define-Fold* transformations as specified by the Determinization Strategy. We will see these subsidiary strategies in action in the examples of Section 7.

During the application of our subsidiary strategies it will be convenient to rewrite every safe clause into its *normal form*. The normal form N of a safe clause can be constructed by performing disequation replacements and disequation promotions, so that the following Properties N1–N5 hold:

(N1) every disequation is of the form: $X \neq t$, with t different from X and unifiable with X,
(N2) every disequation occurs in $bd(N)$ to the left of every atom different from a disequation,
(N3) if $X \neq Y$ occurs in $bd(N)$ and both X and Y are input variables of $hd(N)$, then in $hd(N)$ the leftmost occurrence of X is to the left of the leftmost occurrence of Y,
(N4) for every disequation of the form $X \neq Y$ where Y is an input variable, we have that also X is an input variable, and
(N5) for any pair of disequations d_1 and d_2 in $bd(N)$, it does not exist a substitution ρ which is a bijective mapping from the set of the local variables of d_1 in N onto the set of the local variables of d_2 in N such that $d_1\rho = d_2$.

We have that: (i) the normal form of a safe clause is unique, modulo renaming of variables and disequation promotion, (ii) no two equal disequations occur in the normal form of a safe clause, and (iii) given a program P and a mode M for P such that P is safe w.r.t. M and P satisfies M, if we rewrite a clause of P into its normal form, then the least Herbrand model semantics and the operational semantics are preserved (this fact is a consequence of Theorem 5, Theorem 6, and Proposition 1).

A safe clause for which Properties N1–N5 hold, is said to be *in normal form*. If a clause C is in normal form, then by Property N2, every disequation in $bd(C)$ occurs also in $grd(C)$.

6.2 The Unfold-Simplify Subsidiary Strategy

The Unfold-Simplify strategy first unfolds the clauses in *Cls* w.r.t. the leftmost atom in their body, and then it keeps unfolding the derived clauses as long as input variables are not instantiated. Now, in order to give the formal definition of the Unfold-Simplify strategy we introduce the following concept.

Definition 13 (Consumer Atom) Let P be a program and M a mode for P. A nonbasic atom $q(t_1, \ldots, t_k)$ is said to be a *consumer atom* iff for every nonunit clause in P whose head unifies with that nonbasic atom via an mgu ϑ, we have that for $i = 1, \ldots, k$, if t_i is an input argument of q then $t_i\vartheta$ is a variant of t_i.

The Unfold-Simplify strategy is realized by the following *Unfold-Simplify* procedure, where the expression *Simplify*(S) denotes the set of clauses derived from a given set S of clauses by: (1) first, applying whenever possible the equation elimination rule to the clauses in S, (2) then, rewriting the derived clauses into their normal form, and (3) finally, applying as long as possible the subsumption rule.

Procedure *Unfold-Simplify*(*Cls*, *UnfoldedCls*).
Input: A set *Cls* of clauses in a program P and a mode M_s for P. P is safe w.r.t. M_s and for each $C \in$ *Cls*, the input variables of the leftmost nonbasic atom in the body of C are input variables of the head of C.

Output: A new set *UnfoldedCls* of clauses which are derived from *Cls* by applying the safe unfolding, equation elimination, disequation replacement, and subsumption rules. The clauses in *UnfoldedCls* are safe w.r.t. M_s.

(1) *Unfold w.r.t. Leftmost Non-basic Atom:*

UnfoldedCls := $\{E \mid$ there exists a clause $C \in$ *Cls* and clause E is derived by unfolding C w.r.t. the leftmost nonbasic atom in its body$\}$;
UnfoldedCls := *Simplify*(*UnfoldedCls*)

(2) *Unfold w.r.t. Leftmost Consumer Atom:*

while there exists a clause $C \in$ *UnfoldedCls* whose body has a leftmost consumer atom, say A, such that the unfolding of C w.r.t. A is safe **do**
UnfoldedCls := (*UnfoldedCls* $-$ $\{C\}) \cup \{E \mid E$ is derived by unfolding C w.r.t. $A\}$;
UnfoldedCls := *Simplify*(*UnfoldedCls*)
end-while

Notice that our assumptions on the input program P and clauses *Cls* ensure that the first unfolding step performed by the *Unfold-Simplify* procedure is safe.

Notice also that our Unfold-Simplify strategy may fail to terminate. We will briefly return to this issue in Section 9.

Our Unfold-Simplify strategy differs from usual unfolding strategies for (conjunctive) partial deduction (see, for instance, [8,13,36,41]), because mode information is used. We have found this strategy very effective on several examples as shown in the following Section 7.

6.3 The Partition Subsidiary Strategy

The Partition strategy is realized by the following procedure, where we will write $p(t,u)$ to denote an atom with nonbasic predicate p of arity k (≥ 0), such that: (i) t is an h-tuple of terms, with $0 \leq h \leq k$, denoting the h input arguments of p, and (ii) u is a $(k-h)$-tuple of terms denoting the arguments of p which are *not* input arguments.

Procedure *Partition*(*NonunitCls*, *PartitionedCls*).
Input: A set *NonunitCls* of nonunit clauses in normal form and without variables in common. A mode M_s for *NonunitCls*. The clauses in *NonunitCls* are safe w.r.t. M_s.
Output: A set *PartitionedCls* of clauses which is the union of disjoint packets of clauses such that:
(2a) each packet is a set of clauses of the form (modulo renaming of variables):

$$\begin{cases} H \leftarrow \mathit{Diseqs}, G_1 \\ \quad \dots \\ H \leftarrow \mathit{Diseqs}, G_m \end{cases}$$

where *Diseqs* is a conjunction of disequations and for $k = 1, \dots, m$, no disequation occurs in G_k, and
(2b) for any two clauses C_1 and C_2, if the packet of C_1 is different from the packet of C_2, then C_1 and C_2 are mutually exclusive w.r.t. mode M_s.
The clauses in *PartitionedCls* are in normal form and they are safe w.r.t. M_s.

while there exist in *NonunitCls* two clauses of the form:

C_1. $p(t_1, u_1) \leftarrow \mathit{Body}_1$
C_2. $p(t_2, u_2) \leftarrow \mathit{Body}_2$

such that: (i) C_1 and C_2 are not mutually exclusive w.r.t. mode M_s, and *either*
(ii.1) t_1 is not a variant of t_2 *or*
(ii.2) t_1 is a variant of t_2 via an mgu ϑ such that $t_1\vartheta = t_2$, and for any substitution ρ which is a bijective mapping from the set of local variables of $\mathit{grd}(C_1\vartheta)$ in $C_1\vartheta$ onto the set of

local variables of $grd(C_2)$ in C_2, $grd(C_1\vartheta\rho)$ cannot be made syntactically equal to $grd(C_2)$ by applying disequation promotion

do

We take a binding X/r as follows.

(Case 1) Suppose that t_1 is *not* a variant of t_2. In this case, since C_1 and C_2 are not mutually exclusive, we have that t_1 and t_2 are unifiable and, for some $i, j \in \{1, 2\}$, with $i \neq j$, there exists an mgu ϑ of t_i and t_j and a binding Y/t_a in ϑ such that $t_j\{Y/t_a\}$ is not a variant of t_j. Without loss of generality we may assume that $i=1$ and $j=2$. Then we take the binding X/r to be Y/t_a.

(Case 2) Suppose that t_1 is a variant of t_2 via an mgu ϑ. Now every safe clause whose normal form has a disequation of the form $X \neq t$, where X is a local variable of that disequation in that clause, is mutually exclusive w.r.t. any other safe clause. This is the case because, for any substitution σ which does not bind X, $t\sigma$ is unifiable with X and, thus, $X \neq t\sigma$ is not satisfiable. Thus, for some $i, j \in \{1, 2\}$, with $i \neq j$, there exists a disequation $(Y \neq t_a)\vartheta$ in $grd(C_i\vartheta)$ where $Y\vartheta$ is an input variable of $hd(C_i\vartheta)$, such that for any substitution ρ which is a bijective mapping from the set of local variables of $grd(C_i\vartheta)$ in $C_i\vartheta$ onto the set of local variables of $grd(C_j\vartheta)$ in $C_j\vartheta$ and for every disequation $(Z \neq t_b)\vartheta$ in $grd(C_j\vartheta)$, we have that $(Y \neq t_a)\vartheta\rho$ is different from $(Z \neq t_b)\vartheta$. We also have that $Y\vartheta$ is an input variable of $hd(C_j\vartheta)$. Without loss of generality we may assume that $i=1$, $j=2$, $t_1\vartheta = t_2$, and $C_2\vartheta = C_2$. Then we take the binding X/r to be $(Y/t_a)\vartheta$.

We apply the case split rule to clause C_2 w.r.t. X/r, that is, we derive the two clauses:

C_{21}. $(p(t_2, u_2) \leftarrow Body_2)\{X/r\}$
C_{22}. $p(t_2, u_2) \leftarrow X \neq r, Body_2$

We update the value of *NonunitCls* as follows:

NonunitCls := (*NonunitCls* − $\{C_2\}$) $\cup$ $\{C_{21}, C_{22}\}$
NonunitCls := *Simplify*(*NonunitCls*).

end-while

Now the set *NonunitCls* is partitioned into subsets of clauses and after suitable renaming of variables and disequation promotion, each subset is of the form:

$$\begin{cases} p(t, u_1) \leftarrow Diseqs, Goal_1 \\ \quad \dots \\ p(t, u_m) \leftarrow Diseqs, Goal_m \end{cases}$$

where *Diseqs* is a conjunction of disequations and for $k = 1, \dots, m$, no disequation occurs in $Goal_k$, and any two clauses in different subsets are mutually exclusive w.r.t. mode M_s.

Then we process every subset of clauses we have derived, by applying the safe head generalization rule so to replace the non-input arguments in the heads of the clauses belonging to the same subset by their most specific common generalization. Thus, every subset of clauses will eventually take the form:

$$\begin{cases} p(t, u) \leftarrow Eqs_1, Diseqs, Goal_1 \\ \quad \dots \\ p(t, u) \leftarrow Eqs_m, Diseqs, Goal_m \end{cases}$$

where u is the most specific common generalization of the terms $u_1, \dots, u_m$ and, for $k = 1, \dots, m$, the goal Eqs_k is a conjunction of the equations $V_1 = v_1, \dots, V_r = v_r$ such that $u\{V_1/v_1, \dots, V_r/v_r\} = u_k$.

Finally, we move all disequations to the leftmost positions of the body of every clause, thereby getting the set *PartitionedCls*.

Notice that in the above procedure the application of the case split rule to clause C_2 w.r.t. X/r is safe because: (i) clauses C_1 and C_2 are safe w.r.t. M_s, (ii) X is an input variable of $hd(C_{22})$ (recall that our choice of X/r in Case 2 ensures that X is an input variable of $hd(C_2)$), and (iii) each variable in r is either an input variable of $hd(C_{22})$ or a local variable of $X \neq r$ in C_{22}. Thus, clauses C_{21} and C_{22} are safe w.r.t. mode M_s and they are also mutually exclusive w.r.t. M_s.

The following property is particularly important for the mechanization of our Determinization Strategy.

Theorem 8 *The Partition procedure terminates.*

Proof: See [35, Appendix C]. □

When the Partition procedure terminates, it returns a set *PartitionedCls* of clauses which is the union of packets of clauses enjoying Properties (2a) and (2b) indicated in the Output specification of that procedure. These properties are a straightforward consequence of the termination condition of the while-do statement of that same procedure.

6.4 The Define-Fold Subsidiary Strategy

The Define-Fold strategy is realized by the following procedure.

Procedure *Define-Fold(PartitionedCls, Defs, NewDefs, FoldedCls)*.
Input: (i) A mode M_s, (ii) a set *PartitionedCls* of clauses which are safe w.r.t. M_s, and (iii) a set *Defs* of definition clauses. *PartitionedCls* is the union of the disjoint packets of clauses computed by the Partition subsidiary strategy.
Output: (i) A (possibly empty) set *NewDefs* of definition clauses, together with a mode M_{new} consisting of exactly one mode for each distinct head predicate in *NewDefs*. For each $C \in NewDefs$, the input variables of the leftmost non-basic atom in the body of C are input variables of the head of C. (ii) A set *FoldedCls* of folded clauses.

$NewDefs := \emptyset;\ \ M_{new} := \emptyset;\ \ FoldedCls := \emptyset;$
while there exists in *PartitionedCls* a packet Q of the form:

$$\begin{cases} H \leftarrow\ Diseqs,\ G_1 \\ \quad \cdots \\ H \leftarrow\ Diseqs,\ G_m \end{cases}$$

where *Diseqs* is a conjunction of disequations and for $k = 1, \ldots, m$, no disequation occurs in G_k,
do $PartitionedCls := PartitionedCls - Q$ and apply the definition and safe folding rules as follows.

Case (α) Let us suppose that the set *Defs* of the available definition clauses contains a subset of clauses of the form:

$$\begin{cases} newq(X_1, \ldots, X_h) \ \leftarrow\ G_1 \\ \quad \cdots \\ newq(X_1, \ldots, X_h) \ \leftarrow\ G_m \end{cases}$$

such that: (i) they are all the clauses in *Defs* for predicate *newq*, (ii) $X_1, \ldots, X_h$ include every variable which occurs in one of the goals $G_1, \ldots, G_m$ and also occurs in one of the goals H, *Diseqs* (this property is needed for the correctness of folding, see Section 4.1), and (iii) for $i = 1, \ldots, h$, if X_i is an input argument of *newq* then X_i is either an input variable of H (according to the given mode M_s) or an input variable of the leftmost non-basic atom of one of the goals $G_1, \ldots, G_m$. Then we fold the given packet and we get:
$FoldedCls := FoldedCls \cup \{H \leftarrow Diseqs, newq(X_1, \ldots, X_h)\}$

Case (β) If in *Defs* there is no set of definition clauses satisfying the conditions described in Case (α), then we add to *NewDefs* the following clauses for a new predicate *newr*:

$$\begin{cases} newr(X_1,\ldots,X_h) \leftarrow G_1 \\ \ldots \\ newr(X_1,\ldots,X_h) \leftarrow G_m \end{cases}$$

where, for $i = 1,\ldots,h$, either (i) X_i occurs in one of the goals $G_1,\ldots,G_m$ and also occurs in one of the goals H, *Diseqs*, or (ii) X_i is an input variable of the leftmost nonbasic atom of one of the goals $G_1,\ldots,G_m$. We add to M_{new} the mode $newr(m_1, \ldots, m_h)$ such that for $i = 1,\ldots,h$, $m_i = +$ iff X_i is either an input variable of H or an input variable of the leftmost nonbasic atom of one of the goals $G_1,\ldots,G_m$. We then fold the packet under consideration and we get:
$FoldedCls := FoldedCls \cup \{H \leftarrow Diseqs, newr(X_1,\ldots,X_h)\}$

end-while

Notice that the post-conditions on the set *NewDefs* which is derived by the Define-Fold procedure (see Point (i) of the Output of the procedure), ensure the satisfaction of the pre-conditions on the set *Cls* which is an input of the Unfold-Simplify procedure. Indeed, recall that the set *Cls* is constructed during the Determinization Strategy by the assignment *Cls* := *NewDefs*. Recall also that these pre-conditions are needed to ensure that the first unfolding step performed by the Unfold-Simplify procedure is safe.

Notice also that each application of the folding rule is safe (see Definition 5). This fact is implied in Case (α) by Condition (iii), and in Case (β) by the definition of the mode for *newr*.

Finally, notice that the Define-Fold procedure terminates. However, this procedure does not guarantee the termination of the specialization process, because at each iteration of the while-do loop of the Determinization Strategy, the Define-Fold procedure may introduce a nonempty set of new definition clauses. We will briefly discuss this issue in Section 9.

7 Examples of Application of the Determinization Strategy

In this section we will present some examples of program specialization where we will see in action our Determinization Strategy together with the Unfold-Simplify, Partition, and Define-Fold subsidiary strategies.

7.1 A Complete Derivation: Computing the Occurrences of a Pattern in a String

We consider again the program *Match_Pos* of Section 5.3. The mode M for the program *Match_Pos* is $\{match\_pos(+,+,?),\ append(?,?,+),\ length(+,?)\}$. We leave it to the reader to verify that *Match_Pos* satisfies M.

The derivation we will perform using the Determinization Strategy is more challenging than the ones presented in the literature (see, for instance, [11–13,15,44]) because an occurrence of the pattern P in the string S is specified in the initial program (see clause 1) in a nondeterministic way by stipulating the existence of two substrings L and R such that S is the concatenation of L, P, and R.

We want to specialize the *Match_Pos* program w.r.t. the atom $match\_pos([a,a,b],S,N)$. Thus, we first introduce the definition clause:

6. $match\_pos_s(S,N) \leftarrow match\_pos([a,a,b],S,N)$

The mode of the new predicate is $match\_pos_s(+,?)$ because S is an input argument of *match_pos* and N is not an input argument. Our transformation strategy starts off with the following initial values: $Defs = Cls = \{6\}$, $TransfP = Match\_Pos$, and $M_s = M \cup \{match\_pos_s(+,?)\}$.

First Iteration

Unfold-Simplify. By unfolding clause 6 w.r.t. the leftmost atom in its body we derive:

7. $match\_pos_s(S,N) \leftarrow append(Y,R,S),\ append(L,[a,a,b],Y),\ length(L,N)$

The body of clause 7 has no consumer atoms (notice that, for instance, the mgu of the atom $append(Y,R,S)$ and the head of clause 5 has the binding $S/[A|Z]$ where S is an input variable). Thus, the Unfold-Simplify subsidiary strategy terminates. We have that: $UnfoldedCls = \{7\}$.

Partition. *NonunitCls* is made out of clause 7 only, and thus, the Partition subsidiary strategy immediately terminates and produces a set *PartitionedCls* which consists of a single packet made out of clause 7.

Define-Fold. In order to fold clause 7 in *PartitionedCls*, the Define-Fold subsidiary strategy introduces the following definition clause:

8. $new1(S,N) \leftarrow append(Y,R,S),\ append(L,[a,a,b],Y),\ length(L,N)$

The mode of *new1* is $new1(+,?)$. By folding clause 7 using clause 8 we derive:

9. $match\_pos_s(S,N) \leftarrow new1(S,N)$

Thus, the first iteration of the Determinization Strategy terminates with $Defs = \{6,8\}$, $Cls = \{8\}$, $TransfP = Match\_Pos \cup \{9\}$, and $M_s = M \cup \{match\_pos_s(+,?),\ new1(+,?)\}$.

Second Iteration

Unfold-Simplify. We follow the subsidiary strategy described in Section 6.2 and we first unfold clause 8 in *Cls* w.r.t. the leftmost atom in its body. We get:

10. $new1(S,N) \leftarrow \underline{append(L,[a,a,b],[\,])},\ length(L,N)$
11. $new1([C|S],N) \leftarrow append(Y,R,S),\ \underline{append(L,[a,a,b],[C|Y])},\ length(L,N)$

Now we unfold clauses 10 and 11 w.r.t. the leftmost consumer atom of their bodies (see the underlined atoms). The unfolding of clause 10 amounts to its deletion because the atom $append(L,[a,a,b],[\,])$ is not unifiable with any head in program *Match_Pos*. The unfolding of clause 11 yields two new clauses that are further unfolded according to the Unfold-Simplify subsidiary strategy. After some unfolding steps, we derive the following clauses:

12. $new1([a|S],0) \leftarrow append([a,b],R,S)$
13. $new1([C|S],s(N)) \leftarrow append(Y,R,S),\ append(L,[a,a,b],Y),\ length(L,N)$

Partition. We apply the safe case split rule to clause 13 w.r.t. to the binding C/a, because the input argument in the head of this clause is unifiable with the input argument in the head of clause 12 via the mgu $\{C/a\}$. We derive the following two clauses:

14. $new1([a|S],s(N)) \leftarrow append(Y,R,S),\ append(L,[a,a,b],Y),\ length(L,N)$
15. $new1([C|S],s(N)) \leftarrow C\neq a,\ append(Y,R,S),\ append(L,[a,a,b],Y),\ length(L,N)$

Now, the set of clauses derived so far by the Partition subsidiary strategy can be partitioned into two packets: the first one is made out of clauses 12 and 14, where the input argument of the head predicate is of the form $[a|S]$, and the second one is made out of clause 15 only, where the input argument of the head predicate is of the form $[C|S]$ with $C\neq a$.

The Partition subsidiary strategy terminates by applying the safe head generalization rule to clauses 12 and 14, so to replace the second arguments in their heads by the most specific common generalization of those arguments, that is, a variable. We get the packet:

16. $new1([a|S],M) \leftarrow M=0,\ append([a,b],R,S)$
17. $new1([a|S],M) \leftarrow M=s(N),\ append(Y,R,S),\ append(L,[a,a,b],Y),\ length(L,N)$

For the packet made out of clause 15 only, no application of the safe head generalization rule is performed. Thus, we have derived the set of clauses *PartitionCls* which is the union of the two packets $\{16,17\}$ and $\{15\}$.

Define-Fold. Since there is no set of definition clauses in *Defs* which can be used to fold the packet $\{16, 17\}$, we are in Case (α) of the Define-Fold subsidiary strategy. Thus, we introduce a new predicate $new2$ as follows:

18. $new2(S, M) \leftarrow M=0,\ append([a,b], R, S)$
19. $new2(S, M) \leftarrow M=s(N),\ append(Y, R, S),\ append(L, [a,a,b], Y),\ length(L, N)$

The mode of $new2$ is $new2(+,?)$ because S is an input variable of the head of each clause of the corresponding packet. By folding clauses 16 and 17 using clauses 18 and 19 we derive the following clause:

20. $new1([a|S], M) \leftarrow new2(S, M)$

We then consider the packet made out of clause 15 only. This packet can be folded using clause 8 in *Defs*. Thus, we are in Case (β) of the Define-Fold subsidiary strategy. By folding clause 15 we derive the following clause:

21. $new1([C|S], s(N)) \leftarrow C \neq a,\ new1(S, N)$

Thus, *FoldedCls* is the set $\{20, 21\}$.

After these folding steps we conclude the second iteration of the Determinization Strategy with the following assignments: $\mathit{Defs} := \mathit{Defs} \cup \{18, 19\}$; $\mathit{Cls} := \{18, 19\}$; $\mathit{TransfP} := \mathit{TransfP} \cup \{20, 21\}$; $M_s := M_s \cup \{new2(+,?)\}$.

Third Iteration

Unfold-Simplify. From *Cls*, that is, clauses 18 and 19, we derive the set *UnfoldedCls* made out of the following clauses:

22. $new2([a|S], 0) \leftarrow append([b], R, S)$
23. $new2([a|S], s(0)) \leftarrow append([a,b], R, S)$
24. $new2([C|S], s(s(N))) \leftarrow append(Y, R, S),\ append(L, [a,a,b], Y),\ length(L, N)$

Partition. The set *NonunitCls* is identical to *UnfoldedCls*. From *NonunitCls* we derive the set *PartitionedCls* which is the union of two packets. The first packet consists of the following clauses:

25. $new2([a|S], M) \leftarrow M=0,\ append([b], R, S)$
26. $new2([a|S], M) \leftarrow M=s(0),\ append([a,b], R, S)$
27. $new2([a|S], M) \leftarrow M=s(s(N)), append(Y, R, S), append(L, [a,a,b], Y), length(L, N)$

The second packet consists of the following clause only:

28. $new2([C|S], s(s(N))) \leftarrow C \neq a, append(Y, R, S),\ append(L, [a,a,b], Y),\ length(L, N)$

Define-Fold. We introduce the following definition clauses:

29. $new3(S, M) \leftarrow M=0,\ append([b], R, S)$
30. $new3(S, M) \leftarrow M=s(0),\ append([a,b], R, S)$
31. $new3(S, M) \leftarrow M=s(s(N)),\ append(Y, R, S),\ append(L, [a,a,b], Y),\ length(L, N)$

where the mode for $new3$ is $new3(+,?)$. By folding, from *PartitionedCls* we derive the following two clauses:

32. $new2([a|S], M) \leftarrow new3(S, M)$
33. $new2([C|S], s(s(N))) \leftarrow C \neq a,\ new1(S, N)$

which constitute the set *FoldedCls*.

The third iteration of the Determinization Strategy terminates with the following assignments: $\mathit{Defs} := \mathit{Defs} \cup \{29, 30, 31\}$; $\mathit{Cls} := \{29, 30, 31\}$; $\mathit{TransfP} := \mathit{TransfP} \cup \{32, 33\}$; $M_s := M_s \cup \{new3(+,?)\}$.

Fourth Iteration

Unfold-Simplify. From *Cls* we derive the new set *UnfoldedCls* made out of the following clauses:

34. $new3([b|S],0) \leftarrow append([\,],R,S)$
35. $new3([a|S],s(0)) \leftarrow append([b],R,S)$
36. $new3([a|S],s(s(0))) \leftarrow append([a,b],R,S)$
37. $new3([C|S],s(s(s(N)))) \leftarrow append(Y,R,S),\ append(L,[a,a,b],Y),\ length(L,N)$

Partition. The set *NonunitCls* is identical to *UnfoldedCls*. From *NonunitCls* we derive the new set *PartitionedCls* made out of the following clauses:

38. $new3([a|S],s(M)) \leftarrow M=0,\ append([b],R,S)$
39. $new3([a|S],s(M)) \leftarrow M=s(0),\ append([a,b],R,S)$
40. $new3([a|S],s(M)) \leftarrow M=s(s(N)),\ append(Y,R,S),\ append(L,[a,a,b],Y),\ length(L,N)$
41. $new3([b|S],M) \leftarrow M=0,\ append([\,],R,S)$
42. $new3([b|S],M) \leftarrow M=s(s(s(N))),\ append(Y,R,S),\ append(L,[a,a,b],Y),\ length(L,N)$
43. $new3([C|S],s(s(s(N)))) \leftarrow C\neq a, C\neq b, append(Y,R,S), append(L,[a,a,b],Y), length(L,N)$

PartitionedCls consists of three packets: $\{38,39,40\}$, $\{41,42\}$, and $\{43\}$.

Define-Fold. We introduce two new predicates by means of the following definition clauses:

44. $new4(S,M) \leftarrow M=0,\ append([\,],R,S)$
45. $new4(S,M) \leftarrow M=s(s(s(N))), append(Y,R,S), append(L,[a,a,b],Y), length(L,N)$

We now fold the clauses in *PartitionedCls* and we derive the set *FoldedCls* made out of the following clauses:

46. $new3([a|S],s(M)) \leftarrow new3(R,S)$
47. $new3([b|S],M) \leftarrow new4(R,S)$
48. $new3([C|S],s(s(s(N)))) \leftarrow C\neq a,\ C\neq b,\ new1(S,N)$

The fourth iteration terminates with the following assignments: $Defs := Defs \cup \{44,45\}$; $Cls := \{44,45\}$; $TransfP := TransfP \cup \{46,47,48\}$; $M_s := M_s \cup \{new4(+,?)\}$.

Fifth Iteration

Unfold-Simplify. From *Cls* we derive the new set *UnfoldedCls* made out of the following clauses:

49. $new4(S,0) \leftarrow$
50. $new4([a|S],s(s(s(0)))) \leftarrow append([a,b],R,S)$
51. $new4([C|S],s(s(s(s(N))))) \leftarrow append(Y,R,S),\ append(L,[a,a,b],Y),\ length(L,N)$

Partition. The set *NonunitCls* is made out of clauses 50 and 51. From *NonunitCls* we derive the new set *PartitionedCls* made out of the following clauses:

52. $new4([a|S],s(s(s(M)))) \leftarrow M=0,\ append([a,b],R,S)$
53. $new4([a|S],s(s(s(M)))) \leftarrow M=s(N),\ append(Y,R,S),\ append(L,[a,a,b],Y),\ length(L,N)$
54. $new4([C|S],s(s(s(s(N))))) \leftarrow C\neq a,\ append(Y,R,S),\ append(L,[a,a,b],Y),\ length(L,N)$

PartitionedCls consists of two packets: $\{52,53\}$ and $\{54\}$.

Define-Fold. We are able to perform all required folding steps without introducing new definition clauses (see Case (α) of the Define-Fold procedure). In particular, (i) we fold clauses 52 and 53 using clauses 18 and 19, and (ii) we fold clause 54 using clause 8. Since no new definition is introduced, the set *Cls* is empty and the transformation strategy terminates. Our final specialized program is the program $Match\_Pos_s$ shown in Section 5.3.

The $Match\_Pos_s$ program is semideterministic and it corresponds to the finite automaton with one counter depicted in Figure 1. The predicates correspond to the states of the automaton and the clauses correspond to the transitions. The predicate *new*1 corresponds to the initial state, because the program is intended to be used for goals of the form $match\_pos_s(S, N)$, where S is bound to a list of characters, and by clause 1 $match\_pos_s(S, N)$ calls $new1(S, N)$. Notice that this finite automaton is deterministic except for the state corresponding to the predicate *new*4, where the automaton can either (i) accept the input string by returning the value of N and moving to the final state *true*, even if the input string has not been completely scanned (see clause 49), or (ii) move to the state corresponding to *new*2, if the symbol of the input string which is scanned is a (see clause 55), or (iii) move to the state corresponding to *new*1, if the symbol of the input string which is scanned is different from a (see clause 56).

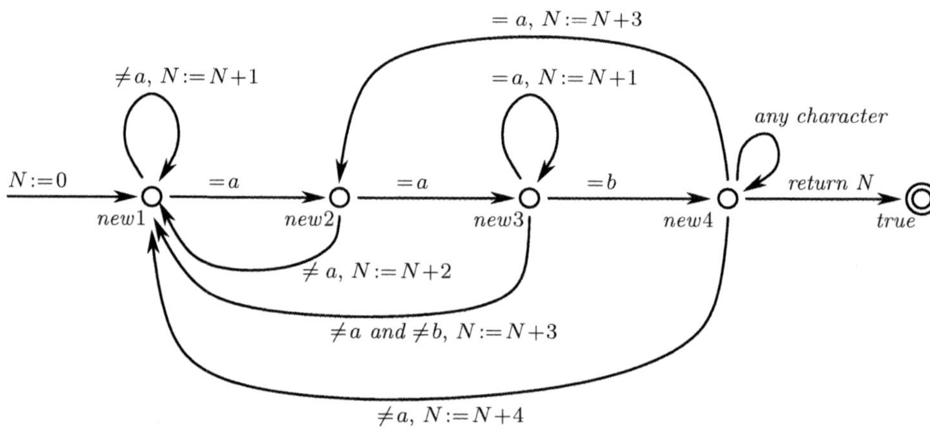

Fig. 1. The finite automaton with the counter N which corresponds to $Match\_Pos_s$.

7.2 Multiple Pattern Matching

Given a list *Ps* of patterns and a string S we want to compute the position, say N, of any occurrence in S of a pattern which is a member of the list *Ps*. For any given *Ps* and S the following program computes N in a nondeterministic way:

Program *Mmatch*	(initial, nondeterministic)
1. $mmatch([P\|Ps], S, N) \leftarrow match\_pos(P, S, N)$	
2. $mmatch([P\|Ps], S, N) \leftarrow mmatch(Ps, S, N)$	

The atom $mmatch(Ps, S, N)$ holds iff there exists a pattern in the list *Ps* of patterns which occurs in the string S at position N. The predicate *match_pos* is defined as in program *Match_Pos* of Section 7.1, and its clauses are not listed here. We consider the following mode for the program *Mmatch*:

$\{mmatch(+, +, ?),\quad match\_pos(+, +, ?),\quad append(?, ?, +),\quad length(+, ?)\}$.

We want to specialize this multipattern matching program w.r.t. the list $[[a, a, a], [a, a, b]]$ of patterns. Thus, we introduce the following definition clause:

3. $mmatch_s(S, N) \leftarrow mmatch([[a, a, a], [a, a, b]], S, N)$

The mode of the new predicate is $mmatch_s(+,?)$ because S is an input argument of *mmatch* and N is not an input argument. Thus, our Determinization Strategy starts off with the following initial values: $Defs = Cls = \{3\}$, $TransfP = Mmatch$, and $M_s = M \cup \{mmatch_s(+,?)\}$.

The output of the Determinization Strategy is the following program $Mmatch_s$:

Program $Mmatch_s$ (specialized, semideterministic)

4. $mmatch_s(S,N) \leftarrow new1(S,N)$
5. $new1([a|S],M) \leftarrow new2(S,M)$
6. $new1([C|S],s(N)) \leftarrow C \neq a,\ new1(S,N)$
7. $new2([a|S],M) \leftarrow new3(S,M)$
8. $new2([C|S],s(s(N))) \leftarrow C \neq a,\ new1(S,N)$
9. $new3([a|S],M) \leftarrow new4(S,M)$
10. $new3([b|S],M) \leftarrow new5(S,M)$
11. $new3([C|S],s(s(s(N)))) \leftarrow C \neq a,\ C \neq b,\ new1(S,N)$
12. $new4(S,0) \leftarrow$
13. $new4([a|S],s(N)) \leftarrow new4(S,N)$
14. $new4([b|S],s(N)) \leftarrow new5(S,N)$
15. $new4([C|S],s(s(s(s(N))))) \leftarrow C \neq a,\ C \neq b,\ new1(S,N)$
16. $new5(S,0) \leftarrow$
17. $new5([a|S],s(s(s(N)))) \leftarrow new2(S,N)$
18. $new5([C|S],s(s(s(s(N))))) \leftarrow C \neq a,\ new1(S,N)$

Similarly to the single-pattern string matching example of the previous Section 7.1, this specialized, semideterministic program corresponds to a finite automaton with counters. This finite automaton is deterministic, except for the states corresponding to the predicates *new4* and *new5* where any remaining portion of the input word is accepted. A similar derivation cannot be performed by usual partial deduction techniques without a prior transformation into *failure continuation passing style* [44].

7.3 From Regular Expressions to Finite Automata

In this example we show the derivation of a deterministic finite automaton by specializing a general parser for regular expressions w.r.t. a given regular expression. The initial program *Reg_Expr* for testing whether or not a string belongs to the language denoted by a regular expression over the alphabet $\{a,b\}$, is the one given below.

Program *Reg_Expr* (initial, nondeterministic)

1. $in\_language(E,S) \leftarrow string(S),\ accepts(E,S)$
2. $string([\,]) \leftarrow$
3. $string([a|S]) \leftarrow string(S)$
4. $string([b|S]) \leftarrow string(S)$
5. $accepts(E,[E]) \leftarrow symbol(E)$
6. $accepts(E_1 E_2,S) \leftarrow append(S_1,S_2,S),\ accepts(E_1,S_1),\ accepts(E_2,S_2)$
7. $accepts(E_1{+}E_2,S) \leftarrow accepts(E_1,S)$
8. $accepts(E_1{+}E_2,S) \leftarrow accepts(E_2,S)$
9. $accepts(E^*,[\,])$
10. $accepts(E^*,S) \leftarrow ne\_append(S_1,S_2,S),\ accepts(E,S_1),\ accepts(E^*,S_2)$
11. $symbol(a) \leftarrow$
12. $symbol(b) \leftarrow$
13. $ne\_append([A],Y,[A|Y]) \leftarrow$
14. $ne\_append([A|X],Y,[A|Z]) \leftarrow ne\_append(X,Y,Z)$

We have that $in\_language(E,S)$ holds iff S is a string in $\{a,b\}^*$ and S belongs to the language denoted by the regular expression E. In this *Reg_Expr* program we have used the predicate $ne\_append(S_1, S_2, S)$ which holds iff the nonempty string S is the concatenation of the *nonempty* string S_1 and the string S_2. The use of the atom $ne\_append(S_1, S_2, S)$ in clause 10 ensures that we have a *terminating* program, that is, a program for which we cannot have an infinite derivation when starting from a ground goal. Indeed, if in clause 10 we replace $ne\_append(S_1, S_2, S)$ by $append(S_1, S_2, S)$, then we may construct an infinite derivation because from a goal of the form $accepts(E^*, S)$ we can derive a new goal of the form $(accepts(E, [\,]),\ accepts(E^*, S))$.

We consider the following mode for the program *Reg_Expr*:

$\{in\_language(+,+),\quad string(+),\quad accepts(+,+),\quad symbol(+),\quad ne\_append(?,?,+),\ append(?,?,+)\}$.

We use our Determinization Strategy to specialize the program *Reg_Expr* w.r.t. the atom $in\_language((aa^*(b{+}bb))^*, S)$. Thus, we begin by introducing the definition clause:

15. $in\_language_s(S) \leftarrow in\_language((aa^*(b{+}bb))^*, S)$

The mode for this new predicate is $in\_language_s(+)$ because S is an input argument of *in_language*. The output of the Determinization Strategy is the following specialized program Reg_Expr_s:

Program Reg_Expr_s	(specialized, semideterministic)
16. $in\_language_s(S) \leftarrow new1(S)$	
17. $new1([\,]) \leftarrow$	
18. $new1([a\|S]) \leftarrow new2(S)$	
19. $new2([a\|S]) \leftarrow new3(S)$	
20. $new2([b\|S]) \leftarrow new4(S)$	
21. $new3([a\|S]) \leftarrow new3(S)$	
22. $new3([b\|S]) \leftarrow new4(S)$	
23. $new4([\,]) \leftarrow$	
24. $new4([a\|S]) \leftarrow new2(S)$	
25. $new4([b\|S]) \leftarrow new1(S)$	

This specialized program corresponds to a deterministic finite automaton.

7.4 Matching Regular Expressions

The following nondeterministic program defines a relation $re\_match(E, S)$, where E is a regular expression and S is a string, which holds iff there exists a substring P of S such that P belongs to the language denoted by E:

Program *Reg_Expr_Match*	(initial, nondeterministic)
1. $re\_match(E, S) \leftarrow append(Y, R, S),\ append(L, P, Y),\ accepts(E, P)$	

The predicates *append* and *accepts* are defined as in the programs *Naive_Match* (see Section 3.3) and *Reg_Expr* (see Section 7.3), respectively, and their clauses are not listed here. We consider the following mode for the program *Reg_Expr_Match*:

$\{append(?,?,+),\ accepts(+,+),\ re\_match(+,+)\}$.

We want to specialize the program *Reg_Expr_Match* w.r.t. the regular expression aa^*b. Thus, we introduce the following definition clause:

2. $re\_match_s(S) \leftarrow re\_match(aa^*b,\ S)$

The mode of this new predicate is $re\_match_s(+)$ because S is an input argument of *re_match*. The output of the Determinization Strategy is the following program:

Program *Reg_Expr_Match*$_s$ (specialized, semideterministic)

3. $re\_match_s(S) \leftarrow new1(S)$
4. $new1([a|S]) \leftarrow new2(S)$
5. $new1([C|S]) \leftarrow C{\neq}a,\ new1(S)$
6. $new2([a|S]) \leftarrow new3(S)$
7. $new2([C|S]) \leftarrow C{\neq}a,\ new1(S)$
8. $new3([a|S]) \leftarrow new4(S)$
9. $new3([b|S]) \leftarrow new3(S)$
10. $new3([C|S]) \leftarrow C{\neq}a,\ C{\neq}b,\ new1(S)$
11. $new4(S) \leftarrow$

Similarly to the single-pattern string matching example of Section 3.3, this specialized, semideterministic program corresponds to a deterministic finite automaton.

7.5 Specializing Context-free Parsers to Regular Grammars

Let us consider the following program for parsing context-free languages:

Program *CF_Parser* (initial, nondeterministic)

1. $string\_parse(G, A, W) \leftarrow string(W),\ parse(G, A, W)$
2. $string([\,]) \leftarrow$
3. $string([0|W]) \leftarrow string(W)$
4. $string([1|W]) \leftarrow string(W)$
5. $parse(G, [\,], [\,]) \leftarrow$
6. $parse(G, [A|X], [A|Y]) \leftarrow terminal(A),\ parse(G, X, Y)$
7. $parse(G, [A|X], Y) \leftarrow nonterminal(A),\ member(A \rightarrow B, G),$
   $append(B, X, Z),\ parse(G, Z, Y)$
8. $member(A, [A|X]) \leftarrow$
9. $member(A, [B|X]) \leftarrow member(A, X)$

together with the clauses for the predicate *append* defined as in program *Match_Pos* (see Section 7.1), and the unit clauses stating that 0 and 1 are terminals and s, u, v, and w are nonterminals. The first argument of *parse* is a context-free grammar, the second argument is a list of terminal and nonterminal symbols, and the third argument is a word represented as a list of terminal symbols. We assume that a context-free grammar is represented as a list of productions of the form $x \rightarrow y$, where x is a nonterminal symbol and y is a list of terminal and nonterminal symbols. We have that $parse(G, [s], W)$ holds iff from the symbol s we can derive the word W using the grammar G.

We consider the following mode for the program *CF_Parser*:

$\{string\_parse(+,+,+),\quad string(+),\quad parse(+,+,+),\quad terminal(+),\quad nonterminal(+),$ $member(?,+),\quad append(+,+,?)\}$.

We want to specialize our parsing program w.r.t. the following regular grammar:

$$
\begin{array}{lll}
s \rightarrow 0\,u & s \rightarrow 0\,v & s \rightarrow 0\,w \\
u \rightarrow 0 & u \rightarrow 0\,u & u \rightarrow 0\,v \\
v \rightarrow 0 & v \rightarrow 0\,v & v \rightarrow 0\,u \\
w \rightarrow 1 & w \rightarrow 0\,w &
\end{array}
$$

To this aim we apply our Determinization Strategy starting from the following definition clause:

$$
\begin{array}{rl}
10.\ string\_parse_s(W) \leftarrow parse([& s \rightarrow [0,u],\ s \rightarrow [0,v],\ s \rightarrow [0,w], \\
& u \rightarrow [0],\quad u \rightarrow [0,u],\ u \rightarrow [0,v], \\
& v \rightarrow [0],\quad v \rightarrow [0,v],\ v \rightarrow [0,u], \\
& w \rightarrow [1],\quad w \rightarrow [0,w] \qquad\qquad],\ [s],\ W)
\end{array}
$$

The mode for this new predicate is *string_parse*$_s$(+). The output of the Determinization Strategy is the following specialized program *CF_Parser*$_s$:

Program *CF_Parser*$_s$ (specialized, semideterministic)

11. *string_parse*$_s$(W) ← *new*1(W)
12. *new*1([0|W]) ← *new*2(W)
13. *new*2([0|W]) ← *new*3(W)
14. *new*2([1|W]) ← *new*4(W)
15. *new*3([ ]) ←
16. *new*3([0|W]) ← *new*5(W)
17. *new*3([1|W]) ← *new*4(W)
18. *new*4([ ]) ←
19. *new*5([ ]) ←
20. *new*5([0|W]) ← *new*3(W)
21. *new*5([1|W]) ← *new*4(W)

This program corresponds to a deterministic finite automaton.

Now, we would like to discuss the improvements we achieved in this example by applying our Determinization Strategy. Let us consider the *derivation tree* T_1 (see Figure 2) generated by the initial program *CF_Parser* starting from the goal *string_parse*$(g, [s], [0^n 1])$, where g denotes the grammar w.r.t. which we have specialized the *CF_Parser* program and $[0^n 1]$ denotes the list $[0, \ldots, 0, 1]$ with n occurrences of 0. The nodes of T_1 are labeled by the goals derived from *string_parse*$(g, [s], [0^n 1])$. In particular, the root of the derivation tree is labeled by *string_parse*$(g, [s], [0^n 1])$ and a node labeled by a goal G has k children labeled by the goals $G_1, \ldots, G_k$ which are derived from G (see Section 2.3). The tree T_1 has a number of nodes which is $O(2^n)$. Thus, by using the initial program *CF_Parser* it takes $O(2^n)$ number of steps to search for a derivation from the root goal *string_parse*$(g, [s], [0^n 1])$ to the goal *true*. (Indeed, this is the case if one uses a Prolog compiler.) In contrast, by using the specialized program *CF_Parser*$_s$, it takes $O(n)$ steps to search for a derivation from the goal *string_parse*$_s([0^n 1])$ to *true*, because the derivation tree T_2 has a number of nodes which is $O(n)$ (see Figure 3).

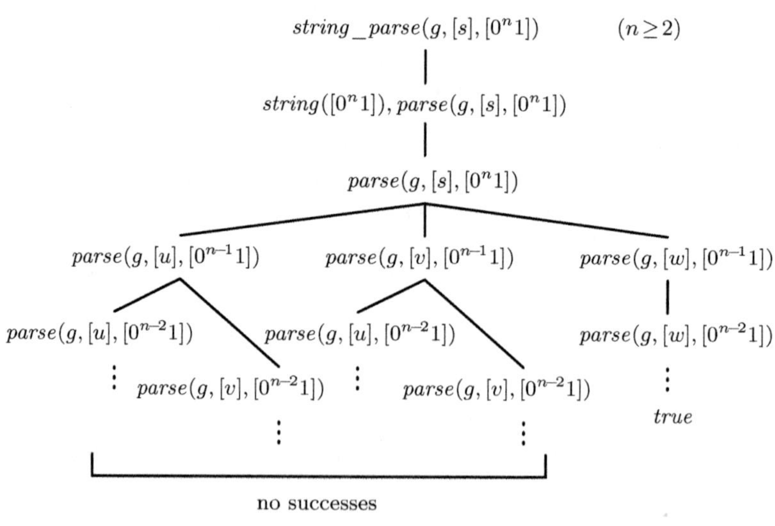

Fig. 2. Derivation tree T_1 for *string_parse*$(g, [s], [0^n 1])$.

The improvement of performance is due to the fact that our Determinization Strategy is able to avoid repeated derivations by introducing new definition clauses whose bodies have goals from which common subgoals are derived. Thus, after performing folding steps which use these definition clauses, we reduce the search space during program execution.

For instance, our strategy introduces the predicate *new*2 defined by the following clauses:

$new2(W) \leftarrow string(W),\ parse(g, [u], W)$
$new2(W) \leftarrow string(W),\ parse(g, [v], W)$
$new2(W) \leftarrow string(W),\ parse(g, [w], W)$

whose bodies are goals from which common subgoals are derived for $W = [0^{n-1}1]$ and $n \geq 2$. Indeed, for instance, $parse(g, [u], [0^{n-2}1])$ can be derived from both $parse(g, [u], [0^{n-1}1])$ and $parse(g, [v], [0^{n-1}1])$ (see Figure 2). The reader may verify that by using the specialized program CF_Parser_s no repeated goal is derived from $string\_parse_s(g, [s], [0^n1])$.

The ability of our Determinization Strategy of putting together the computations performed by the initial program in different branches of the computation tree, so that common repeated subcomputations are avoided, is based on the ideas which motivate the *tupling* strategy [34], first proposed as a transformation technique for functional languages.

$string\_parse_s(g, [s], [0^n1]) \qquad (n \geq 2)$
|
$new1([0^n1])$
|
$new2([0^{n-1}1])$
|
$new3([0^{n-2}1])$
⋮
true

Fig. 3. Derivation tree T_2 for $string\_parse_s([0^n1])$.

8 Experimental Evaluation

The Determinization Strategy has been implemented in the MAP program transformation system [39]. All program specialization examples presented in Sections 3.3, 5.3, and 7 have been worked out in a fully automatic way by the MAP system. We have compared the specialization times and the speedups obtained by the MAP system with those obtained by ECCE, a system for (conjunctive) partial deduction [24]. All experimental results reported in this section have been obtained by using SICStus Prolog 3.8.5 running on a Pentium II under Linux.

In Table 1 we consider the examples of Sections 3.3, 5.3, and 7, and we show the times taken (i) for performing partial deduction by using the ECCE system, (ii) for performing conjunctive partial deduction by using the ECCE system, and (iii) for applying the Determinization Strategy by using the MAP system. The *static input* shown in Column 2 of Table 1 is the goal w.r.t. which we have specialized the programs of Column 1. For running the ECCE system suitable choices among the available unfolding strategies and generalization strategies should be made. We have used the choices suggested by the system itself for partial deduction and conjunctive partial deduction, and we made some changes only when specialization was not performed within a reasonable amount of time. For running the MAP system the only information to be provided by the user is the mode for the program to be specialized. The system assumes that the program satisfies this mode and no mode analysis is performed.

Table 1. Specialization Times (in milliseconds).

Program	Static Input	ECCE (PD)	ECCE (CPD)	MAP (Det)
Naive_Match	$naive\_match([aab], S)$	360	370	70
Naive_Match	$naive\_match([aaaaaaaaab], S)$	420	2120	480
Match_Pos	$match\_pos([aab], S, N)$	540	360	100
Match_Pos	$match\_pos([aaaaaaaaab], S, N)$	650	910	500
Mmatch	$mmatch([[aaa], [aab]], S, N)$	1150	1400	280
Mmatch	$mmatch([[aa], [aaa], [aab]], S, N)$	1740	2040	220
Reg_Expr	$in\_language((aa^*(b{+}bb))^*, S)$	6260	138900	420
Reg_Expr	$in\_language(a^*(b{+}bb{+}bbb), S)$	3460	5430	230
Reg_Expr_Match	$re\_match(aa^*b, S)$	970	5290	210
Reg_Expr_Match	$re\_match(a^*(b + bb), S)$	1970	11200	300
CF_Parser	$string\_parse(g, [s], W)$	23400	32700	1620
CF_Parser	$string\_parse(g_1, [s], W)$	31200	31800	2000

The experimental results of Table 1 show that the MAP implementation of the Determinization Strategy is much faster than the ECCE implementation of both partial deduction and conjunctive partial deduction. We believe that, essentially, this is due to the fact that ECCE employs very sophisticated techniques, such as those based on *homeomorphic embeddings*, for controlling the unfolding and the generalization steps, and ensuring the termination of the specialization process. For a fair comparison, however, we should recall that Determinization may not terminate on examples different from those considered in this paper.

We have already mentioned in Section 3.3 that the performance of the programs derived by the Determinization Strategy may be further improved by applying postprocessing transformations which exploit the semideterminism of the programs. In particular, we may: (i) reorder the clauses so that unit clauses appear before non-unit clauses, and (ii) remove disequations by introducing cuts instead. The reader may verify that these transformations preserve the operational semantics. For a systematic treatment of cut introduction, the reader may refer to [10, 43]. As an example we now show the program obtained from $Match\_Pos_s$ (see Section 5.3) after the above post-processing transformations have been performed.

Program $Match\_Pos_{cut}$ (specialized, with cuts)

$match\_pos_s(S, N) \leftarrow new1(S, N)$
$new1([a|S], M) \leftarrow !,\ new2(S, M)$
$new1([C|S], s(N)) \leftarrow new1(S, N)$
$new2([a|S], M) \leftarrow !,\ new3(S, M)$
$new2([C|S], s(s(N))) \leftarrow new1(S, N)$
$new3([a|S], s(M)) \leftarrow !,\ new3(R, S)$
$new3([b|S], M) \leftarrow !,\ new4(R, S)$
$new3([C|S], s(s(s(N)))) \leftarrow new1(S, N)$
$new4(S, 0) \leftarrow$
$new4([a|S], s(s(s(M)))) \leftarrow !,\ new2(S, M)$
$new4([C|S], s(s(s(s(N))))) \leftarrow new1(S, N)$

In Table 2 below we report the speedups obtained by partial deduction, conjunctive partial deduction, Determinization, and Determinization followed by disequation removal and cut introduction. Every speedup is computed as the ratio between the timing of the initial program and the timing of the specialized program. These timings were obtained by running the various programs several times (up to 10,000) on significantly large input lists (up to 4,000 items).

Table 2. Speedups.

Program	Static Input	Speedup (PD)	Speedup (CPD)	Speedup (Det)	Speedup (Det and Cut)
Naive_Match	*naive_match*([*aab*], *S*)	3.1	5.8×10^3	3.0×10^3	6.8×10^3
Naive_Match	*naive_match*([*aaaaaaaaab*], *S*)	3.3	6.9×10^3	5.8×10^3	12.4×10^3
Match_Pos	*match_pos*([*aab*], *S*, *N*)	1.6	3.6×10^3	1.8×10^3	4.0×10^3
Match_Pos	*match_pos*([*aaaaaaaaab*], *S*, *N*)	2.1	5.3×10^3	2.9×10^3	8.1×10^3
Mmatch	*mmatch*([[*aaa*], [*aab*]], *S*, *N*)	1.7	4.5×10^3	3.5×10^3	6.2×10^3
Mmatch	*mmatch*([[*aa*], [*aaa*], [*aab*]], *S*, *N*)	1.6	2.5×10^3	3.9×10^3	5.4×10^3
Reg_Expr	*in_language*$((aa^*(b+bb))^*, S)$	29.8	6.2×10^3	2.3×10^5	3.9×10^5
Reg_Expr	*in_language*$(a^*(b+bb+bbb), S)$	1.3×10^4	3.3×10^4	4.6×10^4	5.7×10^4
Reg_Expr_Match	*re_match*(aa^*b, S)	5.7×10^2	2.7×10^4	1.5×10^6	3.0×10^6
Reg_Expr_Match	*re_match*$(a^*(b+bb), S)$	2.1×10^2	3.4×10^3	2.5×10^5	4.1×10^5
CF_Parser	*string_parse*$(g, [s], W)$	1.5	1.5	87.1	87.1
CF_Parser	*string_parse*$(g_1, [s], W)$	1.1	1.1	61.3	61.3

To clarify the content of Table 2 let us remark that:

Column 1 shows the names of the initial programs with reference to Sections 3.3, 5.3, and 7.

Column 2 shows the static input. The argument [*aab*] denotes the list $[a, a, b]$. Similar notation has been used for the other static input arguments. The argument g of the first *string_parse* atom denotes the regular grammar considered in Example 7.5. The argument g_1 of the last *string_parse* atom denotes the regular grammar:
$\{s \rightarrow 0\,u,\ \ s \rightarrow 1\,v,\ \ u \rightarrow 0,\ \ u \rightarrow 0\,v,\ \ u \rightarrow 0\,w,\ \ v \rightarrow 1,\ \ v \rightarrow 0\,v,\ \ v \rightarrow 1\,u,\ \ w \rightarrow 1,\ \ w \rightarrow 1\,w\}$.

Column 3, called Speedup (PD), shows the speedups we have obtained after the application of partial deduction.

Column 4, called Speedup (CPD), shows the speedups we have obtained after the application of conjunctive partial deduction.

Column 5, called Speedup (Det), shows the speedups we have obtained after the application of the Determinization Strategy.

Column 6, called Speedup (Det and Cut), shows the speedups we have obtained after the application of the Determinization Strategy followed by the removal of disequations and the introduction of cuts.

Let us now discuss our experimental results of Table 2. In all examples the best speedups are those obtained after the application of the Determinization Strategy followed by the removal of disequations and the introduction of cuts (see column Det and Cut).

As expected, conjunctive partial deduction gives higher speedups than partial deduction.

In some cases, conjunctive partial deduction gives better results than Determinization (see the first five rows of columns CPD and Det). This happens in examples where most nondeterminism is avoided by eliminating intermediate lists (see, for instance, the example of Section 3.3). In those examples, in fact, the Determinization Strategy may be less advantageous than conjunctive partial deduction because it introduces disequations which may be costly to check at runtime. However, as already mentioned, all disequations may be eliminated by introducing cuts (or, equivalently, if-then-else constructs) and the programs derived after disequation removal and cut introduction are indeed more efficient than those derived by conjunctive partial deduction (see column Det and Cut).

For some programs (see, for instance, the entries for *Reg_Expr* and *CF_Parser*) the speedups of the (Det) column are equal to the speedups of the (Det and Cut) column. The reason for this fact is the absence of disequations in the specialized program, so that the introduction of cuts does not improve efficiency.

We would like to notice that further postprocessing techniques are applicable. For instance, similarly to the familiar case of finite automata, we may eliminate clauses corresponding to ε-transitions where no input symbols are consumed (such as clause 9 in program $Match\_Pos_s$), and we may also minimize the number of predicate symbols (this corresponds to the minimization of the number of states). We do not present here these postprocessing techniques because they are outside the scope of the paper.

In summary, the experimental results of Table 2 confirm that in the examples we have considered, the Determinization Strategy followed by the removal of disequations in favour of cuts, achieves greater speedups than (conjunctive) partial deduction. However, it should be noticed that, as already mentioned, Determinization does not guarantee termination, while (conjunctive) partial deduction does, and in order to terminate in all cases, (conjunctive) partial deduction employs generalization techniques that may reduce speedups. In the next section we further discuss the issue of devising a generalization technique that ensures the termination of the Determinization Strategy.

9 Concluding Remarks and Related Work

We have proposed a specialization technique for logic programs based on an automatic strategy, called Determinization Strategy, which makes use of the following transformation rules: (1) definition introduction, (2) definition elimination, (3) unfolding, (4) folding, (5) subsumption, (6) head generalization, (7) case split, (8) equation elimination, and (9) disequation replacement. (Actually, we make use of the safe versions of Rules 4, 6, 7, and 8.) We have also shown that our strategy may reduce the amount of nondeterminism in the specialized programs and it may achieve exponential gains in time complexity.

To get these results, we allow new predicates to be introduced by *one or more* non-recursive definition clauses whose bodies may contain *more than one* atom. We also allow folding steps using these definition clauses. By a folding step several clauses are replaced by a single clause, thereby reducing nondeterminism.

The use of the subsumption rule is motivated by the desire of increasing efficiency by avoiding redundant computations. Head generalizations are used for deriving clauses with equal heads and thus, they allow us to perform folding steps. The case split rule is very important for reducing nondeterminism because it replaces a clause, say C, by several clauses which correspond to exhaustive and mutually exclusive instantiations of the head of C. To get exhaustiveness and mutual exclusion, we allow the introduction of disequalities. To further increase program efficiency, in a postprocessing phase these disequalities may be removed in favour of cuts.

We assume that the initial program to be specialized is associated with a mode of use for its predicates. Our Determinization Strategy makes use of this mode information for directing the various transformation steps, and in particular, the applications of the unfolding and case split rules. Moreover, if our strategy terminates, it derives specialized programs which are semideterministic w.r.t. the given mode. This notion has been formally defined in Section 5.3. Although semideterminism is not in itself a guarantee for efficiency improvement, it is often the case that efficiency is increased because nondeterminism is reduced and redundant computations are avoided.

We have shown that the transformation rules we use for program specialization, are correct w.r.t. the declarative semantics of logic programs based on the least Herbrand model. The proof of this correctness result is similar to the proofs of the correctness results which are presented in [14, 40, 46].

We have also considered an operational semantics for our logic language where a disequation $t_1 \neq t_2$ holds iff t_1 and t_2 are not unifiable. This operational semantics is sound, but not complete w.r.t. the declarative semantics. Indeed, if a goal operationally succeeds in a program, then it is true in the least Herbrand model of the program, but not vice versa. Thus, the proof of correctness of our transformation rules w.r.t. the operational semantics cannot be based on previous results and it is much more elaborate. Indeed, it requires some restrictions, related to the modes of the predicates, both on the programs to be specialized and on the applicability of the transformation rules.

In Section 3 we have extensively discussed the fact that our specialization technique is more powerful than partial deduction [21, 29]. The main reason of the greater power of our technique is that it uses more powerful transformation rules. In particular, partial deduction corresponds to the use the definition introduction, definition elimination, unfolding, and folding transformation rules, with the restriction that we may only fold a single atom at a time in the body of a clause.

Our extended rules allow us to introduce and transform new predicates defined in terms of *disjunctions of conjunctions of atoms* (recall that a set of clauses with the same head is equivalent to a single clause whose premise is the disjunction of the bodies of the clauses in the given set). In this respect, our technique improves over *conjunctive partial deduction* [8], which is a specialization technique where new predicates are defined in terms of conjunctions of atoms.

We have implemented the Determinization Strategy in the MAP transformation system [39] and we have tested this implementation by performing several specializations of string matching and parsing programs. We have also compared the results obtained by using the MAP system with those obtained by using the ECCE system for (conjunctive) partial deduction [24]. Our computer experiments confirm that the Determinization Strategy pays off w.r.t. both partial deduction and conjunctive partial deduction.

Our transformation technique works for programs where the only negative literals which are allowed in the body of a clause, are disequations between terms. The extension of the Determinization Strategy to normal logic programs would require an extension of the transformation rules and, in particular, it would be necessary to use a *negative unfolding* rule, that is, a rule for unfolding a clause w.r.t. a (possibly nonground) negative literal different from a disequation. The correctness of unfold/fold transformation systems which use the negative unfolding rule has been studied in contexts rather different from the one considered here (see, for instance, the work on transformation of *first order programs* [42]) and its use within the Determinization Strategy requires further work.

The Determinization Strategy may fail to terminate for two reasons: (i) the Unfold-Simplify subsidiary strategy may apply the unfolding rule infinitely often, and (ii) the while-do loop of the Determinization Strategy may not terminate, because at each iteration the Define-Fold subsidiary strategy may introduce new predicates.

The termination of the Unfold-Simplify strategy can be guaranteed by applying the techniques for finite unfolding already developed for (conjunctive) partial deduction (see, for instance, [8, 23, 30]). Indeed, the unfolding rule used in this paper is similar to the unfolding rule used in partial deduction.

The introduction of an infinite number of new predicates can be avoided by extending various methods based on *generalization*, such as those used in (conjunctive) partial deduction [8, 13, 25, 37]. Recall that in conjunctive partial deduction we may generalize a predicate definition essentially by means of two techniques: (i) the replacement of a term by a variable, which is then taken as an argument of a new predicate definition, and (ii) the splitting of a conjunction of literals into subconjunctions (together with the introduction of a new predicate for each subconjunction). It has been shown that the use of (i) and (ii) in a suitably controlled way, allows conjunctive partial deduction to terminate in all cases. However, termination is guaranteed at the expense of a possibly incomplete specialization or a possibly incomplete elimination of the intermediate data structures.

In order to avoid the introduction of an infinite number of new predicate definitions while applying the Determinization Strategy, we may follow an approach similar to the one used in the case of conjunctive partial deduction. However, besides the generalization techniques (i) and (ii) mentioned above, we may also need (iii) the splitting of the set of clauses defining a predicate into subsets (together with the introduction of a new predicate for each subset). Similarly to the case of conjunctive partial deduction, it can be shown that suitably controlled applications of the generalization techniques (i), (ii), and (iii) guarantee the termination of the Determinization Strategy at the expense of deriving programs which may fail to be semideterministic.

We leave it for further research the issue of controlling generalization, so that we achieve the termination of the specialization process and at the same time we maximize the reduction of nondeterminism.

In the string matching examples we have worked out, our strategy is able to automatically derive programs which behave like Knuth-Morris-Pratt algorithm, in the sense that they generate a finite automaton from any given pattern and a general pattern matcher. This was done also in the case of programs for matching sets of patterns and programs for matching regular expressions.

In these examples the improvement over similar derivations performed by partial deduction techniques [11,13,44] consists in the fact that we have started from naive, nondeterministic initial programs, while the corresponding derivations by partial deduction described in the literature, use initial programs which are deterministic. Our derivations also improve over the derivations performed by using *supercompilation* with *perfect driving* [15,47] and *generalized partial computation* [12], which start from initial functional programs which already incorporate some ingenuity.

A formal derivation of the Knuth-Morris-Pratt algorithm for pattern matching has also been presented in [3]. This derivation follows the *calculational* approach which consists in applying equivalences of higher order functions. On the one hand the calculational derivation is more general than ours, because it takes into consideration a generic pattern, not a fixed one (the string $[a, a, b]$ in our Example 3.3), on the other hand the calculational derivation is more specific than ours, because it deals with single-pattern string matching only, whereas our strategy is able to automatically derive programs in a much larger class which also includes multi-pattern matching, matching with regular expressions, and parsing.

The use of the case split rule is a form of reasoning *by cases*, which is a very well-known technique in mechanical theorem proving (see, for instance, the Edinburgh LCF theorem prover [17]). Forms of reasoning by cases have been incorporated in program specialization techniques such as the already mentioned supercompilation with perfect driving [15, 47] and generalized partial computation [12]. However, the strategy presented in this paper is the first fully automatic transformation technique which uses case reasoning to reduce nondeterminism of logic programs.

Besides specializing programs and reducing nondeterminism, our strategy is able to eliminate intermediate data structures. Indeed, the initial programs of our examples in Section 7 all have intermediate lists, while the specialized programs do not have them. Thus, our strategy can be regarded as an extension of the transformation strategies for the elimination of intermediate data structures (see the *deforestation* technique [48] for the case of functional programs and the strategy for *eliminating unnecessary variables* [38] for the case of logic programs). Moreover, our strategy derives specialized programs which avoid repeated subcomputations (see the Context-free Parsing example of Section 7.5). In this respect our strategy is similar to the *tupling strategy* for functional programs [34].

Finally, our specialization strategy is related to the program derivation techniques called *finite differencing* [33] and *incrementalization* [27]. These techniques use program invariants to avoid costly, repeated calculations of function calls. Our specialization strategy implicitly discovers and exploits program invariants when using the folding rule. It should be noticed, however, that it is difficult to establish in a rigorous way the formal connection between the basic ideas underlying our specialization strategy and the above mentioned program derivation methods based on program invariants. These methods, in fact, are presented in a very different framework.

Acknowledgments

We would like to thank D. De Schreye, S. Etalle, J. Gallagher, R. Glück, N. D. Jones, M. Leuschel, B. Martens, and M. H. Sørensen for stimulating discussions about partial deduction and logic program specialization. We also acknowledge very constructive and useful comments by the anonymous referees. This work has been partially supported by the EC under the HCM Project "Logic Program Synthesis and Transformation" and the Italian Ministry for Education, University, and Research.

References

1. K. R. Apt: Introduction to logic programming. In J. van Leeuwen, editor, *Handbook of Theoretical Computer Science*, 493–576. Elsevier, 1990.
2. K. R. Apt: *From Logic Programming to Prolog.* Prentice-Hall, London, 1997.
3. R. S. Bird, J. Gibbons, and G. Jones: Formal derivation of a pattern matching algorithm. *Science of Computer Programming*, 12:93–104, 1989.
4. A. Bossi, N. Cocco, and S. Dulli: A method for specializing logic programs. *ACM Transactions on Programming Languages and Systems*, 12(2):253–302, April 1990.
5. A. Bossi, N. Cocco, and S. Etalle: Transforming left-terminating programs. In A. Bossi, editor, *Proceedings of the Ninth International Workshop on Logic-based Program Synthesis, LOPSTR'99, Venezia, Italy, September 22-24, 1999*, Lecture Notes in Computer Science 1817, 156–175. Springer, 2000.
6. R. M. Burstall and J. Darlington: A transformation system for developing recursive programs. *Journal of the ACM*, 24(1):44–67, January 1977.
7. O. Danvy, R. Glück, and P. Thiemann, editors: *Partial Evaluation. International Seminar, Dagstuhl Castle, Germany, February 1996*, Lecture Notes in Computer Science 1110, Springer, 1996.
8. D. De Schreye, R. Glück, J. Jørgensen, M. Leuschel, B. Martens, and M. H. Sørensen: Conjunctive partial deduction: Foundations, control, algorithms, and experiments. *Journal of Logic Programming*, 41(2–3):231–277, 1999.
9. S. K. Debray and D. S. Warren: Automatic mode inference for logic programs. *Journal of Logic Programming*, 5:207–229, 1988.
10. Y. Deville: *Logic Programming: Systematic Program Development.* Addison-Wesley, 1990.
11. H. Fujita: An algorithm for partial evaluation with constraints. Technical Memorandum TM-0367, ICOT, Tokyo, Japan, 1987.
12. Y. Futamura, K. Nogi, and A. Takano: Essence of generalized partial computation. *Theoretical Computer Science*, 90:61–79, 1991.
13. J. P. Gallagher: Tutorial on specialisation of logic programs. In *Proceedings of ACM Sigplan Symposium on Partial Evaluation and Semantics Based Program Manipulation, PEPM '93, Copenhagen, Denmark*, 88–98. ACM Press, 1993.

14. M. Gergatsoulis and M. Katzouraki: Unfold/fold transformations for definite clause programs. In M. Hermenegildo and J. Penjam, editors, *Proceedings Sixth International Symposium on Programming Language Implementation and Logic Programming* (*PLILP '94*), Lecture Notes in Computer Science 844, 340–354. Springer, 1994.
15. R. Glück and A.V. Klimov: Occam's razor in metacomputation: the notion of a perfect process tree. In P. Cousot, M. Falaschi, G. Filé, and A. Rauzy, editors, *3rd International Workshop on Static Analysis, Padova, Italy, September 1993*, Lecture Notes in Computer Science 724, 112–123. Springer, 1993.
16. R. Glück and M. H. Sørensen: A roadmap to metacomputation by supercompilation. In O. Danvy, R. Glück, and P. Thiemann, editors, *Partial Evaluation*, Lecture Notes in Computer Science 1110, 137–160. Springer, 1996.
17. M. J. Gordon, A. J. Milner, and C. P. Wadsworth: *Edinburgh LCF*. Lecture Notes in Computer Science 78. Springer, 1979.
18. F. Henderson, Z. Somogyi, and T. Conway: Determinism analysis in the Mercury compiler. In *Proceedings of the Australian Computer Science Conference, Melbourne, Australia*, 337–346, 1996.
19. M. V. Hermenegildo, F. Bueno, G. Puebla, and P. López: Program analysis, debugging, and optimization using the CIAO system preprocessor. In D. De Schreye, editor, *Proceedings of the 1999 International Conference on Logic Programming, Las Cruces, NM, USA, Nov. 29 – Dec. 4, 1999*, 52–66. MIT Press, 1999.
20. J. Jaffar, M. Maher, K. Marriott, and P. Stuckey: The semantics of constraint logic programming. *Journal of Logic Programming*, 37:1–46, 1998.
21. N. D. Jones, C. K. Gomard, and P. Sestoft: *Partial Evaluation and Automatic Program Generation*. Prentice-Hall, 1993.
22. D. E. Knuth, J. H. Morris, and V. R. Pratt: Fast pattern matching in strings. *SIAM Journal on Computing*, 6(2):323–350, 1977.
23. M. Leuschel: On the power of homeomorphic embedding for online termination. In G. Levi, editor, *Proceedings of the Fifth Static Analysis Symposium, SAS '98, Pisa, Italy*, Lecture Notes in Computer Science 1503, 230–245. Springer, 1998.
24. M. Leuschel: The ECCE partial deduction system and the DPPD library of benchmarks, Release 3, Nov. 2000. Available from `http://www.ecs.soton.ac.uk/~mal`.
25. M. Leuschel, B. Martens, and D. De Schreye: Controlling generalization and polyvariance in partial deduction of normal logic programs. *ACM Transactions on Programming Languages and Systems*, 20(1):208–258, 1998.
26. M. Leuschel, B. Martens, and D. de Schreye: Some achievements and prospects in partial deduction. *ACM Computing Surveys*, 30 (Electronic Section)(3es):4, 1998.
27. Y. A. Liu: Efficiency by incrementalization: An introduction. *Higher-Order and Symbolic Computation*, 13(4):289–313, 2000.
28. J. W. Lloyd: *Foundations of Logic Programming*. Springer, Berlin. Second Edition, 1987.
29. J. W. Lloyd and J. C. Shepherdson: Partial evaluation in logic programming. *Journal of Logic Programming*, 11:217–242, 1991.
30. B. Martens, D. De Schreye, and T. Horváth: Sound and complete partial deduction with unfolding based on well-founded measures. *Theoretical Computer Science*, 122:97–117, 1994.
31. C. S. Mellish: Some global optimizations for a Prolog compiler. *Journal of Logic Programming*, 2(1):43–66, 1985.
32. C. S. Mellish: Abstract interpretation of Prolog programs. In S. Abramsky and C. Hankin, editors, *Abstract Interpretation of Declaratice Languages*, chapter 8, 181–198. Ellis Horwood, 1987.
33. R. Paige and S. Koenig: Finite differencing of computable expressions. *ACM Transactions on Programming Languages and Systems*, 4(3):402–454, 1982.

34. A. Pettorossi: Transformation of programs and use of tupling strategy. In *Proceedings Informatica 77, Bled, Yugoslavia*, 1–6, 1977.
35. A. Pettorossi, M. Proietti, and S. Renault: Derivation of efficient logic programs by specialization and reduction of nondeterminism. *Higher-Order and Symbolic Computation*, 18(1–2):121–210, 2005.
36. S. Prestwich: Online partial deduction of large programs. In *ACM Sigplan Symposium on Partial Evaluation and Semantics-Based Program Manipulation, PEPM '93, Copenhagen, Denmark*, 111–118. ACM Press, 1993.
37. M. Proietti and A. Pettorossi: The loop absorption and the generalization strategies for the development of logic programs and partial deduction. *Journal of Logic Programming*, 16(1–2):123–161, 1993.
38. M. Proietti and A. Pettorossi: Unfolding-definition-folding, in this order, for avoiding unnecessary variables in logic programs. *Theoretical Computer Science*, 142(1):89–124, 1995.
39. S. Renault: A system for transforming logic programs. R 97–04, Department of Computer Science, University of Rome Tor Vergata, Rome, Italy, 1997.
40. A. Roychoudhury, K. Narayan Kumar, C. R. Ramakrishnan, and I. V. Ramakrishnan: A parameterized unfold/fold transformation framework for definite logic programs. In *Proceedings of Principles and Practice of Declarative Programming (PPDP)*, Lecture Notes in Computer Science 1702, 396–413. Springer, 1999.
41. D. Sahlin: Mixtus: An automatic partial evaluator for full Prolog. *New Generation Computing*, 12:7–51, 1993.
42. T. Sato: An equivalence preserving first order unfold/fold transformation system. *Theoretical Computer Science*, 105:57–84, 1992.
43. H. Sawamura and T. Takeshima: Recursive unsolvability of determinacy, solvable cases of determinacy and their application to Prolog optimization. In *Proceedings of the International Symposium on Logic Programming, Boston*, 200–207. IEEE Computer Society Press, 1985.
44. D. A. Smith: Partial evaluation of pattern matching in constraint logic programming languages. In *Proceedings ACM Symposium on Partial Evaluation and Semantics Based Program Manipulation, PEPM '91, New Haven, CT, USA*, Sigplan Notices, 26, 9, 62–71. ACM Press, 1991.
45. Z. Somogyi, F. Henderson, and T. Conway: The execution algorithm of Mercury: an efficient purely declarative logic programming language. *Journal of Logic Programming*, 29(1–3):17–64, 1996.
46. H. Tamaki and T. Sato: Unfold/fold transformation of logic programs. In S.-Å. Tärnlund, editor, *Proceedings of the Second International Conference on Logic Programming*, 127–138, Uppsala University, Uppsala, Sweden, 1984.
47. V. F. Turchin: The concept of a supercompiler. *ACM TOPLAS*, 8(3):292–325, 1986.
48. P. L. Wadler: Deforestation: Transforming programs to eliminate trees. *Theoretical Computer Science*, 73:231–248, 1990.
49. D. H. D. Warren: Implementing Prolog – compiling predicate logic programs. Research Report 39 & 40, Department of Artificial Intelligence, University of Edinburgh, 1977.

Computational Divided Differencing and Divided-Difference Arithmetics

Thomas W. Reps[1] and Louis B. Rall[2] *

[1] Comp. Sci. Dept., Univ. of Wisconsin, 1210 W. Dayton St., Madison, WI 53706, USA. reps@cs.wisc.edu
[2] Dept. of Mathematics, Univ. of Wisconsin, 480 Lincoln Dr., Madison, WI 53706, USA. rall@math.wisc.edu

Summary. Tools for *computational differentiation* transform a program that computes a numerical function $F(x)$ into a related program that computes $F'(x)$ (the derivative of F). This paper describes how techniques similar to those used in computational-differentiation tools can be used to implement other program transformations—in particular, a variety of transformations for *computational divided differencing*. The paper also describes how computational divided differencing relates to the numerical-finite-differencing techniques that motivated Robert Paige's work on finite differencing of set-valued expressions in SETL programs.

Keywords: divided differences, computational differentiation, interpolation, multivariate interpolation, program transformation, round-off error.

1 Introduction

A variety of studies in the field of programming languages have led to useful, high-level transformations that manipulate programs in semantically meaningful ways. In very general terms, these tools transform a program that performs a computation $F(x)$ into a program that performs a related computation $F^\sharp(x)$, for a variety of $F^\sharp$'s of interest.[1] (In some cases, an appropriate preprocessing operation h needs to be applied to the input; in such cases, the transformed program $F^\sharp$ is used to perform a computation of the form $F^\sharp(h(x))$.) Examples of such tools include *partial evaluators* and *program slicers*:

- A *partial evaluator* creates a specialized version of a program when only part of the program's input has been supplied [8, 18, 29]. Partial evaluation is useful for removing interpretive overhead, and can also speed up programs that have two arguments that change value at different rates (such as ray tracing [43]).
- The *slice* of a program with respect to a set of program elements S is a projection of the program that includes only program elements that might affect (either directly or transitively) the values of the variables used at members of S [27,45,56]. Program-slicing tools allow one to find semantically meaningful decompositions of programs, where the decompositions consist of elements that are not textually contiguous. Slicing, and subsequent manipulation of slices, has applications in many software-engineering tools [26].

* Portions of this work appeared in [53]. Supported in part by NSF under grants CCR-{9625667, 9619219, 9986308}, the US-Israel Binational Science Foundation under grant 96-00337, ONR under contract N00014-00-1-0607, the Humboldt Foundation, and the Guggenheim Foundation.

[1] We do not generally make a distinction between *programs* and *procedures*. We use "program" both to refer to the program as a whole, as well as to refer to individual subroutines in a generic sense. We use "procedure" only in places where we wish to emphasize that the focus of interest is an individual subroutine *per se*.

O. Danvy et al. (eds.), Automatic Program Development – A Tribute to Robert Paige, 178–214.

Less well known in the programming-languages community is the work that has been done by numerical analysts on tools for *computational differentiation* (also known as *automatic differentiation* or *algorithmic differentiation*) [3, 22, 23, 48, 57]:

- Given a program that computes a numerical function $F(x)$, a computational-differentiation tool creates a related program that computes $F'(x)$ (the derivative of F).

Applications of computational differentiation include optimization, solving differential equations, curve fitting, and sensitivity analysis.

The work described in this paper expands the set of tools that programmers have at their disposal for performing such high-level, semantically meaningful program manipulations. Because so much scientific, engineering, and graphical software tries to predict and render modeled situations, such software often performs extrapolation and/or interpolation. These operations often involve the computation of divided differences, which can suffer badly from round-off error. In some cases, which motivate this paper, numerically unstable algorithms can be stabilized by computing divided differences in a nonstandard way.

The paper describes how techniques similar to those that have been developed for computational differentiation can be used to transform programs that compute numerical functions into ones that compute divided differences.

- We present a program transformation that, given a numerical function $F(x)$ defined by a program, creates a program that computes $F[x_0, x_1]$, the *first divided difference* of $F(x)$, where

$$F[x_0, x_1] \stackrel{\text{def}}{=} \begin{cases} (F(x_0) - F(x_1)) \;/\; (x_0 - x_1) & \text{if } x_0 \neq x_1 \\ \frac{d}{dz}F(z), \text{ evaluated at } z = x_0 & \text{if } x_0 = x_1 \end{cases} \tag{1}$$

- We show how computational first divided differencing *generalizes* computational differentiation.
- We present a second program transformation that permits the creation of *higher-order divided differences* of a numerical function defined by a program.
- We present a third program transformation that permits higher-order divided differences to be computed more efficiently. This transformation does not apply to all programs; however, we show that there is at least one important situation where this optimization is of use.
- We extend these techniques to handle functions of several variables.
- Finally, we describe how our work on computational differencing relates to the numerical-finite-differencing techniques that motivated Robert Paige's work on finite differencing of set-valued expressions in SETL [47].

Such program transformations can be implemented either as a source-to-source translation, or by means of overloaded operators and reinterpreted operands (in which case the source code is changed very little). The examples in the paper primarily illustrate the latter approach; the paper presents sketches of implementations of the various transformations in the form of C++ class definitions ("divided-difference arithmetics").

The benefits gained from the techniques described in the paper include the following:

- Because divided differences are the basis for a wide variety of numerical techniques, including polynomial interpolation, numerical integration, and solving differential equations [10], this work could lead to more robust programs in scientific and graphics applications, when the function of interest is one that is defined by a program.
- Finite differences on an evenly spaced grid can be used to quickly generate a function's values at any number of points that extend the grid (see [19] and [47, pages 403–404]). Because finite differences on an evenly spaced grid can be obtained from divided differences on an evenly spaced grid, our techniques may be useful in graphics applications for quickly plotting a function, while retaining reasonable accuracy.

- Because the divided-differencing problems that we address can be viewed as generalizations of problems such as differentiation, computation of Taylor coefficients, etc., some of our techniques—in particular, the divided-difference arithmetic presented in Section 7—represent new approaches that, with appropriate simplification, can also be applied in computational-differentiation tools.

Empirical results presented in Sections 5 and 8 provide three concrete demonstrations of some of the benefits that can be gained via our methods.

The remainder of the paper is organized as follows: Section 2 provides a succinct review of the basic principles of computational differentiation. Section 3 discusses the basic principle behind computational divided differencing. Section 4 shows how computational divided differencing generalizes computational differentiation. Section 5 extends the ideas introduced in Section 3 to higher-order computational divided differencing. Section 6 discusses techniques that apply to a useful special case. Section 7 extends the ideas from Sections 3, 5, and 6 to functions of several variables. Section 8 describes how these ideas relate to the numerical-finite-differencing techniques that motivated Robert Paige's work on finite differencing of set-valued expressions in SETL programs. Section 9 discusses other related work.

2 Background on Computational Differentiation

A tool for *computational differentiation* transforms a program that computes a numerical function $F(x)$ into a related program that computes the derivative $F'(x)$. These tools address the following issue: Suppose that you have a program `F(x)` that computes a numerical function $F(x)$. It is a bad idea to try to compute $F'(x_0)$, the value of the derivative of F at x_0, by picking a small value `delta_x` and invoking the following program with the argument $\mathtt{x}_0$:[2]

```
float delta_x = ...⟨some small value⟩ ...;
float F'_naive(float x) {
  return (F(x + delta_x) - F(x))/delta_x;
}
```
(2)

For a small enough value of `delta_x`, the values of `F(x₀+delta_x)` and `F(x₀)` will usually be very close. Round-off errors in the computation of `F(x₀+delta_x)` and `F(x₀)` are magnified by the subtraction of the two quantities, and further amplified by the division by the small quantity `delta_x`, which may cause the overall result to be useless. Computational differentiation sidesteps this problem by computing derivatives in another fashion.

Example 1 [58] Suppose that we have a collection of programs $\mathtt{f}_i$ for the functions f_i, $1 \le i \le k$, together with the program `Prod` shown below, which computes the function $Prod(x) = \prod_{i=1}^{k} f_i(x)$. In addition, suppose that we also have programs $\mathtt{f}'_\mathtt{i}$ for the functions f'_i, $1 \le i \le k$. Finally, suppose that we wish to obtain a program `Prod'` that computes $Prod'(x)$. Column two of the table given below shows mathematical expressions for $Prod(x)$ and $Prod'(x)$. Column three shows two C++ procedures: `Prod` computes $Prod(x)$; `Prod'` is the procedure that a computational-differentiation system would create to compute $Prod'(x)$.

[2] `Courier Font` is used to denote functions defined by programs, whereas *Italic Font* is used to denote mathematical functions. That is, $F(x)$ denotes a function (evaluated over real numbers), whereas `F(x)` denotes a program (evaluated over floating-point numbers). Example programs are all written in C++, although the ideas described apply to other programming languages—including functional programming languages (cf. [33])—as well as to other imperative languages. To emphasize the links between mathematical concepts and their implementations in C++, we take the liberty of sometimes using $'$ and/or subscripts on C++ identifiers.

	Mathematical Notation	Programming Notation

Function	$Prod(x) = \prod_{i=1}^{k} f_i(x)$	`float Prod(float x){`
		`  float ans = 1.0;`
		`  for (int i = 1; i <= k; i++){`
		`    ans = ans * f_i(x);`
		`  }`
		`  return ans;`
		`}`

Derivative	$Prod'(x) = \sum_{i=1}^{k} f_i'(x) * \prod_{j \neq i} f_j(x)$	`float Prod'(float x){`
		`  float ans' = 0.0;`
		`  float ans = 1.0;`
		`  for (int i = 1; i <= k; i++){`
		`    ans' = ans' * f_i(x) + ans * f'_i(x);`
		`    ans = ans * f_i(x);`
		`  }`
		`  return ans';`
		`}`

Notice that `Prod'` resembles `Prod`, as opposed to `F'_naive` (see box (2)). `Prod'` preserves accuracy in its computation of the derivative because, as illustrated below in Example 2, it is based on the rules for the exact computation of derivatives, rather than on the kind of computation performed by `F'_naive`. □

Iteration	Value of `ans'` (as a function of `x`)	Value of `ans` (as a function of `x`)
0	0.0	1.0
1	$f'_1(x)$	$f_1(x)$
2	$f'_1(x) * f_2(x) + f_1(x) * f'_2(x)$	$f_1(x) * f_2(x)$
3	$f'_1(x) * f_2(x) * f_3(x)$ $+ f_1(x) * f'_2(x) * f_3(x)$ $+ f_1(x) * f_2(x) * f'_3(x)$	$f_1(x) * f_2(x) * f_3(x)$
...	...	...
k	$\sum_{i=1}^{k} f'_i(x) * \prod_{j \neq i} f_j(x)$	$\prod_{i=1}^{k} f_i(x)$

Fig. 1. The values of `ans'` and `ans` at the start of each iteration of the for-loop.

The transformation illustrated above is merely one instance of a general transformation that can be applied to any program: Given a program `G` as input, the transformation produces a derivative-computing program `G'`. The method for constructing `G'` is as follows:

- For each variable `v` of type `float` used in `G`, another `float` variable `v'` is introduced.
- Each statement in `G` of the form "`v` = *exp*;", where *exp* is an arithmetic expression, is transformed into "`v'` = *exp'*; `v` = *exp*;", where *exp'* is the expression for the derivative of *exp*. If *exp* involves calls to a procedure `g`, then *exp'* may involve calls to both `g` and `g'`.
- Each statement in `G` of the form "`return v;`" is transformed into "`return v';`".

In general, this transformation can be justified by appealing to the chain rule of differential calculus.

Example 2 For Example 1, we can demonstrate the correctness of the transformation by symbolically executing `Prod'` for a few iterations, comparing the values of `ans'` and `ans` (as functions of `x`) at the start of each iteration of the for-loop, as shown in Figure 1. The loop maintains the invariant that, at the start of each iteration, $\texttt{ans}'(\texttt{x}) = \frac{d}{dx}\texttt{ans}(\texttt{x})$. □

```
enum ArgDesc { CONST, VAR };
class FloatD {
 public:
  float val';
  float val;
  FloatD(ArgDesc,float);
};
// Constructor to convert a constant
// or a value for the independent
// variable to a FloatD
FloatD::FloatD(ArgDesc a, float v){
  switch (a) {
    case CONST:
      val' = 0.0;
      val = v;
    break;
    case VAR:
      val' = 1.0;
      val = v;
    break; }
}
```

```
FloatD operator+(FloatD a, FloatD b){
  FloatD ans;
  ans.val' = a.val' + b.val';
  ans.val = a.val + b.val;
  return ans;
}

FloatD operator*(FloatD a, FloatD b){
  FloatD ans;
  ans.val' = a.val * b.val'
           + a.val' * b.val;
  ans.val = a.val * b.val;
  return ans;
}
```

Fig. 2. A differentiation-arithmetic class.

For the computational-differentiation approach, we did not really need to make the assumption that we were given programs `f'i` for the functions f_i', $1 \leq i \leq k$; instead, the programs `f'i` can be generated from the programs `fi` by applying the same statement-doubling transformation that was applied to `Prod`.

In languages that support operator overloading, such as C++, Ada, and Pascal-XSC, computational differentiation can be carried out by defining a new data type that has fields for both the value and the derivative, and overloading the arithmetic operators to carry out appropriate manipulations of both fields [49], along the lines of the definition of the C++ class `FloatD`, shown in Figure 2. A class such as `FloatD` is called a *differentiation arithmetic* [50].

The transformation then amounts to changing the types of each procedure's formal parameters, local variables, and return value (including those of the `fi`).[3]

Example 3 Using class `FloatD`, the `Prod` program of Example 1 can be handled as follows:

```
float f1(float x){...}           ⇒ FloatD f1(const FloatD &x){...}
         :                                 :
float fk(float x){...}           ⇒ FloatD fk(const FloatD &x){...}
float Prod(float x){               FloatD Prod(const FloatD &x) {
  float ans = 1.0;                   FloatD ans(CONST,1.0); // ans = 1.0
  for (int i = 1; i <= k; i++){      for (int i = 1; i <= k; i++){
    ans = ans * fi(x);           ⇒     ans = ans * fi(x); }
  }                                  return ans;
  return ans;                      }
}
                                   float Prod'(float x) {
                                     FloatD xD(VAR,x);
                                 ⇒   return Prod(xD).val';
                                   }
```

[3] We have referred to both computational differentiation and computational divided differencing as "program transformations", which may conjure up the image of tools that perform source-to-source rewriting fully automatically. Although this is one possible embodiment, in this paper the term "transformation" will also include the use of C++ classes in which the arithmetic operators are overloaded. With the latter approach, rewriting might be carried out by a preprocessor, but might also be performed by hand, since usually only light rewriting of the program source text is required.

By changing the types of the formal parameters, local variables, and the return values of Prod and the f_i (and making a slight change to the initialization of ans in Prod), the program now carries around derivative values (in the val′ field) in addition to performing all of the work performed by the original program. Because of the C++ overload-resolution mechanism, the f_i procedures invoked in the fourth line of the transformed version of Prod are the *transformed* versions of the f_i (i.e., the f_i of type FloatD → FloatD).

The value of Prod's derivative at v is obtained by calling Prod′(v). □

In a differentiation arithmetic, each procedure in the user's program, such as Prod and the f_i in Example 3, can be viewed as a box that maps two inputs to two outputs as depicted below:

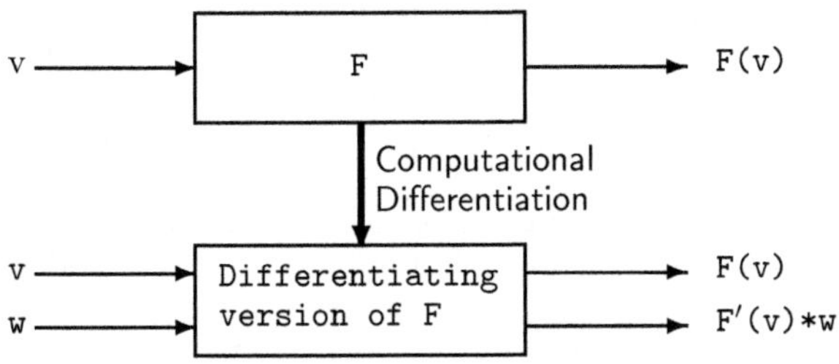

In particular, in each differentiating version of a user-defined or library procedure F, the lower-right-hand output produces the value F′(v)*w. An input value v for the formal parameter is treated as a pair (v,1.0). Such boxes "snap together": when F is composed with G (and the input is v), the output value on the lower-right-hand side is F′(G(v))*G′(v), which agrees with the usual expression for the chain rule for the first-derivative operator:

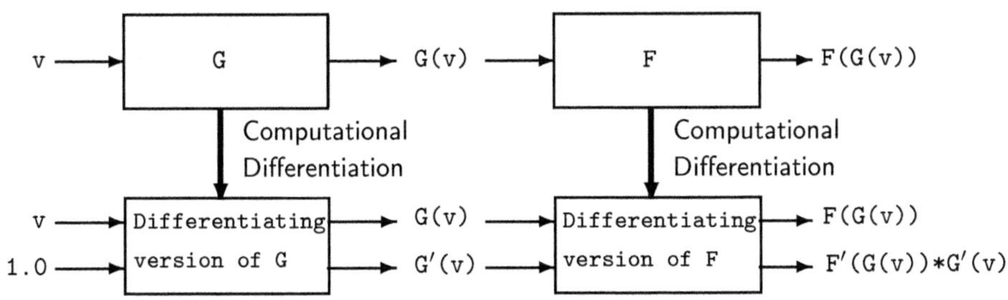

The transformation illustrated in Example 1 is not fully general in that it does not yield a procedure that can be composed with other transformed procedures. To create a composable transformed procedure, the transformation would, in essence, have to make changes that mimic all of the actions of the version created in Example 3 using class FloatD: the procedure would have to take two arguments, x and x′; pass these on to composable transformed versions of the f_i; and return a pair ⟨ans, ans′⟩, instead of ans′ alone.

The computational-differentiation technique summarized above is what is known as *forward-mode* differentiation. When the number of independent variables is much greater than the number of dependent variables, a different computational-differentiation technique, *reverse mode* [20,21,28,36,55], provides theoretically better performance—a greatly reduced number of computational steps—but at the cost of the need to store or recompute intermediate values that affect the final result nonlinearly. Forward mode and reverse mode are the endpoints of a spectrum of algorithmic techniques; in practice, computational-differentiation tools optimize runtime and memory requirements by exploiting associativity properties of the chain rule to permit forward mode and reverse mode to be used in different parts of the computation [7].

The remainder of the paper concerns the generalization of forward-mode computational differentiation to forward-mode computational divided differencing; non-forward-mode techniques are outside the scope of the paper.

The availability of overloading makes it possible to implement (forward-mode) computational differentiation conveniently, by packaging it as a differentiation-arithmetic class, as illustrated above. The alternative to the use of overloading is to build a special-purpose preprocessor to carry out the statement-doubling transformation that was illustrated in Examples 1 and 2. Examples of systems that use the latter approach include ADIFOR [4, 5] and ADIC [6].

2.1 Limitations of Computational Differentiation

This section discusses certain limitations of the computational-differentiation transformation. First, it is worthwhile mentioning that the presence of aliasing (e.g., due to pointers or reference parameters) is *not* a limitation of computational differentiation (nor of computational divided differencing): The transformations presented above (as well as later in Sections 3, 5, 6, and 7) work properly in the presence of aliasing (and are said to be *alias-safe* [22]).

One limitation of computational differentiation comes from the fact that a program `F′(x)` that results from computational differentiation can perform additions and subtractions for which there are no analogues in the original program `F(x)`. For instance, in program `Prod′`, an addition is performed in the statement "`ans′ = ans′ * f_i(x) + ans * f′_i(x);`" whereas no addition is performed in the statement "`ans = ans * f_i(x);`" Consequently, the result of evaluating `F′(x)` can be degraded by round-off error even when `F(x)` is computed accurately. However, the accuracy of the result from evaluating `F′(x)` can be verified by performing the same computation in interval arithmetic [49, 50].

Another problem that arises is that the manner in which a function is programmed influences whether the results obtained from the derivative program are correct. For instance, for programs that use a conditional expression or conditional statement in which the condition depends on the independent variable—i.e., where the function is defined in a piecewise manner—the derivative program may not produce the correct answer.

Suppose that the function $F(x) = x^2$ is programmed using a conditional statement, as shown in the left side of the box below:

```
float F(float x){                          float F′(float x){
  float ans;                                 float ans′;
  if(x == 1.0) { ans = 1.0; }                float ans;
  else { ans = x*x; }             ⇒          if(x == 1.0) { ans′ = 0.0; ans = 1.0; }
  return ans;                                else { ans′ = x+x; ans = x*x; }
}                                            return ans′;
                                           }
```

Computational differentiation would produce the program shown above on the right. With this program, `F′(1.0)` returns `0.0`, rather than the correct value of `2.0` (i.e., correct with respect to the meaning of the program as the mathematical function $F(x) = x^2$) [14].

This phenomenon has been called the *branch problem* or the *if problem* for computational differentiation. A more important example of the branch problem occurs in Gaussian elimination code, where pivoting introduces branches into the program [2, 14, 22]. Some additional problems that can arise with computational differentiation are identified in [14]. A number of different approaches to these problems have been discussed in the literature [2, 14, 22, 34, 54].

Computational divided differencing has some similar (or even worse) problems. All of these issues are outside the scope of the present paper; the problem of finding appropriate ways to generalize the aforementioned techniques to handle the problems that arise with computational divided differencing is left for future work.

3 Computational Divided Differencing

In this paper, we exploit the principle on which computational differentiation is based—namely, that it is possible to differentiate entire *programs*, not just *expressions*—to develop a variety of new *computational divided-differencing* transformations. We develop several transformations that can be applied to numerical programs. One of these corresponds to the *first-divided-difference operator*, denoted by $\cdot\,[x_0, x_1]$ and defined in Equation (1).

As with the differentiation operator, the problem that we face is that because division by a small value and subtraction are both operations that amplify accumulated round-off error, direct use of Equation (1) may lead to highly inaccurate results. In contrast, given a program that computes a numerical function $F(x)$, our technique for computational first divided differencing creates a related program that computes $F[x_0, x_1]$, *but without directly evaluating the right-hand side of Equation* (1).

As we show below, the program transformation that achieves this goal is quite similar to the transformation used in computational-differentiation tools. The transformed program sidesteps the explicit subtraction and division operations that appear in Equation (1), while producing answers that are equivalent (from the standpoint of evaluation in real arithmetic). The program that results thereby avoids many operations that could potentially amplify round-off error, and hence retains accuracy when evaluated in floating-point arithmetic.

Consider the case in which $F(x) = x^2$ and $x_0 \neq x_1$:

$$F[x_0, x_1] = \frac{F(x_0) - F(x_1)}{x_0 - x_1} = \frac{x_0^2 - x_1^2}{x_0 - x_1} = x_0 + x_1.$$

That is, the first divided difference can be obtained by evaluating $x_0 + x_1$. In general, for monomials we have:

$F(x)$	c	x	x^2	x^3	...
$F[x_0, x_1]$	0	1	$x_0 + x_1$	$x_0^2 + x_0x_1 + x_1^2$	...

Turning to programs, suppose that we are given the following program for squaring a number:

```
float Square(float x){
  return x * x;
}
```

To compute the first divided-difference of `Square`, we have our choice between: `Square_1DD_naive` and `Square_1DD`:

```
float Square_1DD_naive(float x0,float x1){
  return (Square(x0) - Square(x1))/(x0 - x1);
}
```

```
float Square_1DD(float x0,float x1){
  return x0 + x1;
}
```

However, the round-off-error characteristics of `Square_1DD` are much better than those of `Square_1DD_naive`.

The basis for creating expressions and programs that compute accurate divided differences is to be found in the basic properties of the first-divided-difference operator [35], which closely resemble those of the first-derivative operator, as shown in Table 1.

The program transformation for performing computational divided differencing can be explained by means of an example.

Example 4 Suppose that we have a C++ class `Poly` that represents polynomials, and a member function `Poly::Eval` that evaluates a polynomial via Horner's rule; i.e., it accumulates the answer by repeatedly multiplying by `x` and adding in the current coefficient, iterating down from the high-order coefficient:[4]

[4] The paper uses the evaluation of a polynomial in `x` via Horner's rule as a running example. It is well known that Horner's rule can return inaccurate results when it is used to evaluate a polynomial in floating-point arithmetic [25, pages 65–67]. Our examples are not meant to illustrate a way to circumvent this shortcoming. Horner's rule is used because it is a simple procedure, which allows various transformations to be illustrated succinctly.

Table 1. Basic properties of the first-derivative and first-divided-difference operators.

First Derivative	First Divided Difference [35]
$c' = 0.0$	$c[x_0, x_1] = 0.0$
$x' = 1.0$	$x[x_0, x_1] = 1.0$
$(c + F)'(x) = F'(x)$	$(c + F)[x_0, x_1] = F[x_0, x_1]$
$(c * F)'(x) = c * F'(x)$	$(c * F)[x_0, x_1] = c * F[x_0, x_1]$
$(F + G)'(x) = F'(x) + G'(x)$	$(F + G)[x_0, x_1] = F[x_0, x_1] + G[x_0, x_1]$
$(F * G)'(x) = F'(x) * G(x) + F(x) * G'(x)$	$(F * G)[x_0, x_1] = [x_0, x_1] * G(x_1) + F(x_0) * G[x_0, x_1]$
$\left(\frac{F}{G}\right)'(x) = \frac{F'(x)*G(x) - F(x)*G'(x)}{G(x)^2}$	$\left(\frac{F}{G}\right)[x_0, x_1] = \frac{F[x_0, x_1]*G(x_1) - F(x_1)*G[x_0, x_1]}{G(x_0) * G(x_1)}$

```
class Poly {
 public:
  float Eval(float);
 private:
  int degree;
  // Array coeff[0..degree]
  float *coeff;
 };
```

```
// Evaluation via Horner's rule
float Poly::Eval(float x){
  float ans = 0.0;
  for (int i = degree; i >= 0; i--) {
    ans = ans * x + coeff[i]; }
  return ans;
}
```

A new member function, `Poly::Eval_1DD`, to compute the first divided difference can be created by transforming `Poly::Eval` as shown below:

```
class Poly {
 public:
  float Eval(float);
  float Eval_1DD(float,float);
 private:
  int degree;
  // Array coeff[0..degree]
  float *coeff;
 };
```

```
float Poly::Eval_1DD(float x0,float x1){
  float ans_1DD = 0.0;
  float ans = 0.0;
  for (int i = degree; i >= 0; i--) {
    ans_1DD = ans_1DD * x1 + ans;
    ans = ans * x0 + coeff[i]; }
  return ans_1DD;
}
```

□

The transformation used to obtain `Eval_1DD` from (the text of) `Eval` is similar to the computational-differentiation transformation that would be used to create a derivative-computing program `Eval'` (`Eval'` appears in Figure 3):

- `Eval_1DD` is supplied with an additional formal parameter (and the two parameters are renamed `x0` and `x1`).
- For each local variable `v` of type `float` used in `Eval`, an additional `float` variable `v_1DD` is introduced in `Eval_1DD`.
- Each statement of the form "`v =` *exp* `;`" in `Eval` is transformed into "`v_1DD =` *exp* `[x0,x1];` `v =` exp_0 `;`", where *exp* `[x0,x1]` is the expression for the divided difference of *exp*, and exp_0 is *exp* with `x0` substituted for all occurrences of `x`.
- Each statement of the form "`return v`" in `Eval` is transformed into "`return v_1DD`".

One caveat concerning the transformation presented above should be noted: the transformation applies only to procedures that have a certain special syntactic structure—namely, the only multiplication operations that depend on the independent variable `x` must be multiplications on the right by `x`. Procedure `Eval` is an example of a procedure that has this property.

A different, but similar, transformation can be used if all of the multiplication operations that depend on the independent variable `x` are multiplications on the left by `x`. (This point is discussed further in Example 8.) It is also possible to give a fully general first-divided-difference transformation; however, this transformation can be viewed as a special case of the material presented in Section 5.

<pre>class Poly {
 public:
  float Eval(float);
  float Eval_1DD(float,float);
  float Eval'(float);
 private:
  int degree;
  // Array coeff[0..degree]
  float *coeff;
 };</pre>
	<pre>float Poly::Eval_1DD(float x0,float x1){
	  float ans_1DD = 0.0;
	  float ans = 0.0;
	  for (int i = degree; i >= 0; i--) {
	    ans_1DD = ans_1DD * x1 + ans;
	    ans = ans * x0 + coeff[i]; }
	  return ans_1DD;
	}
	
	float Poly::Eval'(float x){
	  float ans' = 0.0;
	  float ans = 0.0;
	  for (int i = degree; i >= 0; i--) {
	    ans' = ans' * x + ans;
	    ans = ans * x + coeff[i]; }
	  return ans';
	}</pre>

Fig. 3. The result of applying the computational-differentiation and first-divided-difference transformations to member function `Poly::Eval` of Example 4.

Alternatively, as with computational differentiation, for languages that support operator overloading, computational divided differencing can be carried out with the aid of a new class, say `Float1DD`, for which the arithmetic operators are appropriately redefined. (We will call such a class a *divided-difference arithmetic.*) Computational divided differencing is then carried out by making appropriate changes to the types of each procedure's formal parameters, local variables, and return value. Again, definitions of first-divided-difference arithmetic classes—both for the case of general first divided differences, as well as for the special case that covers programs like `Eval`—can be viewed as special cases of the divided-difference arithmetic classes `FloatDD` and `FloatDDR1` discussed in Sections 5 and 6, respectively.

4 Computational Divided Differencing as a Generalization of Computational Differentiation

In this section, we explain the sense in which computational divided differencing generalizes computational differentiation. First, observe that over real numbers,

$$\lim_{x_1 \to x_0} F[x_0, x_1] = \lim_{x_1 \to x_0} \frac{F(x_0) - F(x_1)}{x_0 - x_1} = F'(x_0). \tag{3}$$

However, although Equation (3) holds over reals, it does not hold over floating-point numbers: as $\mathtt{x}_1$ approaches $\mathtt{x}_0$, because of accumulated round-off error, the quantity $\frac{\mathtt{F}(\mathtt{x}_0) - \mathtt{F}(\mathtt{x}_1)}{\mathtt{x}_0 - \mathtt{x}_1}$ *does not*, in general, approach $\mathtt{F}'(\mathtt{x}_0)$. (Note the use of `Courier Font` here; this is a statement about quantities computed by *programs*.) This is why derivatives cannot be computed accurately by procedure `F'_naive` (see box (2)). In contrast, for the *programs* $\mathtt{F}[\mathtt{x}_0, \mathtt{x}_1]$ and $\mathtt{F}'(\mathtt{x}_0)$, we have: $\lim_{\mathtt{x}_1 \to \mathtt{x}_0} \mathtt{F}[\mathtt{x}_0, \mathtt{x}_1] = \mathtt{F}'(\mathtt{x}_0)$.

More precisely, we have equality when $\mathtt{x}_1$ equals $\mathtt{x}_0$:

$$\mathtt{F}[\mathtt{x}_0, \mathtt{x}_0] = \mathtt{F}'(\mathtt{x}_0). \tag{4}$$

Example 5 To illustrate Equation (4), consider applying the two transformations to member function `Poly::Eval` of Example 4; the result is shown in Figure 3. When formal parameters `x0`, `x1`, and `x` all have the same value—say `v`—then *exactly the same operations* are performed by `Eval_1DD(v,v)` and `Eval'(v)`. □

Because computations are carried out over floating-point numbers, the programs $\mathtt{F}[\mathtt{x}_0, \mathtt{x}_1]$ and $\mathtt{F}'(\mathtt{x}_0)$ are only approximations to the functions that we actually desire. That is, $\mathtt{F}[\mathtt{x}_0, \mathtt{x}_1]$ approximates the function $F[x_0, x_1]$, and $\mathtt{F}'(\mathtt{x}_0)$ approximates $F'(x_0)$. The relationships among these functions and programs are depicted below:

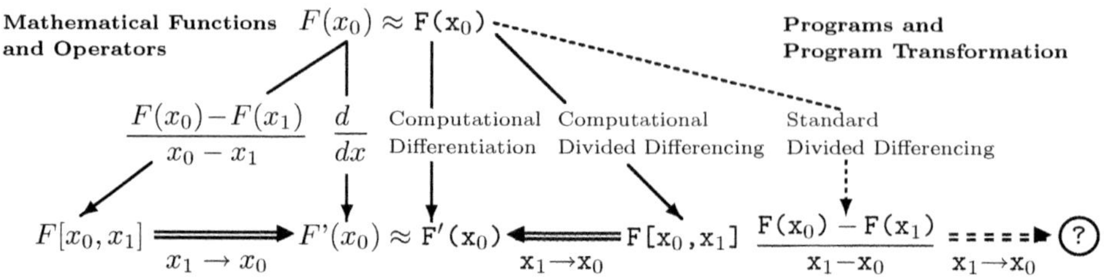

In particular, as $\mathtt{x}_1$ approaches $\mathtt{x}_0$, $\mathtt{F}[\mathtt{x}_0, \mathtt{x}_1]$ approaches $\mathtt{F}'(\mathtt{x}_0)$. Consequently, a program produced by a system for computational divided differencing can be used to compute values of derivatives (in addition to divided differences) by feeding it duplicate actual parameters (e.g., `Eval_1DD(v,v)`). In this sense, computational divided differencing can be said to generalize computational differentiation.

Computational divided differencing suffers from one of the same problems that arises with computational differentiation—namely that the program $\mathtt{F}[\mathtt{x}_0, \mathtt{x}_1]$ that results from the transformation can perform additions and subtractions that have no analogues in the original program $\mathtt{F}(\mathtt{x})$ (see the discussion in Section 2.1). Consequently, the result of evaluating $\mathtt{F}[\mathtt{x}_0, \mathtt{x}_1]$ can be degraded by round-off error even when $\mathtt{F}(\mathtt{x})$ is computed accurately. However, computational divided differencing is no worse in this regard than computational differentiation. Moreover, because of the fact that $\mathtt{F}[\mathtt{x}_0, \mathtt{x}_1]$ converges to $\mathtt{F}'(\mathtt{x}_0)$ as $\mathtt{x}_1$ approaches $\mathtt{x}_0$, if $\mathtt{F}'(\mathtt{x}_0)$ returns a result of sufficient accuracy, then $\mathtt{F}[\mathtt{x}_0, \mathtt{x}_1]$ will return a result of sufficient accuracy when $|\mathtt{x}_0 - \mathtt{x}_1|$ is small.

5 Higher-Order Computational Divided Differencing

In this section, we show that the idea from Section 3 can be generalized to define a transformation for *higher-order computational divided differencing*. To do so, we define a divided-difference arithmetic that manipulates *divided-difference tables*. Higher-order divided differences are divided differences of divided differences, defined recursively as follows:

$$F[x_i] \stackrel{\text{def}}{=} F(x_i) \tag{5}$$

$$F[x_0,x_1,\ldots,x_{n-1},x_n] \stackrel{\text{def}}{=} \begin{cases} (F[x_0,x_1,\ldots,x_{n-1}]-F[x_1,\ldots,x_{n-1},x_n])/(x_0-x_n) & \text{if } x_0 \neq x_n \\ \frac{\partial}{\partial z}F[z,x_1,\ldots,x_{n-1}]\big|_{z=x_0} & \text{if } x_0 = x_n \end{cases} \tag{6}$$

Higher-order divided differences have numerous applications in interpolation and approximation of functions [10].

In our context, a divided-difference table for a function F is an upper-triangular matrix whose entries are divided differences of different orders, as indicated below:

$$\begin{pmatrix} F(x_0) & F[x_0,x_1] & F[x_0,x_1,x_2] & F[x_0,x_1,x_2,x_3] \\ 0 & F(x_1) & F[x_1,x_2] & F[x_1,x_2,x_3] \\ 0 & 0 & F(x_2) & F[x_2,x_3] \\ 0 & 0 & 0 & F(x_3) \end{pmatrix} \tag{7}$$

Other arrangements of F's higher-order divided differences into matrix form are possible. For example, one could have a lower triangular matrix with the $F(x_i)$ running down the first column. However, the use of the arrangement shown in Equation (7) is key to being able to use simple notation—i.e., ordinary matrix operations—to describe our methods [44].

It is worthwhile mentioning here that higher-order divided differences are symmetric in the x_k; that is, for any permutation π of the sequence $[0, \ldots, n]$,

$$F[x_0, x_1, ..., x_n] = F[x_{\pi(0)}, x_{\pi(1)}, ..., x_{\pi(n)}].$$

Notation that emphasized the order-independence of the x_k, such as $F\{x_0, x_1, ..., x_n\}$, might have been employed. However, the arrangement of higher-order divided differences into the form shown in Equation (7)—together with the use of ordinary matrix operations—imposes some order on the diagonal elements, say, $F(x_0)$, $F(x_1)$, ..., $F(x_n)$. Our notation, therefore, will always reflect this order: for $0 \leq i \leq j \leq n$, the (i, j) element of a divided-difference table whose diagonal elements are $F(x_0)$, $F(x_1)$, ..., $F(x_n)$ is denoted by $F[x_i, x_{i+1}, ..., x_j]$, where the sequence $[i, i+1, \ldots, j]$ is a contiguous subsequence of $[0, \ldots, n]$.

We occasionally use $[x_{i,j}]$ as an abbreviation for $[x_i, ..., x_j]$. However, the reader should note that $F[x_{0,2}]$ is not the same as $F[x_0, x_2]$. We use $\cdot^{\triangledown[x_0, \ldots, x_n]}$ to denote the operator that yields the divided-difference table for a function with respect to points $x_0, \ldots, x_n$. (We use $\cdot^{\triangledown}$ if the points $x_0, \ldots, x_n$ are clear from the context.)

Table 2. Basic properties of two divided-difference operators.

First Divided Difference [35]	Divided-Difference Table [44]
$c[x_0, x_1] = 0.0$	$c^{\triangledown[x_0, ..., x_n]} = c * I$
$x[x_0, x_1] = 1.0$	$x^{\triangledown[x_0, ..., x_n]} = A_{[x_0, \ldots, x_n]}$
$(c+F)[x_0,x_1] = F[x_0, x_1]$	$(c+F)^{\triangledown[x_0, ..., x_n]} = c * I + F^{\triangledown[x_0, ..., x_n]}$
$(c*F)[x_0,x_1] = c * F[x_0, x_1]$	$(c*F)^{\triangledown[x_0, ..., x_n]} = c * F^{\triangledown[x_0, ..., x_n]}$
$(F+G)[x_0,x_1] = F[x_0, x_1] + G[x_0, x_1]$	$(F+G)^{\triangledown[x_0,...,x_n]} = F^{\triangledown[x_0,...,x_n]} + G^{\triangledown[x_0,...,x_n]}$
$(F*G)[x_0,x_1] = F[x_0,x_1]*G(x_1) + F(x_0)*G[x_0,x_1]$	$(F*G)^{\triangledown[x_0,...,x_n]} = F^{\triangledown[x_0,...,x_n]} * G^{\triangledown[x_0,...,x_n]}$
$\left(\frac{F}{G}\right)[x_0,x_1] = \frac{F[x_0, x_1]*G(x_1) - F(x_1)*G[x_0, x_1]}{G(x_0) * G(x_1)}$	$\left(\frac{F}{G}\right)^{\triangledown[x_0, ..., x_n]} = \frac{F^{\triangledown[x_0, ..., x_n]}}{G^{\triangledown[x_0, ..., x_n]}}$

A method for creating accurate divided-difference tables for rational expressions is found in Opitz [44]. This method is based on the properties of $\cdot^{\triangledown[x_0,\ldots,x_n]}$ given in the right-hand column of Table 2, where

- I denotes the identity matrix.
- $A_{[x_0,\ldots,x_n]}$ denotes the matrix $\begin{pmatrix} x_0 & 1 & 0 & \cdots & 0 \\ 0 & x_1 & 1 & \cdots & 0 \\ \vdots & \ddots & \ddots & \ddots & \vdots \\ 0 & \cdots & \ddots & x_{n-1} & 1 \\ 0 & \cdots & \cdots & 0 & x_n \end{pmatrix}$
- In the entry for $(F * G)^{\triangledown[x_0,\ldots,x_n]}$, the multiplication operation in $F^{\triangledown[x_0,\ldots,x_n]} * G^{\triangledown[x_0,\ldots,x_n]}$ is matrix multiplication.
- In the entry for $\left(\frac{F}{G}\right)^{\triangledown[x_0,\ldots,x_n]}$, the division operation in $\frac{F^{\triangledown[x_0,\ldots,x_n]}}{G^{\triangledown[x_0,\ldots,x_n]}}$ is matrix division (i.e., $\frac{P}{Q} = P * Q^{-1}$).

The two columns of Table 2 can be read as recursive definitions for the operations $\cdot[x_0, x_1]$ and $\cdot^{\triangledown[x_0,\ldots,x_n]}$, respectively. These have straightforward implementations as recursive programs that walk over an expression tree.

It is easy to verify (by means of induction) that the first and second columns of Table 2 are consistent with each other: in each case, the quantity $e[x_0, x_1]$ represents the (0,1) entry of the matrix $e^{\triangledown[x_0,\ldots,x_n]}$.

The second column of Table 2 has another interpretation:

Observation 1 [Reinterpretation Principle]. The divided-difference table of an arithmetic expression $e(x)$ with respect to the $n+1$ points $x_0, \ldots, x_n$ can be obtained by reinterpreting $e(x)$ as a *matrix expression*, where the matrix $A_{[x_0,\ldots,x_n]}$ is used at each occurrence of the variable x, and $c * I$ is used at each occurrence of a constant c.

That is, the expression tree for $e(x)$ is unchanged—except at its leaves, where $A_{[x_0,\ldots,x_n]}$ is used in place of x, and $c * I$ is used in place of c—but the operators at all internal nodes are reinterpreted as denoting matrix operations. This observation is due to Opitz [44]. With only a slight abuse of notation, we can express this as $e^{\triangledown[x_0,\ldots,x_n]} = e(A_{[x_0,\ldots,x_n]})$.

Using this notation, we can show that the chain rule for the divided-difference operator $\cdot^{\triangledown[x_0,\ldots,x_n]}$ has the following particularly simple form:

$$(F \circ G)^{\triangledown[x_0,\ldots,x_n]} = (F \circ G)(A_{[x_0,\ldots,x_n]}) = F(G(A_{[x_0,\ldots,x_n]})) = F(G^{\triangledown[x_0,\ldots,x_n]}).$$

Opitz's idea can be extended to the creation of accurate divided-difference tables for functions defined by *programs* by overloading the arithmetic operators used in the program to be matrix operators—i.e., by defining a divided-difference arithmetic that manipulates divided-difference tables:

Observation 2 [Computational Divided-Differencing Principle]. Rather than computing a divided-difference table with respect to the points x_0, x_1, …, x_n by invoking the program $n+1$ times and then applying Equations (5) and (6), we may instead evaluate the program (once) using a divided-difference arithmetic that overloads arithmetic operations as matrix operations, substituting $A_{[x_0,\ldots,x_n]}$ for each occurrence of the formal parameter x, and $c * I$ for each occurrence of a constant c.

The single invocation of the program using the divided-difference arithmetic will actually be more expensive than the $n + 1$ ordinary invocations of the program. The advantage of using divided-difference arithmetic is *not* that execution is speeded up because the program is only invoked once (in fact, execution is slower); the advantage is that the result computed using divided-difference arithmetic is much more *accurate*.

Because higher-order divided differences are defined recursively in terms of divided differences of lower order (cf. Equations (5) and (6)), it would be possible to define an algorithm for higher-order computational-divided-differencing using repeated applications of lower-order computational-divided-differencing transformations. However, with each application of the transformation for computational first divided differencing, the program that results performs (roughly) three times the number of operations that are performed by the program the transformation starts with. Consequently, this approach has a significant drawback: the final program that would be created for computing k^{th} divided differences could be $O(3^k)$ times slower than the original program. In contrast, the slow-down factor with the approach based on Observation 2 is $O(k^3)$.

We now sketch how a version of higher-order computational divided differencing based on Observation 2 can be implemented in C++. Below, we present highlights of a divided-difference arithmetic class, named `FloatDD`. We actually make use of two classes:

(i) class `FloatDD`, the divided-difference arithmetic proper, and

(ii) class `FloatV`, vectors of $\mathtt{x_i}$ values.

These classes are defined as follows:

```
class FloatDD {
 public:
  int numPts; // Size is numPts-by-numPts
  float **divDiffTable; // Two-dimensional upper-triangular array
  FloatDD(ArgDesc ad, int N, float v); // N-by-N constant or variable
  FloatDD(const FloatV &); // Construct a FloatDD from a FloatV
  FloatDD(int N); // Construct a zero-valued FloatDD of size N-by-N
  FloatDD& operator+ (const FloatDD &) const; // binary addition
  FloatDD& operator- (const FloatDD &) const; // binary subtraction
  FloatDD& operator* (const FloatDD &) const; // binary multiplication
  FloatDD& operator/ (const FloatDD &) const; // binary division
};
```

```
class FloatV {
 public:
  int numPts;
  float *val; // An array of values: val[0]..val[numPts-1]
  FloatV(int N, ...); // N points
  FloatV(float start, int N, float incr); // N equally spaced points
};
```

The constructor `FloatDD(const FloatV &)` plays the role of generating a matrix $A_{[x_0,\ldots,x_n]}$ from a vector $[x_0, \ldots, x_n]$ of values for the independent variable.

It is defined as follows:

```
// Construct a FloatDD from a FloatV
FloatDD::FloatDD(const FloatV &fv) :
  numPts(fv.numPts),
  divDiffTable(calloc_ut(numPts)) // allocate upper-triangular matrix of zeros
{
  for (int i = 0; i < numPts; i++) {
    divDiffTable[i][i] = fv.val[i];
    if (i < numPts-1) divDiffTable[i][i+1] = 1.0; }
}
```

The procedure `calloc_ut` allocates an upper-triangular matrix in such a way that ordinary array-indexing operations can be used to access the elements; all elements of the matrix are initialized to zero.

```
// Allocate upper-triangular matrix of zeros
float **calloc_ut(int n){
  int size = (n * (n+1)) / 2;
  float *arr = new float[size];
  float **a = new float*[n];
  for (float *p = arr; p < &arr[size]; p++) *p = 0.0;
  float *q = arr;
  for (int i = 0; i < n; i++) {
    a[i] = q;
    q += (n-(i+1)); }
  return a;
}
```

The constructor `FloatDD(ArgDesc ad, int N, float v)` generates either the matrix $A_{[\mathtt{v},\ldots,\mathtt{v}]}$ of size `N`-by-`N` or the matrix $\mathtt{v} * I$ of size `N`-by-`N`, depending on the value of parameter `ad`. It is defined as follows:

```
FloatDD::FloatDD(ArgDesc ad, int N, float v):
  numPts(N),
  divDiffTable(calloc_ut(numPts))
{
  int i;
  switch (ad) {
    case CONST:
      for (i = 0; i < numPts; i++) {
        divDiffTable[i][i] = v; }
      break;
    case VAR:
      for (i = 0; i < numPts; i++) {
        divDiffTable[i][i] = v;
        if (i < numPts - 1)
          divDiffTable[i][i+1] = 1.0; }
      break; }
}
```

The multiplication operator of class `FloatDD` performs matrix multiplication:

```
FloatDD& FloatDD::operator* (const FloatDD &fdd) const{
  assert(numPts == fdd.numPts);
  FloatDD *ans = new FloatDD(numPts);
  for (int r = 0; r < numPts; r++) {
    for (int c = r; c < numPts; c++) {
      float temp = 0.0;
      for (int k = r; k <= c; k++) {
        temp += divDiffTable[r][k] * fdd.divDiffTable[k][c]; }
      ans->divDiffTable[r][c] = temp; }
  }
  return *ans;
}
```

The division operator of class `FloatDD` is implemented using back substitution. That is, suppose we wish to find the value of `A/B` (call this value `X`). `X` can be found by solving the system `X * B = A`. Because the divided-difference tables `A` and `B` are both upper-triangular matrices, this can be done using back substitution:

```
// Use back substitution
FloatDD& FloatDD::operator/ (const FloatDD &fdd) const{
  assert(numPts == fdd.numPts);
  assert(NonZeroDiagonal(fdd));
  FloatDD *ans = new FloatDD(numPts);
  for (int r = 0; r < numPts; r++) {
    for (int c = r; c < numPts; c++) {
      float temp = 0.0;
      for (int k = r; k < c; k++) {
        temp += ans->divDiffTable[r][k] * fdd.divDiffTable[k][c]; }
      ans->divDiffTable[r][c] =
           (divDiffTable[r][c] - temp) / fdd.divDiffTable[c][c]; }
  }
  return *ans;
}
```
(8)

Example 6 To illustrate these definitions, consider again procedure `Poly::Eval`, which evaluates a polynomial via Horner's rule.

Computational divided differencing is carried out by changing the types of `Eval`'s formal parameters, local variables, and return value from `float` to `FloatDD`:

```
// Evaluation via Horner's rule
float Poly::Eval(float x){
  float ans = 0.0;
  for (int i = degree; i >= 0; i--){
    ans = ans * x + coeff[i];
  }
  return ans;
}
```

⇒

```
// Evaluation via Horner's rule
FloatDD Poly::Eval(const FloatDD &x){
  FloatDD ans(x.numPts); // ans = 0.0
  for (int i = degree; i >= 0; i--){
    ans = ans * x + coeff[i];
  }
  return ans;
}
```

The transformed procedure can be used to generate the divided-difference table for the polynomial $P(x) = 2.1 * x^3 - 1.4 * x^2 - .6 * x + 1.1$ with respect to the (unevenly spaced) points 3.0, 3.01, 3.02, 3.05 by performing the following operations:

```
Poly *P = new Poly(4,2.1,-1.4,-0.6,1.1);
FloatV x(4,3.0,3.01,3.02,3.05);
FloatDD A(x); // Corresponds to A[3.0,3.01,3.02,3.05]
FloatDD fdd = P->Eval(A);
```

(9)

We now present some empirical results that illustrate the advantages of the computational-divided-differencing method. In this experiment, we worked with the polynomial $P(x) = 2.1 * x^3 - 1.4 * x^2 - .6 * x + 1.1$, and performed computations on a Sun SPARCstation 20/61 running SunOS 5.6. Programs were compiled with the egcs-2.91.66 version of `g++` (egcs-1.1.2 release). The experiment compared the standard method for generating divided-difference tables—namely, the recursive definition given by Equation (5) and the first line of Equation (6)—against the overloaded version of procedure `Poly::Eval` from Example 6 (which was invoked using code like the fragment shown in box (9)) using single-precision and double-precision floating-point arithmetic.

In each of the examples shown below, the values used for the (unevenly spaced) points x_0, x_1, x_2, and x_3 are shown on the left. (Differences are indicated in boldface.)

Computational Divided Differencing (single-precision arithmetic) | Standard Divided Differencing (single-precision arithmetic)

$$\begin{array}{ll} x_0: & 3.0 \\ x_1: & 4.0 \\ x_2: & 5.0 \\ x_3: & 7.0 \end{array} \begin{pmatrix} 43.4 & 67.3 & 23.8 & \mathbf{2.1} \\ & 110.7 & 114.9 & 32.2 \\ & & 225.6 & 211.5 \\ & & & 648.6 \end{pmatrix} \begin{pmatrix} 43.4 & 67.3 & 23.8 & 2.\mathbf{09999} \\ & 110.7 & 114.9 & 32.2 \\ & & 225.6 & 211.5 \\ & & & 648.6 \end{pmatrix}$$

$$\begin{array}{ll} x_0: & 3.0 \\ x_1: & 3.01 \\ x_2: & 3.02 \\ x_3: & 3.05 \end{array} \begin{pmatrix} 43.4 & 47.87\mathbf{52} & 17.5\mathbf{63} & \mathbf{2.1} \\ & 43.8787 & 48.226\mathbf{5} & 17.66\mathbf{8} \\ & & 44.361 & 48.9332 \\ & & & 45.829 \end{pmatrix} \begin{pmatrix} 43.4 & 47.87\mathbf{49} & 17.5\mathbf{858} & \mathbf{1.59073} \\ & 43.8787 & 48.226\mathbf{6} & 17.66\mathbf{53} \\ & & 44.361 & 48.9332 \\ & & & 45.829 \end{pmatrix}$$

$$\begin{array}{ll} x_0: & 3.0 \\ x_1: & 3.001 \\ x_2: & 3.002 \\ x_3: & 3.005 \end{array} \begin{pmatrix} 43.4 & 47.717\mathbf{5} & \mathbf{17.5063} & \mathbf{2.1} \\ & 43.477 & 47.7\mathbf{525} & \mathbf{17.5168} \\ & & 43.4955 & 47.82\mathbf{26} \\ & & & 43.6389 \end{pmatrix} \begin{pmatrix} 43.4 & 47.717\mathbf{7} & \mathbf{15.2886} & \mathbf{754.685} \\ & 43.4477 & 47.7\mathbf{483} & \mathbf{19.0621} \\ & & 43.4955 & 47.82\mathbf{45} \\ & & & 43.6389 \end{pmatrix}$$

$$\begin{array}{ll} x_0: & 3.0 \\ x_1: & 3.0001 \\ x_2: & 3.0002 \\ x_3: & 3.0005 \end{array} \begin{pmatrix} 43.4 & 47.\mathbf{7017} & \mathbf{17.5006} & \mathbf{2.1} \\ & 43.4048 & 47.\mathbf{7052} & \mathbf{17.5017} \\ & & 43.4095 & 47.7\mathbf{122} \\ & & & 43.4238 \end{pmatrix} \begin{pmatrix} 43.4 & 47.\mathbf{6945} & \mathbf{3.62336} & \mathbf{117520} \\ & 43.4048 & 47.\mathbf{6952} & \mathbf{62.379} \\ & & 43.4095 & 47.7\mathbf{202} \\ & & & 43.4238 \end{pmatrix}$$

In particular, because P is a cubic polynomial whose high-order coefficient is 2.1, the proper value of $P[x_0, x_1, x_2, x_3]$—the third divided difference of P—is 2.1, not 117,520! (Compare the entries that appear in the upper-right-hand corners of the fourth pair of divided-difference tables shown above.)

Switching to double-precision arithmetic, and continuing to move the points closer together, we obtain

Computational Divided Differencing (double-precision arithmetic) — Standard Divided Differencing (double-precision arithmetic)

$$\begin{array}{ll} x_0: & 3.0 \\ x_1: & 3.0001 \\ x_2: & 3.0002 \\ x_3: & 3.0005 \end{array} \begin{pmatrix} 43.4 & 47.7018 & 17.5006 & 2.1 \\ & 43.4048 & 47.7053 & 17.5017 \\ & & 43.4095 & 47.7123 \\ & & & 43.4239 \end{pmatrix} \quad \begin{pmatrix} 43.4 & 47.7018 & 17.5006 & 2.10\mathbf{121} \\ & 43.4048 & 47.7053 & 17.5017 \\ & & 43.4095 & 47.7123 \\ & & & 43.4239 \end{pmatrix}$$

$$\begin{array}{ll} x_0: & 3.0 \\ x_1: & 3.00001 \\ x_2: & 3.00002 \\ x_3: & 3.00005 \end{array} \begin{pmatrix} 43.4 & 47.7002 & 17.5001 & \mathbf{2.1} \\ & 43.4005 & 47.7005 & 17.5002 \\ & & 43.4001 & 47.7012 \\ & & & 43.4024 \end{pmatrix} \quad \begin{pmatrix} 43.4 & 47.7002 & 17.5001 & \mathbf{1.89257} \\ & 43.4005 & 47.7005 & 17.5002 \\ & & 43.4001 & 47.7012 \\ & & & 43.4024 \end{pmatrix}$$

$$\begin{array}{ll} x_0: & 3.0 \\ x_1: & 3.000001 \\ x_2: & 3.000002 \\ x_3: & 3.000005 \end{array} \begin{pmatrix} 43.4 & 47.7 & 17.5 & \mathbf{2.1} \\ & 43.4 & 47.7001 & \mathbf{17.5} \\ & & 43.4001 & 47.7001 \\ & & & 43.4002 \end{pmatrix} \quad \begin{pmatrix} 43.4 & 47.7 & 17.50\mathbf{77} & \mathbf{-1640.17} \\ & 43.4 & 47.7001 & 17.\mathbf{4995} \\ & & 43.4001 & 47.7001 \\ & & & 43.4002 \end{pmatrix}$$

Finally, with either single-precision or double-precision arithmetic, when we set all of the input values to 3.0, we obtain

Computational Divided Differencing — Standard Divided Differencing

$$\begin{array}{l} x_0: 3.0 \\ x_1: 3.0 \\ x_2: 3.0 \\ x_3: 3.0 \end{array} \qquad \begin{pmatrix} 43.4 & \mathbf{47.7} & \mathbf{17.5} & \mathbf{2.1} \\ & 43.4 & \mathbf{47.7} & \mathbf{17.5} \\ & & 43.4 & \mathbf{47.7} \\ & & & 43.4 \end{pmatrix} \quad \begin{pmatrix} 43.4 & \mathbf{NaN} & \mathbf{NaN} & \mathbf{NaN} \\ & 43.4 & \mathbf{NaN} & \mathbf{NaN} \\ & & 43.4 & \mathbf{NaN} \\ & & & 43.4 \end{pmatrix}$$

With the standard divided-differencing method, division by 0 occurs and yields the exceptional value NaN. In contrast, computational divided differencing produces values for P's first, second, and third derivatives. More precisely, each kth divided-difference entry in the computational-divided-differencing table equals

$$\left. \frac{1}{k!} \frac{d^k P(x)}{dx^k} \right|_{x = 3.0} \tag{10}$$

The $k = 1$ case was already discussed in Section 4, where we observed that computational first divided differencing could be used to compute first derivatives. □

Example 7 [30]. Suppose that we wish to compute the future value of n monthly payments, each of 1 unit, paid at the end of each month into a savings account that compounds interest at the rate of α per month (where α is a small positive value and n is a positive integer). This answers the question "How many dollars are accumulated after n months, when you deposit \$1 per month for n months, into a savings account that pays annual interest at the rate of $(12 \times \alpha \times 100)\%$, compounded monthly?" Future value can be computed by the function

$$\mathit{FutureValue}(\alpha, n) \stackrel{\text{def}}{=} \frac{((1+\alpha)^n - 1)}{\alpha}. \tag{11}$$

However, this can also be written as $\mathit{FutureValue}(\alpha, n) = \dfrac{((1+\alpha)^n - 1^n)}{(1+\alpha) - 1}$, and thus is equal to the following first-divided difference of the power function:

$\mathit{FutureValue}(\alpha, n) = (x^n)[1+\alpha, 1]$.

The latter quantity can be computed to nearly full accuracy using computational divided differencing by computing $(x^n)^{\triangledown[1+\alpha,1]}$, and then extracting the $(0, 1)$ entry. For instance, we can start with the following procedure `power`, which computes `x`$^{\mathtt{n}}$ via repeated squaring and multiplication by `x`, according to the bits of argument `n`:

```
const unsigned int num_bits = sizeof(unsigned int)*8;
float power(float x, unsigned int n) {
  unsigned int mask = 1 << (num_bits - 1);
  float ans = 1.0;
  for (unsigned int i = 0; i < num_bits; i++) {
    ans = ans * ans;
    if (mask & n)
      ans = ans * x;
    mask >>= 1;
  }
  return ans;
}
```

By changing the types of `power`'s formal parameters, local variables, and return value, we create a version that computes a divided-difference table:

```
FloatDD power(FloatV &x, unsigned int n) {
  unsigned int mask = 1 << (num_bits - 1);
  FloatDD ans(CONST, 2, 1.0); // ans = 1.0
  for (unsigned int i = 0; i < num_bits; i++) {
    ans = ans * ans;
    if (mask & n)
      ans = ans * x;
    mask >>= 1;
  }
  return ans;
}
```

The transformed procedure can be used to compute the desired computation to nearly full accuracy by calling the procedure `FutureValue` that is defined below:

```
float FutureValue(float alpha, unsigned int n) {
  float w[2] = { 1+alpha, 1 };
  FloatV v(2, w);
  return power(v,n).divDiffTable[0][1];
}
```

We now present some empirical results that illustrate the advantages of this approach. In this experiment, we performed computations using single-precision floating-point arithmetic on a Sony VAIO PCG-Z505JSK (650 MHz Intel Pentium III processor) running Windows 2000. Programs were compiled with Microsoft Visual C++ 6.0. Table 3 shows the future values of \$1 deposits made for 360 months, as computed via Equation (11) versus procedure `FutureValue`, for a variety of interest rates. □

6 A Special Case

A divided-difference table for a function F can be thought of as a (redundant) representation of an interpolating polynomial for F. For instance, if you have a divided-difference table T (and also know the appropriate vector of values x_0, x_1, ..., x_n), you can explicitly construct the Newton form of the interpolating polynomial for F according to the following definition [10, page 197]:

$$p_n(x) = \sum_{i=0}^{n} F[x_0, \ldots, x_i] * \prod_{j=0}^{i-1} (x - x_j) \tag{12}$$

Note that to be able to create the Newton form of the interpolating polynomial for F via Equation (12), only the first row of divided-difference table T is required to be at hand—i.e., the values $F[x_0, \ldots, x_i]$, for $0 \leq i \leq n$—together with the values of x_0, x_1, ..., x_n. This observation suggests that we should develop an alternative divided-difference arithmetic that builds up and manipulates only first rows of divided-difference tables. We call this divided-difference arithmetic `FloatDDR1` (for **D**ivided-**D**ifference **R**ow **1**). The motivation

Rate	Equation (11)	FutureValue
8%	1490.36	1490.36
4%	694.0**45**	694.0**48**
2%	492.7**13**	492.7**22**
0.8%	406.**692**	406.**718**
0.4%	382.4	382.4**22**
0.2%	370.**934**	370.**986**
0.08%	364.**142**	364.**34**
0.04%	362.**595**	362.**165**
0.02%	361.**505**	361.**08**
0.008%	360.**882**	360.**432**
0.004%	360.**668**	360.**216**
0.002%	360.**56**	360.**108**
0.0008%	**386.238**	**360.046**
0.0004%	**386.238**	**360.023**
0.0002%	**257.492**	**360.008**

Table 3. The future values of $1 deposits made for 360 months, as computed via Equation (11) versus procedure `FutureValue`. (Differences are indicated in boldface.)

for this approach is that `FloatDDR1` operations will be much faster than `FloatDD` ones, because `FloatDD` operations must manipulate upper-triangular matrices, whereas `FloatDDR1` operations involve only simple vectors.

To achieve this, we define class `FloatDDR1` as follows:

```
class FloatDDR1 {
  friend FloatDDR1& operator+ (const FloatDDR1 &, const float);
  friend FloatDDR1& operator+ (const FloatDDR1 &, const FloatV &);
  friend FloatDDR1& operator+ (const float, const FloatDDR1 &);
  friend FloatDDR1& operator+ (const FloatV &, const FloatDDR1 &);
  friend FloatDDR1& operator- (const FloatDDR1 &, const float);
  friend FloatDDR1& operator- (const FloatDDR1 &, const FloatV &);
  friend FloatDDR1& operator- (const float, const FloatDDR1 &);
  friend FloatDDR1& operator- (const FloatV &, const FloatDDR1 &);
  friend FloatDDR1& operator* (const FloatDDR1 &, const float);
  friend FloatDDR1& operator* (const FloatDDR1 &, const FloatV &);
  friend FloatDDR1& operator* (const float, const FloatDDR1 &);
  friend FloatDDR1& operator* (const FloatV &, const FloatDDR1 &);
  friend FloatDDR1& operator/ (const FloatDDR1 &, const float);
  friend FloatDDR1& operator/ (const FloatDDR1 &, const FloatV &);
  public:
  int numPts;
  float *divDiffTable; // One-dimensional array of divided differences
  FloatDDR1(int N) // Construct a zero-valued FloatDDR1 of length N
    : numPts(N), divDiffTable(new float[numPts])
    { }
  FloatDDR1& operator+ (const FloatDDR1 &) const; // binary addition
  FloatDDR1& operator- (const FloatDDR1 &) const; // binary subtraction
};
```

Compared with class `FloatDD`, class `FloatDDR1` is somewhat impoverished: we can add or subtract two arbitrary `FloatDDR1`'s; however, because we do not have full divided-difference tables available, we cannot multiply two arbitrary `FloatDDR1`'s; nor do we have the full $A_{[x_0,\ldots,x_n]}$ matrices that are used at each occurrence of the independent variable. We finesse these difficulties by limiting the other operations of class `FloatDDR1` to those defined by the friend functions indicated in the class definition given above: (i) addition, subtraction, and multiplication on either side by a `float` or a `FloatV`; (ii) division on the right by a `float` or a `FloatV`.

The operations that involve a `float` argument `c` have their "obvious" meanings, if one bears in mind that a `float` value `c` serves as a stand-in for a full matrix `c*I`. For the addition (subtraction) operations, `c` is only added to (subtracted from) the `divDiffTable[0]` entry of the `FloatDDR1` argument. For the multiplication (division) operations, all of the `divDiffTable` entries are multiplied by (divided by) `c`.

In the operations that involve a `FloatV` argument, the `FloatV` value serves as a stand-in for a full $A_{[x_0,\ldots,x_n]}$ matrix. For instance, the operator for multiplication on the right by a `FloatV` can be thought of as performing a form of matrix multiplication—but specialized to produce only the first row of the output divided-difference table (and to use only values that are available in the given `FloatDDR1` and `FloatV` arguments):

```
FloatDDR1& operator* (const FloatDDR1 &fddr1, const FloatV &fv){
  FloatDDR1 *ans = new FloatDDR1(fddr1.numPts);
  ans->divDiffTable[0] = fddr1.divDiffTable[0] * fv.val[0];
  for (int c = 1; c < fddr1.numPts; c++) {
   ans->divDiffTable[c] = fddr1.divDiffTable[c-1] + fddr1.divDiffTable[c] * fv.val[c];
  }
  return *ans;
}
```

It might be thought that the operator for multiplication on the left by a `FloatV` does not have the proper values available in the given `FloatV` and `FloatDDR1` arguments to produce the first row of the product divided-difference table as output. (In particular, the second argument, which is of type `FloatDDR1`, is a row vector, yet we want to produce a row vector as the result.) However, it is easy to show that divided-difference matrices are commutative:

$$F^{\triangledown} * G^{\triangledown} = (F * G)^{\triangledown} = (G * F)^{\triangledown} = G^{\triangledown} * F^{\triangledown} \tag{13}$$

Consequently, the operator for multiplication on the left by a `FloatV` can be treated as if the `FloatV` were on the right:

```
FloatDDR1& operator* (const FloatV &fv, const FloatDDR1 &fddr1){
  return fddr1 * fv;
}
```

As with class `FloatDD`, the division operator is implemented using a form of back substitution — specialized here to compute just what is needed for the first row of the divided-difference table:

```
FloatDDR1& operator/ (const FloatDDR1 &fddr1, const FloatV &fv){
  FloatDDR1 *ans = new FloatDDR1(fddr1.numPts);
  ans->divDiffTable[0] = fddr1.divDiffTable[0] / fv.val[0];
  for (int c = 1; c < fddr1.numPts; c++) {
    ans->divDiffTable[c] = (fddr1.divDiffTable[c] - ans->divDiffTable[c-1])
                           / fv.val[c]; }
  return *ans;
}
```

Because only a limited set of arithmetic operations is available for objects of class `FloatDDR1`, this divided-difference arithmetic can only be applied to procedures that have a certain special syntactic structure, namely ones that are "accumulative" in the independent variable (with only "right-accumulative" quotients). In other words, the procedure must never perform arithmetic operations (other than addition or subtraction) on two local variables that both depend on the independent variable.

Example 8 The procedure `Poly::Eval` for evaluating a polynomial via Horner's rule is an example of a procedure of the right form. Consequently, an overloaded version of `Poly::Eval` that uses `FloatDDR1` arithmetic can be written as shown below:

<pre>// Evaluation via Horner's rule
float Poly::Eval(float x){
  float ans = 0.0;
  for (int i = degree; i >= 0; i--){
    ans = ans * x + coeff[i];
  }
  return ans;
}</pre>
	<pre>// Evaluation via Horner's rule
	FloatDDR1 Poly::Eval(const FloatV &x){
	  FloatDDR1 ans(x.numPts); // ans = 0.0
	  for (int i = degree; i >= 0; i--){
	    ans = ans * x + coeff[i];
	  }
	  return ans;
	}</pre>

In Section 3, Example 4 discussed the procedure `Poly::Eval_1DD`, a transformed version of `Poly::Eval` that computes the value of the first divided difference of a polynomial with respect to two values, x_0 and x_1. With the way that the overloaded operations are defined for class `FloatDDR1`, when the actual parameter supplied for `x` is a `FloatV` of length two consisting of x_0 and x_1, the procedure `FloatDDR1 Poly::Eval(const FloatV &x)` performs essentially the same steps as `Poly::Eval_1DD`. One slight difference is that, in addition to returning the value of the first divided difference, the `FloatDDR1` version also returns the result of evaluating the polynomial on x_0. Another difference is that with class `FloatDDR1`, because of our trick for handling multiplication by a `FloatV` on the left (cf. Equation (13) and the discussion that follows), `FloatDDR1` arithmetic can be used with programs that contain multiplications by the independent variable `x` on the left *as well as* on the right. (The transformation used in Example 4 could also be enhanced in this fashion.) □

Some empirical results that illustrate the advantages of `FloatDDR1` arithmetic in a useful application are presented at the end of Section 8.

As with the methods discussed in Sections 3 and 5, `FloatDDR1` arithmetic can be used to produce values of interest for computational differentiation. For instance, suppose we have transformed procedure `F`:

```
float F(float x); ⇒ FloatDDR1 F(const FloatV &x);
```

When all of the x_i values in the actual parameter supplied for `FloatV x` are the same value, say $\overline{x}$, then the `FloatDDR1` value returned as the output holds the Taylor coefficients for the expansion of `F` at $\overline{x}$ (cf. formula (10)). Thus, the `FloatV` divided-difference arithmetic generalizes previously known techniques for producing accurate Taylor coefficients for functions defined by programs [49].

If we attempt to use `FloatDDR1` arithmetic in a procedure that is not "accumulative" in the independent variable, with only "right-accumulative" quotients, the overload-resolution mechanism of the C++ compiler will detect and report a problem.

In the future-value calculation performed in Example 7, we cannot apply `FloatDDR1` arithmetic to procedure `power` because the statement "`ans = ans * ans;`" multiplies two local variables that both depend on the independent variable `x`. Consequently, `power` is not accumulative in the independent variable. In the case of Microsoft Visual C++ 6.0, the following error message is issued: `binary '*' : no operator defined which takes a left-hand operand of type 'class FloatDDR1'`.

7 Multidimensional Computational Divided Differencing

In this section, we define a third divided-differencing arithmetic that allows one to perform computational divided differencing of functions of several variables.

7.1 Background Discussion

As background to the material that will be discussed in Section 7.2, let us reiterate a few points concerning the divided-difference tables that result from computational divided differencing of functions of a single variable. In the following discussion, we assume that the divided-difference table in question has been constructed with respect to some known collection of values x_0, x_1, ..., x_n.

As mentioned at the beginning of Section 6, a divided-difference table can be thought of as a (redundant) representation of an interpolating polynomial. For instance, if you have a divided-difference table T (and know the appropriate vector of values x_0, x_1, ..., x_n, as well), you can explicitly construct the interpolating polynomial in Newton form by using the values in the first row of T in accordance with Equation (12). One of the consequences of this point is so central to what follows in Section 7.2 that it is worthwhile to state it explicitly and to introduce some helpful notation:

Observation 3 [Representation Principle]. A divided-difference table T is a finite representation of a function $Func[\![T]\!]$ defined by Equation (12). (Note that if $F = Func[\![T]\!]$, then $T = F^{\triangledown}$.) Given two divided-difference tables, T_1 and T_2, that are defined with respect to the same set of points x_0, x_1, ..., x_n, the operations of matrix addition, subtraction, multiplication, and division applied to T_1 and T_2 yield representations of the sum, difference, product, and quotient, respectively, of $Func[\![T_1]\!]$ and $Func[\![T_2]\!]$.

In other words, the operations of class `FloatDD` provide ways to (i) instantiate representations of functions of one variable (by evaluating programs in which `floats` have been replaced by `FloatDDs`), and (ii) perform operations on function representations (i.e., by addition, multiplication, etc. of `FloatDD` values).

We also restate the Computational Divided-Differencing Principle (Observation 2), adding the additional remark given in the second paragraph:

Observation 4 [Computational Divided-Differencing Principle Redux]. Rather than computing a divided-difference table with respect to the points x_0, x_1, ..., x_n by invoking the program $n+1$ times and then applying Equations (5) and (6), we may instead evaluate the program (once) using a divided-difference arithmetic that overloads arithmetic operations as matrix operations, substituting $A_{[x_0,\ldots,x_n]}$ for each occurrence of the formal parameter x, and $c * I$ for each occurrence of a constant c.

Furthermore, this principle can be applied to divided-difference tables for functions on any field (because addition, subtraction, multiplication, and division operations are required, together with additive and multiplicative identity elements).

7.2 Computational Divided Differencing of Functions of Several Variables

We now consider the problem of defining an appropriate notion of divided differencing for a function F of several variables. Observation 3 provides some guidance, as it suggests that the generalized divided-difference table for F that we are trying to create should also be thought of as a representation of a function of several variables that *interpolates* F. Such a generalized computational divided-differencing technique will be based on the combination of Observations 3 and 4.

Because we have already used the term *higher-order* to refer generically to 2nd, 3rd, ..., n^{th} divided differences, we use the term *higher-kind* to refer to the generalized divided-difference tables that arise with functions of several variables. In the remainder of this section, we make use of an alternative notation for the divided-difference operator $\cdot^{\triangledown[x_0,\ldots,x_n]}$:

$$\mathbf{DD}^1_{[x_0,\ldots,x_n]}[\![F]\!] \stackrel{\text{def}}{=} F^{\triangledown[x_0,\ldots,x_n]}.$$

We use $\mathbf{DD}^1[\![F]\!]$ when the x_i are understood, and abbreviate ranges of variables in the usual way, e.g., $\mathbf{DD}^1_{[x_{0,3}]}[\![F]\!] = \mathbf{DD}^1_{[x_0,x_1,x_2,x_3]}[\![F]\!] = F^{\triangledown[x_0,x_1,x_2,x_3]}$. The notation $\mathbf{DD}^1[\![\cdot]\!]$ refers to divided-difference tables of kind 1 (the kind we are already familiar with from Section 5, namely `FloatDD` values). Below, we use $\mathbf{DD}^2[\![\cdot]\!]$ to refer to divided-difference tables of kind 2; in general, we use $\mathbf{DD}^k[\![\cdot]\!]$ to refer to divided-difference tables of kind k.

To understand the basic principle that underlies our approach, consider the problem of creating a surface that interpolates a two-variable function $F(x, y)$ with respect to a grid formed by three coordinate values x_0, x_1, x_2 in the x-dimension, and four coordinate values y_0, y_1, y_2, y_3 in the y-dimension.

The clearest way to explain the technique in common programming-language terminology involves currying F. That is, instead of working with F : *float* $\times$ *float* $\rightarrow$ *float*, we work with F : *float* $\rightarrow$ *float* $\rightarrow$ *float*. We can create (a representation of) an interpolating surface for F (i.e., a divided-difference table of kind 2, denoted by $\mathbf{DD}^2_{[x_{0,2}],[y_{0,3}]}[\![F]\!]$) by building a divided-difference table of kind 1 using the functions $F(x_0)$, $F(x_1)$, and $F(x_2)$, each of which is of type *float* $\rightarrow$ *float*, as the "interpolation points". (This idea is a specific instance of the very general approach to surface approximation via the tensor-product construction given in [11, Chapter XVII].)

Note that this process requires that we be capable of performing addition, subtraction, multiplication, and division of *functions*. However, each of the functions $F(x_0)$, $F(x_1)$, and $F(x_2)$ is itself a one-argument function for which we can create a representation, namely by building the divided-difference tables $\mathbf{DD}^1_{[y_{0,3}]}[\![F(x_0)]\!]$, $\mathbf{DD}^1_{[y_{0,3}]}[\![F(x_1)]\!]$, and $\mathbf{DD}^1_{[y_{0,3}]}[\![F(x_2)]\!]$ (with respect to the coordinate values y_0, y_1, y_2, and y_3). By Observation 4, the arithmetic operations on functions $F(x_0)$, $F(x_1)$, and $F(x_2)$ needed to create $\mathbf{DD}^2_{[x_{0,2}],[y_{0,3}]}[\![F]\!]$ can be carried out by performing matrix operations on the matrices $\mathbf{DD}^1_{[y_{0,3}]}[\![F(x_0)]\!]$, $\mathbf{DD}^1_{[y_{0,3}]}[\![F(x_1)]\!]$, and $\mathbf{DD}^1_{[y_{0,3}]}[\![F(x_2)]\!]$. For instance,

$$\mathbf{DD}^1_{[y_{0,3}]}[\![F[x_0, x_1]]\!] = \mathbf{DD}^1_{[y_{0,3}]}[\![\frac{F(x_0)-F(x_1)}{x_0-x_1}]\!] = \frac{\mathbf{DD}^1_{[y_{0,3}]}[\![F(x_0)]\!]-\mathbf{DD}^1_{[y_{0,3}]}[\![F(x_1)]\!]}{x_0 - x_1} \tag{14}$$

In what follows, it is convenient to express functions using lambda notation (i.e., in $\lambda z.exp$, z is the name of the formal parameter, and exp is the function body). For instance, $\lambda x.\lambda y.x$ denotes the curried two-argument function (of type *float* $\rightarrow$ *float* $\rightarrow$ *float*) that merely returns its first argument. For our purposes, the advantage of lambda notation is that it provides a way to express the anonymous one-argument function that is returned when a curried two-argument function is supplied a value for its first argument (e.g., $(\lambda x.\lambda y.x)(x_0)$ returns $\lambda y.x_0$).

In short, the idea is that a divided-difference table of kind 2 for function F is a matrix of matrices (see also [53, Figure 4]):

$$\begin{aligned}\mathbf{DD}^2_{[x_{0,2}],[y_{0,3}]}[\![F]\!] &= \mathbf{DD}^1_{[x_{0,2}]}[\![\lambda x.\mathbf{DD}^1_{[y_{0,3}]}[\![F(x)]\!]]\!] \\ &= \begin{pmatrix} \mathbf{DD}^1_{[y_{0,3}]}[\![F(x_0)]\!] & \mathbf{DD}^1_{[y_{0,3}]}[\![F[x_0,x_1]]\!] & \mathbf{DD}^1_{[y_{0,3}]}[\![F[x_0,x_1,x_2]]\!] \\ & \mathbf{DD}^1_{[y_{0,3}]}[\![F(x_1)]\!] & \mathbf{DD}^1_{[y_{0,3}]}[\![F[x_1,x_2]]\!] \\ & & \mathbf{DD}^1_{[y_{0,3}]}[\![F(x_2)]\!] \end{pmatrix}\end{aligned}$$

It is instructive to consider some instances of $\mathbf{DD}^2_{[x_{0,2}],[y_{0,3}]}[\![F]\!]$ for various F's:

Example 9 Consider the function $\lambda x.\lambda y.x$. For $0 \leq i \leq 2$, we have

$$\mathbf{DD}^1_{[y_{0,3}]}[\![(\lambda x.\lambda y.x)(x_i)]\!] = \mathbf{DD}^1_{[y_{0,3}]}[\![\lambda y.x_i]\!] = \begin{pmatrix} x_i & 0 & 0 & 0 \\ & x_i & 0 & 0 \\ & & x_i & 0 \\ & & & x_i \end{pmatrix}$$

and, for $0 \leq i \leq 1$, we have

$$\mathbf{DD}^1_{[y_{0,3}]}[\![(\lambda x.\lambda y.x)[x_i, x_{i+1}]]\!] = \mathbf{DD}^1_{[y_{0,3}]}[\![\frac{(\lambda x.\lambda y.x)(x_i) - (\lambda x.\lambda y.x)(x_{i+1})}{x_i - x_{i+1}}]\!]$$

$$= \mathbf{DD}^1_{[y_{0,3}]}[\![\frac{\lambda y.x_i - \lambda y.x_{i+1}}{x_i - x_{i+1}}]\!] = \mathbf{DD}^1_{[y_{0,3}]}[\![\lambda y.1]\!] = \begin{pmatrix} 1 & 0 & 0 & 0 \\ & 1 & 0 & 0 \\ & & 1 & 0 \\ & & & 1 \end{pmatrix}$$

Consequently, we have that: $\mathbf{DD}^2_{[x_{0,2}],[y_{0,3}]}[\![\lambda x.\lambda y.x]\!] =$

$$= \begin{pmatrix} \begin{pmatrix} x_0 & 0 & 0 & 0 \\ & x_0 & 0 & 0 \\ & & x_0 & 0 \\ & & & x_0 \end{pmatrix} & \begin{pmatrix} 1 & 0 & 0 & 0 \\ & 1 & 0 & 0 \\ & & 1 & 0 \\ & & & 1 \end{pmatrix} & \begin{pmatrix} 0 & 0 & 0 & 0 \\ & 0 & 0 & 0 \\ & & 0 & 0 \\ & & & 0 \end{pmatrix} \\ \begin{pmatrix} \\ \\ \\ \\ \end{pmatrix} & \begin{pmatrix} x_1 & 0 & 0 & 0 \\ & x_1 & 0 & 0 \\ & & x_1 & 0 \\ & & & x_1 \end{pmatrix} & \begin{pmatrix} 1 & 0 & 0 & 0 \\ & 1 & 0 & 0 \\ & & 1 & 0 \\ & & & 1 \end{pmatrix} \\ \begin{pmatrix} \\ \\ \\ \\ \end{pmatrix} & \begin{pmatrix} \\ \\ \\ \\ \end{pmatrix} & \begin{pmatrix} x_2 & 0 & 0 & 0 \\ & x_2 & 0 & 0 \\ & & x_2 & 0 \\ & & & x_2 \end{pmatrix} \end{pmatrix} \tag{15}$$

□

Example 10 Consider the function $\lambda x.\lambda y.y$. For $0 \leq i \leq 2$, we have

$$\mathbf{DD}^1_{[y_{0,3}]}[\![(\lambda x.\lambda y.y)(x_i)]\!] = \mathbf{DD}^1_{[y_{0,3}]}[\![\lambda y.y]\!] = \begin{pmatrix} y_0 & 1 & 0 & 0 \\ & y_1 & 1 & 0 \\ & & y_2 & 1 \\ & & & y_3 \end{pmatrix}$$

and, for $0 \leq i \leq 1$, we have

$$\mathbf{DD}^1_{[y_{0,3}]}[\![(\lambda x.\lambda y.y)[x_i, x_{i+1}]]\!] = \mathbf{DD}^1_{[y_{0,3}]}[\![\frac{(\lambda x.\lambda y.y)(x_i) - (\lambda x.\lambda y.y)(x_{i+1})}{x_i - x_{i+1}}]\!]$$

$$= \mathbf{DD}^1_{[y_{0,3}]}[\![\frac{\lambda y.y - \lambda y.y}{x_i - x_{i+1}}]\!] = \mathbf{DD}^1_{[y_{0,3}]}[\![\lambda y.0]\!] = \begin{pmatrix} 0 & 0 & 0 & 0 \\ & 0 & 0 & 0 \\ & & 0 & 0 \\ & & & 0 \end{pmatrix}$$

Consequently, we have that: $\mathbf{DD}^2_{[x_{0,2}],[y_{0,3}]}[\![\lambda x.\lambda y.y]\!] =$

$$= \begin{pmatrix} \begin{pmatrix} y_0 & 1 & 0 & 0 \\ & y_1 & 1 & 0 \\ & & y_2 & 1 \\ & & & y_3 \end{pmatrix} & \begin{pmatrix} 0 & 0 & 0 & 0 \\ & 0 & 0 & 0 \\ & & 0 & 0 \\ & & & 0 \end{pmatrix} & \begin{pmatrix} 0 & 0 & 0 & 0 \\ & 0 & 0 & 0 \\ & & 0 & 0 \\ & & & 0 \end{pmatrix} \\ \begin{pmatrix} \\ \\ \\ \\ \end{pmatrix} & \begin{pmatrix} y_0 & 1 & 0 & 0 \\ & y_1 & 1 & 0 \\ & & y_2 & 1 \\ & & & y_3 \end{pmatrix} & \begin{pmatrix} 0 & 0 & 0 & 0 \\ & 0 & 0 & 0 \\ & & 0 & 0 \\ & & & 0 \end{pmatrix} \\ \begin{pmatrix} \\ \\ \\ \\ \end{pmatrix} & \begin{pmatrix} \\ \\ \\ \\ \end{pmatrix} & \begin{pmatrix} y_0 & 1 & 0 & 0 \\ & y_1 & 1 & 0 \\ & & y_2 & 1 \\ & & & y_3 \end{pmatrix} \end{pmatrix} \tag{16}$$

□

Moreover, Observation 4 tells us that divided-difference tables for functions of two variables can be built up by means of a divided-difference arithmetic that operates on matrices of matrices. That is, we can build up divided-difference tables of kind 2 for more complex functions of x and y by using operations on matrices of matrices, substituting $\mathbf{DD}^2[\![\lambda x.\lambda y.x]\!]$ for each occurrence of the formal parameter x in the function, and $\mathbf{DD}^2[\![\lambda x.\lambda y.y]\!]$ for each occurrence of the formal parameter y.

Example 11 For the function $\lambda x.\lambda y.(x \times y)$, $\mathbf{DD}^2[\![\lambda x.\lambda y.(x \times y)]\!]$ can be created by *multiplying the matrices* $\mathbf{DD}^2[\![\lambda x.\lambda y.x]\!]$ and $\mathbf{DD}^2[\![\lambda x.\lambda y.y]\!]$ from Equations (15) and (16), respectively:

$$\mathbf{DD}^2[\![\lambda x.\lambda y.(x \times y)]\!] = \mathbf{DD}^2[\![(\lambda x.\lambda y.x) \times (\lambda x.\lambda y.y)]\!] = \mathbf{DD}^2[\![\lambda x.\lambda y.x]\!] \times \mathbf{DD}^2[\![\lambda x.\lambda y.y]\!] =$$

$$= \begin{pmatrix} \begin{pmatrix} x_0y_0 & x_0 & 0 & 0 \\ & x_0y_1 & x_0 & 0 \\ & & x_0y_2 & x_0 \\ & & & x_0y_3 \end{pmatrix} & \begin{pmatrix} y_0 & 1 & 0 & 0 \\ & y_1 & 1 & 0 \\ & & y_2 & 1 \\ & & & y_3 \end{pmatrix} & \begin{pmatrix} 0 & 0 & 0 & 0 \\ & 0 & 0 & 0 \\ & & 0 & 0 \\ & & & 0 \end{pmatrix} \\ \begin{pmatrix} \; \end{pmatrix} & \begin{pmatrix} x_1y_0 & x_1 & 0 & 0 \\ & x_1y_1 & x_1 & 0 \\ & & x_1y_2 & x_1 \\ & & & x_1y_3 \end{pmatrix} & \begin{pmatrix} y_0 & 1 & 0 & 0 \\ & y_1 & 1 & 0 \\ & & y_2 & 1 \\ & & & y_3 \end{pmatrix} \\ \begin{pmatrix} \; \end{pmatrix} & \begin{pmatrix} \; \end{pmatrix} & \begin{pmatrix} x_2y_0 & x_2 & 0 & 0 \\ & x_2y_1 & x_2 & 0 \\ & & x_2y_2 & x_0 \\ & & & x_2y_3 \end{pmatrix} \end{pmatrix}$$

Note, for example, that the (0,1) entry in the above matrix, namely

$$\begin{pmatrix} y_0 & 1 & 0 & 0 \\ & y_1 & 1 & 0 \\ & & y_2 & 1 \\ & & & y_3 \end{pmatrix}$$

was obtained via the calculation

$$\begin{pmatrix} x_0 & 0 & 0 & 0 \\ & x_0 & 0 & 0 \\ & & x_0 & 0 \\ & & & x_0 \end{pmatrix} \times \begin{pmatrix} 0 & 0 & 0 & 0 \\ & 0 & 0 & 0 \\ & & 0 & 0 \\ & & & 0 \end{pmatrix} + \begin{pmatrix} 1 & 0 & 0 & 0 \\ & 1 & 0 & 0 \\ & & 1 & 0 \\ & & & 1 \end{pmatrix} \times \begin{pmatrix} y_0 & 1 & 0 & 0 \\ & y_1 & 1 & 0 \\ & & y_2 & 1 \\ & & & y_3 \end{pmatrix} + \begin{pmatrix} 0 & 0 & 0 & 0 \\ & 0 & 0 & 0 \\ & & 0 & 0 \\ & & & 0 \end{pmatrix} \times \begin{pmatrix} 0 & 0 & 0 & 0 \\ & 0 & 0 & 0 \\ & & 0 & 0 \\ & & & 0 \end{pmatrix} \tag{17}$$

and *not* by the use of Equation (14), which involves a matrix subtraction, a scalar subtraction, and a scalar division. By sidestepping the explicit subtraction and division operations, expression (17) avoids the potentially disastrous magnification of round-off error that can occur with floating-point arithmetic. □

The principle illustrated in Example 11 gives us the machinery that we need to perform computational divided differencing for bivariate functions defined by programs. As usual, computational divided differencing is performed by changing the types of formal parameters, local variables, and return values to the type of an appropriate divided-difference arithmetic.

Furthermore, these ideas can be applied to a function F with an arbitrary number of variables: when F has k variables, $\mathbf{DD}^k[\![F]\!]$, F's divided-difference table of kind k, is a matrix of matrices of ... of matrices nested to depth k. Currying with respect to the first parameter of F "peels off" one dimension; $\mathbf{DD}^k[\![F]\!]$ is a matrix whose entries are divided-difference tables of kind $k-1$ (i.e., matrices of matrices of ... of matrices nested to depth $k-1$). For instance, the diagonal entries are the divided-difference tables of kind $k-1$ for the $(k-1)$-parameter functions $F(x_0)$, $F(x_1)$, ..., $F(x_n)$ (i.e., $\mathbf{DD}^{k-1}[\![F(x_0)]\!]$, $\mathbf{DD}^{k-1}[\![F(x_1)]\!]$, ..., $\mathbf{DD}^{k-1}[\![F(x_n)]\!]$).

To implement this approach in C++, we define two classes and one class template:

- Class template `template <int k> class DivDiffArith` can be instantiated with a positive value `k` to represent divided-difference tables of kind `k`. Each object of class `DivDiffArith<k>` has links to sub-objects of class `DivDiffArith<k-1>`.
- Class `DivDiffArith<0>` represents the base case; `DivDiffArith<0>` objects simply hold a single `float`.
- Class `IntVector`, a vector of `int`'s, is used to describe the number of points in each dimension of the grid of coordinate points.

Excerpts from the definitions of these classes are shown below:

```
template <int k> class DivDiffArith {
 public:
  int numPts;
  DivDiffArith<k-1> **divDiffTable; // Two-dimensional upper-triangular array
  DivDiffArith(const FloatV &v, const IntVector &grid, int d);
  DivDiffArith(float, const IntVector &grid); //constant; shape conforms to grid
  DivDiffArith(float, const DivDiffArith<k-1> &dda);//constant; shape conforms to dda
  DivDiffArith<k>& operator+ (const DivDiffArith<k> &) const;//binary addition
  DivDiffArith<k>& operator- (const DivDiffArith<k> &) const;//binary subtraction
  DivDiffArith<k>& operator* (const DivDiffArith<k> &) const;//binary multiplication
  DivDiffArith<k>& operator/ (const DivDiffArith<k> &) const;//binary division
};
```

```
class DivDiffArith<0> {
 public:
  float value;
  DivDiffArith(float v = 0.0); // Default constructor
  DivDiffArith(const FloatV &v, const IntVector &grid, int d);
  DivDiffArith<0>& operator+ (const DivDiffArith<0> &) const;// binary addition
  DivDiffArith<0>& operator- (const DivDiffArith<0> &) const;// binary subtraction
  DivDiffArith<0>& operator* (const DivDiffArith<0> &) const;// binary multiplication
  DivDiffArith<0>& operator/ (const DivDiffArith<0> &) const;// binary division
};
```

```
class IntVector {
 public:
  int numPts;
  int *val; // An array of values: val[0]..val[numPts-1]
  IntVector();
  IntVector(int N, ...); // Construct IntVector given N values
  IntVector& operator<< (const int i); // left shift
 };
```

The operations of class `DivDiffArith<k>` are overloaded in a fashion similar to those of `FloatDD`. (`FloatDD` is essentially identical to `DivDiffArith<1>`.) For instance, the overloaded multiplication operator performs matrix multiplication:

```
template <int k>
DivDiffArith<k>& DivDiffArith<k>::operator* (const DivDiffArith<k> &dda) const{
  assert(numPts == dda.numPts);
  DivDiffArith<k> *ans = new DivDiffArith<k>(numPts);
  for (int r = 0; r < numPts; r++) {
    for (int c = r; c < numPts; c++) {
      DivDiffArith<k-1> temp((float)0.0, divDiffTable[r][c]); // temp = 0.0
      for (int j = r; j <= c; j++) {
        temp += divDiffTable[r][j] * dda.divDiffTable[j][c];
      }
      ans->divDiffTable[r][c] = temp;
    }
  }
  return *ans;
}
```

Class `DivDiffArith<k>` has two constructors for creating a `DivDiffArith<k>` object from a `float` constant. They differ only in their second arguments (an `IntVector` versus a `DivDiffArith<k>`), which are used to determine the appropriate dimensions to use at each level in the nesting of matrices.

Suppose that in the procedure on which computational divided differencing is to be carried out, variable `z` is the independent variable associated with argument position `d+1`. To generate an appropriate `DivDiffArith<k>` object for `z` for a given set of grid values z_0, ..., z_m, a `FloatV` with the values z_0, ..., z_m is created, and then passed to the following `DivDiffArith<k>` constructor:

```
template <int k>
DivDiffArith<k>::DivDiffArith(const FloatV &v, const IntVector &grid, int d):
  numPts(grid.val[0]),
  divDiffTable(calloc_ut< k >(numPts))
{
  assert(grid.val[d] == v.numPts);
  IntVector tail = grid << 1;
  for (int r = 0; r < numPts; r++) {
    for (int c = r; c < numPts; c++) {
      divDiffTable[r][c] = DivDiffArith<k-1>((float)0.0,tail); }
  }
  if(d == 0) {
    DivDiffArith<k-1> one((float)1.0, tail);
    for (int i = 0; i < numPts; i++) {
      divDiffTable[i][i] = DivDiffArith<k-1>(v.val[i],tail);
      if(i < numPts - 1) { divDiffTable[i][i+1] = one; } }
  }
  else {
    for (int i = 0; i < numPts; i++) {
      divDiffTable[i][i] = DivDiffArith<k-1>(v,tail,d-1); } }
}
```

Example 12 The following code fragment generates two `DivDiffArith<2>` values, `x` and `y`, which correspond to the matrices shown in Equations (15) and (16), respectively:

```
IntVector grid(2,3,4);
FloatV fv_x(3,x0,x1,x2);
DivDiffArith<2> x(fv_x,grid,0); // argument position 1
FloatV fv_y(4,y0,y1,y2,y3);
DivDiffArith<2> y(fv_y,grid,1); // argument position 2
```

□

Example 13 Consider a C++ class `BivariatePoly` that represents bivariate polynomials, and a member function `BivariatePoly::Eval` that evaluates a polynomial via a bivariate version of Horner's rule:

```
class BivariatePoly {
 public:
  float Eval(float,float);
 private:
  int degree1,degree2;
  //Array coeff[0..degree1][0..degree2]
  float **coeff;
 };
```

```
//Evaluation via bivariate Horner's rule
float BivariatePoly::Eval(float x, float y){
  float ans = 0.0;
  for (int i = degree1; i >= 0; i--){
    float temp = 0.0;
    for (int j = degree2; j >= 0; j--){
      temp = temp * y + coeff[i][j]; }
    ans = ans * x + temp;
  }
  return ans;
}
```

Similar to what has been done in Examples 4, 5, 6, and 8, computational divided differencing is carried out on this version of `Eval` by changing the types of its formal parameters, local variables, and return value from `float` to `DivDiffArith<2>`.

```
// Evaluation via bivariate Horner's rule
DivDiffArith<2> BivariatePoly::Eval(const DivDiffArith<2>, const DivDiffArith<2> &y)
{ DivDiffArith<2> ans(0.0,x); // ans = 0.0
  for (int i = degree1; i >= 0; i--){
    DivDiffArith<2> temp(0.0,y); // temp = 0.0
    for (int j = degree2; j >= 0; j--){
      temp = temp * y + coeff[i][j]; }
    ans = ans * x + temp;
  }
  return ans;
}
```

To use this procedure to create a divided-difference table of kind 2 for a given variable `P` of type `BivariatePoly*`, with respect to the 3-by-4 grid {x_0,x_1,x_2} × {y_0,y_1,y_2,y_3}, we would generate the `IntVector grid` and `DivDiffArith<2>` values `x` and `y` as shown in Example 12, and then invoke "`P->Eval(x,y);`" □

In general, if there are k independent variables (i.e., k dimensions), and v_i is the number of sample coordinate values for the ith dimension, where $1 \leq i \leq k$, each value of type `DivDiffArith<k>` will use space $\prod_{i=1}^{k} \frac{v_i(v_i+1)}{2}$. Compared with the time required for the original program, the slow-down factor for the `DivDiffArith<k>` version is bounded by $O(\prod_{i=1}^{k} v_i^3)$.

One final point concerning costs: by generalizing the grid descriptors slightly, it is possible to devise an even more general divided-differencing arithmetic that is *heterogeneous* in shape with respect to different argument positions. By "heterogeneous", we mean that full two-dimensional (upper-triangular) divided-difference tables could be provided for some argument positions, while other argument positions could just provide a single row of divided differences (i.e., one-dimensional, `FloatDDR1`-like tables). By this means, when a procedure body is "accumulative" in certain of its formal parameters but not others, it would be possible to tailor the divided-differencing version of the procedure to improve its efficiency. (In the case of `DivDiffArith<2> BivariatePoly::Eval`, it would be possible to specify that both argument positions provide `FloatDDR1`-like tables.)

8 Paige's Work on Finite Differencing of Computable Expressions

Starting in the mid-1970s, Paige studied how finite-differencing transformations of applicative set-former expressions could be exploited to optimize loops in very-high-level languages, such as SETL [47]. These ideas were implemented in his RAPTS system [46]. Some of the techniques that Paige explored have their roots in earlier work by Earley [12, 13].

Independently of and contemporaneously with Paige, similar loop-optimization methods targeted toward very-high-level set-theoretic languages were investigated by Fong and Ullman [15–17]. More recently, Liu and Stoller have used some extensions of these ideas to optimize array computations [37] and recursive programs [38]. Liu et al. have also shown how such transformations can be applied to derive algorithms for incremental-computation problems (i.e., problems in which the goal is to maintain the value of some function $F(x)$ as the input x undergoes small changes) [39, 40].

The basic idea for optimizing SETL loops is described as a generalization of strength reduction, an idea attributed to John Cocke from the 1960s, whereby a loop is transformed so that a multiplication operation in the loop is eliminated in favor of an addition, as shown below:

```
                           i = ...;
i = ...;                   T = i*c; // T depends on i
while (...) {              deltaT = delta*c;
  ... i*c ...;       ⇒     while (...) {
  i = i + delta;             ... T ...; // T replaces i*c
}                            i = i + delta; // change to i
                             T = T + deltaT // update of T
                           }
```

This transformation improves the running time of the loop if the cost of the additions performed by the transformed loop are less than the cost of the multiplications performed in the original loop. In [9], Cocke and Schwartz presented a variety of strength-reduction transformations for use in optimizing compilers.

Paige's work on loop optimization in SETL was based on the observation that a similar transformation could be applied to loops that involve set-former expressions. In this transformation, an expensive set-former expression in a loop is replaced by a set-initialization statement (placed before the loop) together with a set-update operation (placed inside the loop):

```
                                   A = ...;
A = ...;                           T = {x ∈ A | x%2 == 0}; // T depends on A
while (...) {                      while (...) {
  ... {x ∈ A | x%2 == 0} ...;  ⇒     ... T ...; // T replaces the set former
  d = ...;                           d = ...;
  A = A ∪ {d};                       A = A ∪ {d}; // change to A
}                                    if (d%2 == 0) T = T ∪ {d}; // update of T
                                   }
```

In the transformed program shown above on the right, the expression `{x ∈ A | x%2 == 0}` in the loop is replaced by a use of `T`. Because the statement "`A = A ∪ {d};`" may alter the value of variable `A`, just after this statement a new statement is introduced: "`if (d%2 == 0) T = T ∪ {d};`". The latter statement updates the value of variable `T` to have the same value that the expression `{x ∈ A | x%2 == 0}` has when evaluated with the new value of variable `A`. In Paige's terminology, the code fragment `if (d%2 == 0) T = T ∪ {d};` is called a *post-derivative* of `T = {x ∈ A | x%2 == 0};` with respect to the change `A = A ∪ {d};`. Similarly, a code fragment for updating variable `T` that is placed before the change `A = A ∪ {d};` is called a *prederivative*.

Of more direct relevance to the topic of the present paper is the discussion in Paige's papers in which he points out affinities between his work, on the one hand, and differentiation and finite differencing of numerical functions, on the other hand. In particular, the (forward) *finite difference* of a function F with respect to h is defined as follows:

$$\Delta_h F(x) \stackrel{\text{def}}{=} F(x+h) - F(x).$$

Paige and Koenig describe the relationship between their SETL finite-differencing methods and numerical finite-difference methods as follows [47, pages 403–404]:

«It is interesting to note that the origins of our method may be traced back to the finite difference techniques introduced by the English mathematician Henry Briggs in the sixteenth century. His method, which can be used to generate a sequence of polynomial values $p(x_0)$, $p(x_0+h)$, $p(x_0+2h)$, ..., hinges on the following idea. For a given polynomial $p(x)$ of degree n and an increment h, the first difference polynomial

$$p_1(x) = p(x+h) - p(x)$$

is of degree $n-1$ or less, the second difference polynomial

$$p_2(x) = p_1(x+h) - p_1(x)$$

is of degree $n-2$ or less, ..., and, finally, $p_n(x)$ must be a constant. Thus, to tabulate successive values of $p(x)$ starting with $x = x_0$, we can perform these two steps:

1. Calculate initial values for $p(x_0)$, $p_1(x_0)$, ..., $p_n(x_0)$ and store them in $t(1)$, $t(2)$, ..., $t(n+1)$.
2. Generate the desired polynomial table by iterating over the following code block:

```
print x, t(1);                   $ print x and p(x)
t(1) := t(1) + t(2);             $ place new values for
t(2) := t(2) + t(3);             $ p(x), p1(x), ..., pn-1(x)
...                              $ into
t(n) := t(n) + t(n + 1);         $ t(1), t(2), ..., t(n).
x := x + h;                      $
```

Note that Briggs's method requires only n additions in step 2 to compute each new polynomial value, while Horner's rule to compute a fresh polynomial value costs n additions and n multiplications.»

They relate Briggs's method to strength reduction in the following passage [47, pages 404–405]:

«Although Cocke's technique does not treat polynomials as special objects, strength reduction is sufficiently powerful to transform a program involving repeated calculations of a polynomial according to Horner's rule into an equivalent program that essentially uses the more efficient finite difference method of Briggs. Indeed, this is a surprising and important result that demonstrates that the success of polynomial evaluation by differencing results from properties of the elementary operations used to form polynomials rather than from properties exclusive to polynomials. In other words, Cocke's method works because the following distributive and associative laws hold for sums and products:

$$(i \pm delta) * c \quad \Rightarrow \quad i * c \pm delta * c;$$
$$(i \pm delta) + c \quad \Rightarrow \quad (i + c) \pm delta.$$

In [9] Cocke and Schwartz extend this idea to show how reduction in strength (which we call finite differencing) applies to a wide range of arithmetic operations that exhibit appropriate distributive properties.»

Later in the paper, after Paige and Koenig have introduced their rules for finite differencing of set-former expressions with respect to changes in argument values (an operation that they sometimes call "differentiation"), they return to the discussion of Briggs's method [47, page 421]:

«Profitable differentiation of an expression f can sometimes be supported by differentiating f together with a chain of auxiliary expressions (as in Briggs's first, second, ..., difference polynomials . . .). Thus, the prederivative $\nabla^- E\langle x \mathrel{+}:= delta;\rangle$ of the nth degree polynomial $E = P(x)$ is

$$E \mathrel{+}:= P_1(x)$$

where $P_1(x)$ is the first difference polynomial. However, for the prederivative code above to be inexpensive, we must also differentiate the second, third, ..., n^{th} difference polynomials, denoted $E_i = P_i(x)$, $i = 2..n$. To realize Briggs's efficient technique, we consider the extended prederivative (of expressions ordered carefully into a "differentiable chain") $\nabla^- E_{n-1}, \ldots, E_1, E\langle x \mathrel{+}:= delta;\rangle$ that expands into

$$E \mathrel{+}:= E_1; \quad E_1 \mathrel{+}:= E_2; \quad \ldots \quad E_{n-1} \mathrel{+}:= E_n;$$»

Essentially all of the material that Paige and Koenig present in their paper to relate their work to Briggs's method has been quoted above. However, a few details about the derivation of Briggs's method were not spelled out in their treatment, which we now attempt to rectify. We will show below that computational divided differencing supplies a clean way to handle an important step in the derivation for which what Paige and Koenig say is ambiguous.

The initial program for tabulating a polynomial at a collection of equally spaced points can be written in C++ as follows:[5]

```
void Poly::Tabulate(float start, int numPoints, float h){
  float x = start;
  float y;
  for (int i = 1; i <= numPoints; i++) { // Tabulation loop
    y = Eval(x);
    cout << x << ": " << y << endl;
    x += h;
  }
}
```

In the program shown above, `Eval` is the member function of class `Poly` that evaluates a polynomial via Horner's rule (see Section 3), of type `float Poly::Eval(float x);`

The intention of Paige and Koenig is to transform procedure `Poly::Tabulate` into something like the following version:

```
void Poly::Tabulate(float start, int numPoints, float h){
  float x = start;
  float y;
  float E = Eval(start); // Call on Eval hoisted out of the loop
  float E_1 = ???; // Unspecified initialization
          ...
  float E_{n-1} = ???;
  float E_n = ???;
  for (int i = 1; i <= numPoints; i++) { // Tabulation loop
    y = E;
    cout << x << ": " << y << endl;
    E += E_1; // Extended prederivative w.r.t. x += h;
    E_1 += E_2;
          ...
    E_{n-1} += E_n;
    x += h;
  }
}
```

However, as indicated by the question marks in the above code, Paige and Koenig do not state explicitly how they plan to arrive at the proper initialization code that is to be placed just before the loop in the transformed program. It is unclear whether they intend to generate the values E_1, E_2, …, E_{n-1}, E_n by evaluating polynomial `P` at `start`, `start+h`, …, `start+(n-1)*h`, `start+n*h` and then create the E_i via subtraction operations, or whether they intend to generate the finite-difference polynomials P_1, P_2, …, P_{n-1}, P_n symbolically and then apply each of them to `start`. The former method can lead to very inaccurate results (see below), whereas the latter method requires that a substantial amount of symbolic manipulation be performed to generate the P_i. However, the divided-difference arithmetic `FloatDDR1` gives us an easy way to create suitable initialization code that produces accurate values for the E_i. This initialization code consists of three steps:

- Create a `FloatV` of equally spaced points, starting at `start` and separated by `h`, where the number of points is one more than the degree of the polynomial.
- Introduce a single call on the member function

  `FloatDDR1 Poly::Eval(const FloatV &x);`

  to create the first row of the divided-difference table for the polynomial with respect to the given `FloatV`.

[5] It should be pointed out that in practice it is better to code the statement in the loop that changes the value of `x` as "`x = start + i * h;`", rather than as "`x += h;`", so that small errors in `h` do not accumulate in `x` due to repeated addition. We have chosen to use the latter form to emphasize the similarities between `Tabulate` and the two earlier strength-reduction examples.

- Convert the resulting `FloatDDR1` (which holds *divided-difference* values) into the first row of a *finite-difference* table for the polynomial by multiplying each entry by an appropriate adjustment factor (i.e., the ith entry of the finite-difference table, where $0 \le i \le n$, is $\mathtt{i!} * \mathtt{h^i} * \mathtt{P}[\mathtt{x_0}, \mathtt{x_1}, \ldots, \mathtt{x_i}]$ [10, Lemma 4.1]).

This initialization method is used in the version of `Tabulate` shown below. (In this version of `Tabulate`, the $\mathtt{E_i}$ are renamed `diffTable[i]`.)

```
void Poly::Tabulate(float start, int numPoints, float h){
 float x = start;
 float y;
 // Create accurate divided-difference table
 FloatV fv(x, degree+1, h);
 FloatDDR1 fddr1 = Eval(fv); // Calls FloatDDR1 Poly::Eval(const FloatV &);
 // Convert divided-difference entries to finite-difference entries
 float *diffTable(new float[degree+1]);
 float adjustment = 1.0;
 for (int i = 0; i <= degree; i++) {
   diffTable[i] = fddr1.divDiffTable[i] * adjustment;
   adjustment *= (h * (i+1));
 }
 for (int i = 1; i <= numPoints; i++) { // Tabulation loop
   y = diffTable[0];
   cout << x << ": " << y << endl;
   for (int j = 0; j < degree; j++) { // Prederivative w.r.t. x += h;
     diffTable[j] += diffTable[j+1]; }
   x += h;
 }
}
```

Empirical Results: Tabulation of a Polynomial via Briggs's Method.
We now present some empirical results that illustrate the advantages of the final version of `Poly::Tabulate` presented above. Again, we work with the polynomial $P(x) = 2.1 * x^3 - 1.4 * x^2 - .6 * x + 1.1$, and perform computations using single-precision floating-point arithmetic on a Sun SPARCstation 20/61 running SunOS 5.6. Programs were compiled with the egcs-2.91.66 version of `g++` (egcs-1.1.2 release) with optimization at the `-O1` level.

The final version of `Poly::Tabulate` uses the divided-difference arithmetic `FloatDDR1` in the initialization step that creates the initial finite-difference vector `diffTable`. An alternative way to generate `diffTable` is to evaluate polynomial `P` at `start`, `start+h`, ..., `start+(n-1)*h`, `start+n*h` and then create `diffTable` via subtraction operations, according to the standard definition [10, page 214]. However, the latter way of generating `diffTable` involves subtraction operations, and hence may magnify any round-off errors in the $n + 1$ values computed for `P`. In contrast, the method using computational divided differencing yields a way to create a more accurate initial finite-difference table.

	Evaluate via Horner	Comp. Div. Diff. + Briggs	Standard FD + Briggs
x	P(x)	P(x)	P(x)
0.0000	1.10000	1.10000	1.10000
0.0001	1.09994	1.09994	1.09994
0.0002	1.09988	1.09988	1.09988
...	...	...	...
0.9998	1.1994**2**	1.1994**1**	**19844.3**
0.9999	1.1997**1**	1.1997**0**	**19850.3**
1.0000	1.**20000**	1.**19999**	**19856.2**
Time (milliseconds)	7.64	5.62	5.49

To give a concrete illustration of the benefits, the table on the right shows what happens when $P(x)$ is evaluated at the 10,001 points in the interval $[0.0, 1.0]$ with a grid spacing of .0001. (Differences are indicated in boldface.) The numbers that appear in the rightmost column for `P(.9998)`, `P(.9999)`, and `P(1.0000)` are *not* typographical errors. What happens is that round-off errors in the computation of the initial finite-difference table via the standard method causes the table to be initialized to 1.1, −5.99623*e*-05, −1.19209*e*-07, and 1.19209*e*-07. In contrast, the initial values produced via the method based on computational divided differencing are 1.1, −6.0014*e*-05, −2.79874*e*-08, and 1.26*e*-11. After 10,000 iterations of the Briggs calculation, accumulated errors have caused the values in the rightmost column to diverge widely from the correct ones.

Overall, the method based on computational divided differencing is far more accurate than the one in which the vector needed for Briggs's method is obtained by subtraction operations (and only 2% slower). Furthermore, the results from the method based on computational divided differencing are nearly as accurate as those obtained by reevaluating the polynomial at each point, but the reevaluation method is 36% slower.

9 Other Related Work

Computational Differentiation. Computational differentiation is a well-established area of numerical analysis, with its own substantial literature [3, 22, 23, 48, 57].

As discussed in Section 4, computational divided differencing is a *generalization* of computational differentiation: a program resulting from computational divided differencing can be used to obtain derivatives (as well as divided differences), whereas a program resulting from computational differentiation can only produce derivatives (and not divided differences).

As already mentioned in Section 2, the computational-differentiation technique that was summarized there is what is known as *forward-mode* (first-order) differentiation. A different computational-differentiation technique for first-order differentiation, *reverse mode* [20, 21, 28, 36, 55], provides theoretically better performance—a greatly reduced number of computational steps—but at the cost of the need to store or recompute intermediate values that affect the final result nonlinearly. It is possible to develop a reverse-mode version of (first-order) computational divided differencing, although it is only in special circumstances that it offers the same potential savings in operations performed that reverse mode achieves for (first-order) computational differentiation:

- Just as forward-mode computational divided differencing generalizes forward-mode computational differentiation, reverse-mode computational divided differencing generalizes reverse-mode computational differentiation: when divided differencing involves coinciding pairs (e.g., x_0 coincides with x_1, y_0 coincides with y_1, etc.), then reverse-mode computational divided differencing computes the first derivative.
- Viewing computational differentiation from the more general perspective of computational divided differencing provides some insight into the source of the advantage of reverse-mode computational differentiation. Without going into details, the advantage can be said to come from the fact that the "degenerate case" of reverse-mode computational divided differencing—reverse-mode computational differentiation—has a number of common subexpressions (and such common subexpressions do not arise in the case of forward-mode computational differentiation). Thus, the advantage of reverse mode comes from exploiting these symmetries, which lowers the execution cost.
- In the case of reverse-mode computational divided differencing when the divided differencing involves *no* coinciding pairs, then there are no common subexpressions to take advantage of. Forward mode has no symmetries to exploit either, so in this case, reverse-mode first-order computational divided differencing does not offer the potential savings in operations performed that reverse-mode computational differentiation offers.

- However, in the case of reverse-mode computational divided differencing when there is at least *one* coinciding pair (i.e., x_0 coincides with x_1, or y_0 coincides with y_1, etc.) then there are symmetries that can be exploited to create savings in operations performed, compared with forward mode (which still has no symmetries to exploit). Thus, in principle, one could build a reverse-mode computational divided differencing tool that is theoretically faster than a forward-mode computational divided differencing tool, when divided differencing involves one or more coinciding pair.

Other Work on Accurate Divided Differencing. The program-transformation techniques for creating accurate divided differences described in this paper are based on a 1964 result of Opitz's [44], which was later rediscovered in 1980 by McCurdy [42] and again in 1998 by one of us (Reps). However, Opitz and McCurdy both discuss how to create accurate divided differences only for *expressions*. In this paper, the idea has been applied to the creation of accurate divided differences for functions defined by *programs*.

McCurdy, and later Kahan and Fateman [31] and Rall and Reps [51], have looked at ways to compute accurate divided differences for library functions (i.e., sin, cos, exp, etc.).

Kahan and Fateman have also investigated how similar techniques can be used to avoid unsatisfactory numerical answers when evaluating formulas returned by symbolic-algebra systems. In particular, their work was motivated by the observation that naive evaluation of a definite integral $\int_a^b f(x)\,dx$ can sometimes produce meaningless answers: When a symbolic-algebra system produces a closed-form solution for the indefinite integral $\int f(x)\,dx$, say $G(x)$, the result of the computation $G(b) - G(a)$ may have no significant digits. Kahan and Fateman show that divided differences can be used to develop accurate numerical formulas that sidestep this problem.

One of the techniques developed by McCurdy for computing accurate divided-difference tables involved first computing just the first row of the table and then generating the rest of the entries by a backfilling algorithm. He studied the conditions under which this technique maintained sufficient accuracy. However, his algorithm for accurately computing the first row of the divided-difference table was based on a series expansion of the function, rather than a divided-difference arithmetic, such as the `FloatDDR1` arithmetic developed in Section 6.

Divided-difference arithmetic for first divided differences has also been called *slope arithmetic*, and an interval version of it has been investigated previously as a way to obtain an interval enclosure for the range of a function evaluated over an interval [35, 52, 59]. It has been shown that interval slope arithmetic can yield tighter enclosures than methods based on derivatives [35, 52, 59]. Zuhe and Wolfe have also shown that a form of interval divided-difference arithmetic involving second divided differences can provide even tighter interval enclosures [59]. In the present paper, we have confined ourselves to point (non-interval) divided-difference arithmetics, but have explored the use of a divided-difference arithmetic for divided differences of arbitrary order (originally due to Opitz [44]). We have shown how to extend the latter divided-difference arithmetic to the case of multivariate functions involving a fixed, but arbitrary, number of variables, and have also proposed various specializations of it, which would improve runtime efficiency in certain situations.

Other Work on Controlling Round-Off Error in Numerical Computations. Computational differentiation and computational divided differencing are methods for controlling round-off error that can arise in two types of numerical computations. Other work aimed at controlling round-off error in numerical computations includes interval-arithmetic methods for verifying the accuracy of computed results, which have been developed for many basic numerical computations [24, 25], as well as the work on remainder differential algebra, which can be viewed as a way of extending higher-order computational differentiation so that the resulting program also computes an interval remainder term [41].

Other Work that Exploits Operator Overloading. Representing particular sequences (such as arithmetic progressions) by compact terms, and using them as mathematical entities with appropriate overloaded arithmetic operations has been generalized to geometric progressions, descending factorials, etc. [1,32]. In Section 8, we showed how a divided-difference arithmetic provides a method for accurately initializing a finite-difference table for a polynomial. The methods described in [1, 32] provide alternative initialization methods; for instance, the example from Section 8 could be handled using Karczmarczuk's approach by evaluating the polynomial $P(x) = 2.1 * x^3 - 1.4 * x^2 - .6 * x + 1.1$ in a "sequence arithmetic" in which the argument x is an arithmetic sequence represented by a pair $x = (x_0, Dx)$, standing for $[x_0, x_0 + Dx, x_0 + 2Dx, \ldots]$, where Dx is some constant. However, in a more complicated sequence $y = (y_0, Dy)$, Dy is itself allowed to be a sequence (z_0, Dz), and the overall top-level sequence y can be reconstructed through partial sums. For a sequence generated by a polynomial, Dy is allowed to have deeper nested structure.

For instance, the evaluation of $y = P(x)$ using sequence arithmetic, with $x = (x_0, Dx) = (0, 0.0001)$, would yield $y = (1.1, Dy)$, where $Dy = (-6.0014e\text{-}05, DDy)$, where $DDy = (-2.79874e\text{-}08, DDDy)$, where $DDDy = 1.26e\text{-}11$ //a constant, at last...

The four values 1.1, −6.0014*e*-05, −2.79874*e*-08, and 1.26*e*-11 are the four values of the accurate initial finite-difference table.

Acknowledgments

We are grateful for discussions with C. Bischof, C. de Boor, J. O'Brien, W. Kahan, and R. Zippel about various aspects of the work. Example 1 is due to R. Zippel [58]; Example 7 was suggested by W. Kahan [30]. The observation in Section 9 that the methods from [1,32] provide an alternative to the use of a divided-difference arithmetic for accurately initializing a finite-difference table for a polynomial (cf. Section 8) is due to one of the referees of [53].

References

1. O. Bachmann, P. S. Wang, and E. Zima: Chains of recurrences — a method to expedite the evaluation of closed form functions. In *Proc. of ISSAC 1994: Int. Symp. on Symb. and Alg. Comp.*, 242–249, New York, ACM Press, 1994.
2. T. Beck and H. Fischer: The if-problem in automatic differentiation. *J. Comp. and Appl. Math.*, 50:119–131, 1994.
3. M. Berz, C. Bischof, G. F. Corliss, and A. Griewank, editors: *Computational Differentiation: Techniques, Applications, and Tools.* Soc. for Indust. and Appl. Math., Philadelphia, PA, 1996.
4. C. Bischof, A. Carle, G. Corliss, A. Griewank, and P. Hovland: ADIFOR: Generating derivative codes from Fortran programs. *Scientific Programming*, 1(1):11–29, 1992.
5. C. Bischof, A. Carle, P. Khademi, and A. Mauer: ADIFOR 2.0: Automatic differentiation of Fortran 77 programs. *IEEE Comp. Sci. and Eng.*, 3:18–32, 1996.
6. C. Bischof, L. Roh, and A. Mauer: ADIC: An extensible automatic differentiation tool for ANSI-C. *Software — Practice and Experience*, 27(12):1427–1456, 1997.
7. C. H. Bischof and M. R. Haghighat: Hierarchical approaches to automatic differentiation. In Berz et al. [3], 83–94, 1996.
8. D. Bjørner, A. P. Ershov, and N. D. Jones, editors: *Partial Evaluation and Mixed Computation: Proc. of the IFIP TC2 Workshop on Partial Evaluation and Mixed Computation.* North-Holland, New York, 1988.
9. J. Cocke and J. T. Schwartz: *Programming Languages and Their Compilers: Preliminary Notes, 2nd Rev. Version.* Courant Inst. of Math. Sci., New York Univ., New York, 1970.
10. S. D. Conte and C. de Boor: *Elementary Numerical Analysis: An Algorithmic Approach, 2nd. Ed.* McGraw-Hill, New York, 1972.
11. C. de Boor: *A Practical Guide to Splines*, volume 27 of *Appl. Math. Sciences.* Springer-Verlag, New York, 1978.

12. J. Earley: High-level operations in automatic programming. In *Symp. on Very High Level Lang.*, New York, ACM Press, April 1974.
13. J. Earley: High-level iterators and a method for automatically designing data structure representation. *J. Comp. Lang.*, 1(4):321–342, 1976.
14. H. Fischer: Special problems in automatic differentiation. In Griewank and Corliss [23], 43–50, 1992.
15. A. Fong: Elimination of common subexpressions in very high level languages. In *Symp. on Princ. of Prog. Lang.*, 48–57, January 1977.
16. A. Fong: Inductively computable constructs in very high level languages. In *Symp. on Princ. of Prog. Lang.*, 21–28, January 1979.
17. A. Fong and J. Ullman: Induction variables in very high level languages. In *Symp. on Princ. of Prog. Lang.*, 104–112, January 1976.
18. Y. Futamura: Partial evaluation of computation process – an approach to a compiler-compiler. *Higher-Order and Symb. Comp.*, 12(4), 1999.
Reprinted from Systems • Computers • Controls 2(5), 1971.
19. H. H. Goldstine: *A History of Numerical Analysis.* Springer-Verlag, 1977.
20. A. Griewank: On automatic differentiation. In M. Iri and K. Tanabe, editors, *Mathematical Programming: Recent Developments and Applications*, 83–108. Kluwer Academic Press, Boston, MA, 1989.
21. A. Griewank: The chain rule revisited in scientific computing. *SIAM News*, 24, 1991.
22. A. Griewank: *Evaluating Derivatives: Principles and Techniques of Algorithmic Differentiation*, volume 19 of *Frontiers in Applied Mathematics.* Soc. for Indust. and Appl. Math., Philadelphia, PA, 2000.
23. A. Griewank and G. F. Corliss, editors: *Automatic Differentiation of Algorithms: Theory, Implementation, and Application.* Soc. for Indust. and Appl. Math., Philadelphia, PA, 1992.
24. R. Hammer, M. Hocks, U. Kulisch, and D. Ratz: *Numerical Toolbox for Verified Computing I: Basic Numerical Problems*, volume 21 of *Springer Ser. in Comp. Math.* Springer, New York, 1993.
25. R. Hammer, M. Hocks, U. Kulisch, and D. Ratz: *C++ Toolbox for Verified Computing I: Basic Numerical Problems.* Springer, New York, 1995.
26. S. Horwitz and T. Reps: The use of program dependence graphs in software engineering. In *Int. Conf. on Softw. Eng.*, 392–411, May 1992.
27. S. Horwitz, T. Reps, and D. Binkley: Interprocedural slicing using dependence graphs. *Trans. on Prog. Lang. and Syst.*, 12(1):26–60, January 1990.
28. M. Iri: Simultaneous computation of functions, partial derivatives and estimates of rounding errors: Complexity and practicality. *Japan J. Appl. Math.*, 1(2):223–252, 1984.
29. N. D. Jones, C. K. Gomard, and P. Sestoft: *Partial Evaluation and Automatic Program Generation.* Prentice-Hall International, 1993.
30. W. Kahan: Personal communication to Thomas Reps and Louis Rall. September 2000.
31. W. Kahan and R. J. Fateman: Symbolic computation of divided differences. Unpublished; see http://www.cs.berkeley.edu/~fateman/papers/divdiff.pdf, 1985.
32. J. Karczmarczuk: Traitement paresseux et optimisation des suites numeriques. In *Proc. of JFLA 2000*, 17–30, 2000.
33. J. Karczmarczuk: Functional differentiation of computer programs. *Higher-Order and Symb. Comp.*, 14(1):35–57, 2001.
34. R. B. Kearfott: Automatic differentiation of conditional branches in an operator overloading context. In Berz et al. [3], 75–81, 1996.
35. R. Krawczyk and A. Neumaier: Interval slopes for rational functions and associated centered forms. *SIAM J. Numer. Anal.*, 22(5):604–616, June 1985.
36. S. Linnainmaa: Taylor expansion of the accumulated rounding error. *BIT*, 16(1):146–160, 1976.

37. Y. A. Liu and S. D. Stoller: Loop optimization for aggregate array computations. In *Int. Conf. on Comp. Lang.*, May 1998.
38. Y. A. Liu and S. D. Stoller: From recursion to iteration: What are the optimizations? In *Workshop on Part. Eval. and Sem.-Based Prog. Manip.*, 73–82, January 2000.
39. Y. A. Liu, S. D. Stoller, and T. Teitelbaum: Discovering auxiliary information for incremental computation. In *Symp. on Princ. of Prog. Lang.*, 157–170, January 1996.
40. Y. A. Liu and T. Teitelbaum: Systematic derivation of incremental programs. *Sci. of Comp. Program.*, 24:1–39, 1995.
41. K. Makino and M. Berz: Remainder differential algebras and their applications. In Berz et al. [3], 63–74, 1996.
42. A. C. McCurdy: Accurate computation of divided differences. Ph.D. diss. and Tech. Rep. UCB/ERL M80/28, Univ. of Calif.–Berkeley, CA, 1980.
43. T. Mogensen: The application of partial evaluation to ray-tracing. Master's thesis, Datalogisk Institut, Univ. of Copenhagen, Copenhagen, Denmark, 1986.
44. G. Opitz: Steigungsmatrizen. *Zeitschrift für Angewandte Mathematik und Mechanik*, 44:T52–T54, 1964. In German. In English at: http://www.cs.wisc.edu/wpis/papers/opitz.zamm64.ps.
45. K. J. Ottenstein and L. M. Ottenstein: The program dependence graph in a software development environment. In *Softw. Eng. Symp. on Practical Softw. Dev. Environments*, 177–184, 1984.
46. R. Paige: Transformational programming – Applications to algorithms and systems. In *Symp. on Princ. of Prog. Lang.*, 73–87, New York, ACM Press, January 1983.
47. R. Paige and S. Koenig: Finite differencing of computable expressions. *Trans. on Prog. Lang. and Syst.*, 4(3):402–454, July 1982.
48. L. Rall: *Automatic Differentiation: Techniques and Applications, Lecture Notes in Computer Science, Vol. 120.* Springer, 1981.
49. L. B. Rall: Differentiation and generation of Taylor coefficients in Pascal-SC. In U.W. Kulisch and W.L. Miranker, editors, *A New Approach to Scientific Computation*, 291–309. Academic Press, New York, 1983.
50. L. B. Rall: Point and interval differentiation arithmetics. In Griewank and Corliss [23], 17–24, 1992.
51. L. B. Rall and T. W. Reps: Algorithmic differencing. In U. Kulisch, R. Lohner, and A. Facius, editors, *Perspectives in Enclosure Methods*, 133–147. Springer, Vienna, 2001.
52. D. Ratz: An optimized interval slope arithmetic and its application. Bericht 4/1996, Institut für Angewandte Mathematik, Universität Karlsruhe, Karlsruhe, Germany, 1996.
53. T. W. Reps and L. B. Rall: Computational divided differencing and divided-difference arithmetics. *Higher-Order and Symb. Comp.*, 16:93–149, 2003.
54. K. Shamseddine and M. Berz: Exception handling in derivative computation with nonarchimedean calculus. In Berz et al. [3], 37–51, 1996.
55. B. Speelpenning: *Compiling Fast Partial Derivatives of Functions Given by Algorithms.* Ph.D. thesis, Dept. of Comp. Sci., Univ. of Illinois, Urbana, IL, January 1980.
56. M. Weiser: Program slicing. *Trans. on Softw. Eng.*, SE-10(4):352–357, July 1984.
57. R. E. Wengert: A simple automatic derivative evaluation program. *Commun. ACM*, 7(8):463–464, 1964.
58. R. Zippel: Personal communication to Thomas Reps. July 1996.
59. S. Zuhe and M. A. Wolfe: On interval enclosures using slope arithmetic. *Applied Mathematics and Computation*, 39(1):89–105, 1990.

Least Reflexive Points of Relations

Jules Desharnais[1]* and Bernhard Möller[2]

[1]Département d'Informatique, Université Laval, Québec, QC, G1K 7P4 Canada
Jules.Desharnais@ift.ulaval.ca
[2]Institut für Informatik, Universität Augsburg, D-86135 Augsburg, Germany
Bernhard.Moeller@informatik.uni-augsburg.de

Summary. Assume a partially ordered set $(S, \leq)$ and a relation R on S. We consider various sets of conditions in order to determine whether they ensure the existence of a least reflexive point, that is, a least x such that xRx. This is a generalization of the problem of determining the least fixed point of a function and the conditions under which it exists. To motivate the investigation we first present a theorem by Cai and Paige giving conditions under which iterating R from the bottom element necessarily leads to a minimal reflexive point; the proof is by a concise relation-algebraic calculation. Then, we assume a complete lattice and exhibit sufficient conditions, depending on whether R is partial or not, for the existence of a least reflexive point. Further results concern the structure of the set of all reflexive points; among other results we give a sufficient condition for these to form a complete lattice, thus generalizing Tarski's classical result to the nondeterministic case.

Keywords: partial order, fixed point, least reflexive point, greatest reflexive point, lattice, relation, inflationary relation.

1 Introduction

Iterative and recursive processes are at the center of computer science. The mathematical background is the theory of (least) fixed points and is well understood in the case where the iteration can be described by a (total) *function* [4, 10].

Much less is known about fixed points of *relations*. The problem *Find the least x related to itself* was stated to the first author in these terms by Robert Paige in 1992, at the 44th meeting of the IFIP Working Group 2.1 (*Algorithmic Languages and Calculi*), which was held in Augsburg, Germany, and was organized by the second author. This problem has its origin in the work presented in [3]. There, the authors are concerned with the construction of efficient algorithms expressed in a language using set-theoretic queries augmented with nondeterministic minimal and maximal fixed point queries (the deterministic case is treated in an earlier paper [2]).

Let us state the problem more precisely. Consider a partially ordered set $(S, \leq)$ and a binary relation R on S. A *reflexive point of* R is an element $x \in S$ such that xRx. In the sequel we give conditions under which R has a least reflexive point and investigate the structure of the set of all reflexive points of R. It turns out that under a suitable relational generalization of the property of monotonicity, the set of reflexive points even forms a complete lattice, so that Tarski's classical results generalize nicely to the nondeterministic case.

As a starting point, we present in Section 2 a theorem by Cai and Paige [3] giving conditions under which iterating R from the bottom element of the partial order necessarily leads to a minimal reflexive point; this theorem is based on the notion of an *inflationary relation*.

* This research was supported by a grant from NSERC (Natural Sciences and Engineering Research Council of Canada).

O. Danvy et al. (eds.), Automatic Program Development – A Tribute to Robert Paige, 215–228.

In Section 3, we first define four conditions generalizing monotonicity. Then, we examine which combinations of these, if any, are sufficient to ensure the existence of a least reflexive point; total and partial relations are tackled separately. In Section 4, we uncover some additional structure on the set of reflexive points. In Section 5, we use duality principles to present analogous results about greatest reflexive points. In Section 6, we study another set of four possible generalizations of monotonicity and we explain their relationship with the four conditions of Section 3.

We conclude with an evaluation of the results achieved and directions for future research. There are also two appendices. Appendix A gives graphical representations of the lattices and relations used as examples in the paper. They are grouped together to facilitate comparisons. The diagrams are labelled alphabetically. In the text, we refer to these diagrams by "Appendix A(a)", "Appendix A(b)", etc. Appendix B contains examples for all possible combinations of the conditions from Section 3 and all possible combinations of those of Section 6. Most proofs are omitted. They can be found in [5,6].

2 Reflexive Points of Inflationary Relations

As stated in the introduction, Cai and Paige [3] are concerned with the construction of efficient algorithms expressed in a language using set-theoretic queries augmented with nondeterministic minimal and maximal fixed point queries. A typical algorithm is one that finds a maximal independent set of vertices of an undirected graph. A *maximal independent set* of an undirected graph (V, E) is a subset $U \subseteq V$ such that for any $(u, v) \in E$, at most one of u and v is in U and, for any $u \in V - U$, there is a vertex $v \in U$ such that $(u, v) \in E$.

Consider the following graph. It has two maximal independent sets of vertices, namely $\{1, 3\}$ and $\{2, 4\}$.

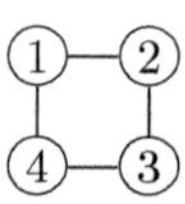

An algorithm incrementally building a maximal independent set would initially choose any vertex and add new vertices, provided that this preserves independence, until a fixed point is reached where no more vertices can be added. Here is such an algorithm:[1]

$U := \emptyset$;
while $\exists(v : v \in V - U : U \cup \{v\}$ is independent$)$ **do**
  $U := U \cup \ni\{v \mid U \cup \{v\}$ is independent$\}$

The expression $\ni S$ denotes an arbitrary element from the nonempty set S. The following relation R on the powerset $\mathcal{P}\{1, 2, 3, 4\}$ is the relation computed by the body of the nondeterministic loop, i.e., the set of pairs (U_1, U_2) such that U_2 is a possible value of variable U after one execution of the body of the loop if U_1 is the value of U before the execution.

$$R := \left\{ \begin{array}{lll} (\emptyset, \{1\}) & (\{1,2\}, \{1,2\}) & (\{1,2,3\}, \{1,2,3\}) \\ (\emptyset, \{2\}) & (\{1,3\}, \{1,3\}) & (\{1,2,4\}, \{1,2,4\}) \\ (\emptyset, \{3\}) & (\{1,4\}, \{1,4\}) & (\{1,3,4\}, \{1,3,4\}) \\ (\emptyset, \{4\}) & (\{2,3\}, \{2,3\}) & (\{2,3,4\}, \{2,3,4\}) \\ (\{1\}, \{1,3\}) & (\{2,4\}, \{2,4\}) & (\{1,2,3,4\}, \{1,2,3,4\}) \\ (\{2\}, \{2,4\}) & (\{3,4\}, \{3,4\}) & \\ (\{3\}, \{1,3\}) & & \\ (\{4\}, \{2,4\}) & & \end{array} \right\} \quad (1)$$

[1] Quantifiers have three arguments: a list of variables, the domain over which the quantification applies, and the quantified expression; for instance, $\forall(x : P : Q)$ is read "for all x satisfying P, Q holds", or "for all x, $P \Rightarrow Q$", while $\exists(x : P : Q)$ is read "there exists an x satisfying P and Q". When the second argument is true, it is omitted.

This relation has no least reflexive point, but it has many minimal ones, namely all the subsets of $\{1, 2, 3, 4\}$ with exactly two elements.

One interesting property of relation R in (1) is that any path starting at $\emptyset$ necessarily leads to a minimal reflexive point. Hence, one can build a minimal reflexive point iteratively starting from $\emptyset$ — this is what the above algorithm does.

We now give sufficient conditions that ensure this property.

By V and I we denote the universal and identity relations, respectively. The complement of a set or relation R is denoted by $\overline{R}$. The *composition* (or *relative product*) of two relations Q and R is defined by $Q;R := \{(s,u) \mid \exists(t :: sQt \text{ and } tRu)\}$. As usual, R^* denotes the reflexive and transitive closure of R. The *converse* of a relation R is defined by $R^{\smile} := \{(s,t) \mid tRs\}$.

Definition 1 Let $(S, \leq, \bot)$ be a partial order with least element $\bot$. We say that a relation R on S is *inflationary* [3] iff R is total and included in $\leq$, i.e., $R;V = V$ and $R \subseteq \leq$. In elementwise terms this means $\forall(x :: \exists(y :: xRy))$ and $\forall(x :: \forall(y : xRy : x \leq y))$.

A relation R is *progressively finite* iff there is no infinite chain $s_0, s_1, s_2, \ldots$, such that $(s_i, s_{i+1}) \in R$, for all $i \geq 0$ [9].

Because the notion of *well-foundedness* is often used to characterize relations that do not have infinite chains, we remark that R is progressively finite iff its converse is well-founded.

Proposition 3 *Let Q and R be relations and $f(X) := R;X \cup Q$.*
1. *If R is progressively finite, then f has a unique fixed point, viz. $R^*;Q$* [1].
2. *If $Q \subseteq R$ and R is progressively finite, then Q is progressively finite.*

Is is shown in [3] that, for an inflationary relation R on a progressively finite order, iteration from an arbitrary element necessarily leads to a reflexive point. To state this in relation-algebraic terms we first observe that $R \cap I$ is a partial identity relation characterizing the set of reflexive points of R. Hence we have $x\, R^*; (R \cap I)\, y$ iff from x we can reach a reflexive point y by iterating R. The claim follows if we can show that this relation is total, which is expressed by $R^*; (R \cap I); V = V$.

Theorem 1 *Let $(S, \leq)$ be a partial order such that $<$ is progressively finite. Let R be a relation on S. If R is inflationary, then $R^*; (R \cap I); V = V$.*

If $(S, \leq)$ has a least element $\bot$ then $\bot$ is a natural starting point for the iteration of R.

The relation R in (1) is inflationary, using the ordering $\subseteq$ on $\mathcal{P}\{1, 2, 3, 4\}$, and $\subset$ is progressively finite. This is why from any subset of $\{1, 2, 3, 4\}$ there is a path by R to a reflexive subset of $\{1, 2, 3, 4\}$.

We now present a different example, where relation R still satisfies the preconditions of Theorem 1, and also has a least reflexive point. The lattice is $\mathcal{P}\{1, 2, 3\}$ with the inclusion ordering. The relation R is the Hasse diagram of $\subset$ plus the pair $(\{1, 2, 3\}, \{1, 2, 3\})$, that is,

$$R := \left\{ \begin{array}{lll} (\emptyset, \{1\}) & (\{1\}, \{1, 2\}) & (\{1, 2\}, \{1, 2, 3\}) \\ (\emptyset, \{2\}) & (\{1\}, \{1, 3\}) & (\{1, 3\}, \{1, 2, 3\}) \\ (\emptyset, \{3\}) & (\{2\}, \{1, 2\}) & (\{2, 3\}, \{1, 2, 3\}) \\ & (\{2\}, \{2, 3\}) & (\{1, 2, 3\}, \{1, 2, 3\}) \\ & (\{3\}, \{1, 3\}) & \\ & (\{3\}, \{2, 3\}) & \end{array} \right\} \qquad (2)$$

This relation could correspond to an (extremely simple) algorithm that, given a set T, adds to T an element not in T, if there is any. Note that R is inflationary and that $\subset$ is progressively finite. Thus Theorem 1 applies and explains why from any subset of $\{1, 2, 3\}$ there is a path to the unique reflexive point $\{1, 2, 3\}$.

3 Four Conditions Generalizing Monotonicity

In this section, we generalize the classical fixed-point theory of monotonic functions on complete lattices to the relational case. Therefore, we assume the partial order to be a complete lattice $(S, \sqcap, \sqcup, \bot, \top, \leq)$. Letting R be a binary relation on S, we give in Theorem 2 a sufficient condition implying the existence of a least reflexive point for R when R is a total relation. A consequence of this lemma is Theorem 3 in Section 4, which shows that under the same condition the set of reflexive points of total relations is a complete lattice. We deal with partial relations in Section 3.4.

We denote, for relation R and element x, by $xR := \{y \mid xRy\}$ the set of images of x, and, for any $T \subseteq S$, by $\bigsqcap T$ and $\bigsqcup T$ the meet and join, respectively, of the elements in T. With these conventions we define the following notations:

$$\begin{array}{ll} \text{(a)} & m_\sqcap := \bigsqcap \{x \mid \bigsqcap xR \leq x\}, \\ \text{(b)} & m_\sqcup := \bigsqcap \{x \mid \bigsqcup xR \leq x\}. \end{array} \tag{3}$$

The elements $m_\sqcap$ and $m_\sqcup$ generalize the notion of least prefixed point of a function; indeed, for a total function R, we have $\bigsqcap xR = \bigsqcup xR =$ the unique image of x, so that $m_\sqcap = m_\sqcup =$ the least prefixed point of R.

3.1 Monotonicity of Relations

In the theory of fixed points of functions, monotonic functions play a major role and we seek generalizations of this notion to the case of relations. The following are four natural conditions that can be imposed on R:

$$\begin{array}{ll} \text{(a)} & \forall(x, y : x < y : \bigsqcap xR \leq \bigsqcap yR), \\ \text{(b)} & \forall(x, y : x < y : \bigsqcap xR \leq \bigsqcup yR), \\ \text{(c)} & \forall(x, y : x < y : \bigsqcup xR \leq \bigsqcap yR), \\ \text{(d)} & \forall(x, y : x < y : \bigsqcup xR \leq \bigsqcup yR). \end{array} \tag{4}$$

These are natural conditions because they all constrain in some way how the "packet" of images of x increases with increasing x, by saying how the lower and upper bounds of these images increase. When R is a total function, all are equivalent and they all state that R is monotonic (due to $\bigsqcap xR = \bigsqcup xR =$ the unique image of x).

A total function is an extreme case of a relation. There are two more relaxed cases: that of total relations and that of partial functions (*functions* for short, in the sequel). If relation R is total, then $\bigsqcap xR \leq \bigsqcup xR$, for all x. If R is *functional* (i.e., is a partial function), then $\bigsqcup xR \leq \bigsqcap xR$. This is why we obtain the following implications between Conditions 4(a,b,c,d).

$$R \text{ total: } \begin{array}{ccc} 4(c) & \Rightarrow & 4(a) \\ \Downarrow & & \Downarrow \\ 4(d) & \Rightarrow & 4(b) \end{array} \qquad R \text{ function: } \begin{array}{ccc} 4(c) & \Leftarrow & 4(a) \\ \Uparrow & & \Uparrow \\ 4(d) & \Leftarrow & 4(b) \end{array} \tag{5}$$

From these diagrams, one can deduce that if R is a total function, all four Conditions 4(a,b,c,d) are equivalent, as mentioned above.

Although Conditions 4(a,b,c,d) are not independent for the special kinds of relations mentioned above, in the general case they are, as the following examples show.

1. Condition 4(a) does not follow from (the conjunction of) 4(b,c,d). Take S to be the lattice $\{\bot, \mathsf{a}, \mathsf{b}, \top\}$ with ordering $\bot < \mathsf{a} < \top$ and $\bot < \mathsf{b} < \top$ (see Appendix A(a)), and
$$R := \{(\mathsf{a}, \mathsf{a}), (\mathsf{a}, \mathsf{b}), (\mathsf{b}, \mathsf{a}), (\mathsf{b}, \mathsf{b}), (\top, \top)\}. \tag{6}$$
2. Condition 4(b) does not follow from 4(a,c,d). This can be seen by taking the lattice $\{\bot, \top\}$ with ordering $\bot < \top$ (see Appendix A(b)), and the empty relation
$$R := \emptyset. \tag{7}$$

3. Condition 4(c) does not follow from 4(a,b,d). Take S to be the lattice $\{\bot, \mathsf{a}, \mathsf{b}, \top\}$ with ordering $\bot < \mathsf{a} < \top$ and $\bot < \mathsf{b} < \top$ (see Appendix A(c)), and
$$R := \{(\bot, \mathsf{a}), (\bot, \mathsf{b}), (\mathsf{a}, \bot), (\mathsf{a}, \top), (\mathsf{b}, \bot), (\mathsf{b}, \top), (\top, \mathsf{a}), (\top, \mathsf{b})\}. \tag{8}$$
4. Condition 4(d) does not follow from 4(a,b,c). Take S to be the lattice $\{\bot, \top\}$ with ordering $\bot < \top$ (see Appendix A(d)), and
$$R := \{(\bot, \bot), (\bot, \top)\}. \tag{9}$$

In fact, the independence is even more "complete": there are examples for all 16 possible combinations of 4(a,b,c,d) (see Appendix B).

As is easily seen, 4(a) is equivalent to $x \leq y \Rightarrow \bigsqcap xR \leq \bigsqcap yR$, for all x and y; this means that the function $(x :: \bigsqcap xR)$ is monotonic. Similarly, from 4(d), we get that $(x :: \bigsqcup xR)$ is monotonic. Because we assume a complete lattice, $m_\sqcap$ and $m_\sqcup$ are the least fixed points of $(x :: \bigsqcap xR)$ and $(x :: \bigsqcup xR)$, respectively:
$$\begin{array}{ll} \text{(a)} & \text{Condition 4(a) implies } \bigsqcap m_\sqcap R = m_\sqcap, \\ \text{(b)} & \text{Condition 4(d) implies } \bigsqcup m_\sqcup R = m_\sqcup. \end{array} \tag{10}$$

3.2 A Stronger Set of Conditions

One may wonder why in 4(a,b,c,d) we did not use $x \leq y$ instead of $x < y$, which would give
$$\begin{array}{llll} \text{(a)} & \forall(x, y : x \leq y : \bigsqcap xR \leq \bigsqcap yR), & \text{(c)} & \forall(x, y : x \leq y : \bigsqcup xR \leq \bigsqcap yR), \\ \text{(b)} & \forall(x, y : x \leq y : \bigsqcap xR \leq \bigsqcup yR), & \text{(d)} & \forall(x, y : x \leq y : \bigsqcup xR \leq \bigsqcup yR). \end{array} \tag{11}$$
This is because of the following relationship between these properties:
$$\begin{array}{llll} 11(\text{a}) \Leftrightarrow 4(\text{a}), & & 11(\text{c}) \Leftrightarrow 4(\text{c}) \text{ and } R \text{ functional}, \\ 11(\text{b}) \Leftrightarrow 4(\text{b}) \text{ and } R \text{ total} \quad (\text{if } \top \neq \bot), & & 11(\text{d}) \Leftrightarrow 4(\text{d}). \end{array} \tag{12}$$
The first and last equivalences are easy to see. The proof of the second one is based on the observation that if the set of images of x is empty, then $\bigsqcap xR \leq \bigsqcup xR \Leftrightarrow \top \leq \bot$, and the third equivalence is due to the fact that $\bigsqcup xR \leq \bigsqcap xR$ holds only if the set of images of x contains at most one element.

Since Conditions 11(b,c) are too strong, it is better to use 4(a,b,c,d) and add totality or functionality only as needed.

3.3 Reflexive Points of Total Relations

In this and the following subsection, we deal separately with total relations and partial relations, because this gives a clearer picture of the problem. We begin with the simpler case and assume in this section that relation R is total. Before showing the main result (Theorem 2), we need a lemma.

Lemma 1 *Let R be a total relation satisfying Condition* 4(c). *Then $m_\sqcup R = \{m_\sqcup\}$. In particular, $m_\sqcup$ is a reflexive point of R.*

One may wonder whether $m_\sqcap$ is also a reflexive point under the same conditions (totality + 4(c)). The following relation, on the lattice $\{\bot, \mathsf{a}, \mathsf{b}, \mathsf{c}, \top\}$ with ordering $\bot < \mathsf{a} < \mathsf{c} < \top$ and $\bot < \mathsf{b} < \mathsf{c} < \top$, shows that this is not the case (Appendix A(e)). For this relation, $m_\sqcap = \bot$.
$$R := \{(\bot, \mathsf{a}), (\bot, \mathsf{b}), (\mathsf{a}, \mathsf{c}), (\mathsf{b}, \mathsf{c}), (\mathsf{c}, \mathsf{c}), (\top, \top)\} \tag{13}$$
And here is the main result of this section.

Theorem 2 *Let R be a total relation and assume Condition* 4(c). *Then R has a least reflexive point, viz., $l := \bigsqcap C$, where $C := \{x \mid xRx\}$.*

The relation given in (13) is an example of a total relation that satisfies the precondition of Theorem 2 (i.e., Condition 4(c)). For another, less trivial, example of a relation satisfying 4(c), see (18). Condition 4(c) is very strong, since it implies 4(a,b,c,d) (see (5)).

One can check that the relation (1) of Section 2 satisfies none of the Conditions 4(a,b,c,d) (for instance, note that $\{1\} \subseteq \{1,2\}$ while $\bigcap\{1\}R = \bigcup\{1\}R = \{1,3\}$ and $\bigcap\{1,2\}R = \bigcup\{1,2\}R = \{1,2\}$). Hence, it is not too surprising that there is no least reflexive point.

We conclude this subsection on the case of total relations with the remark that Condition 4(c) in Theorem 2 cannot be relaxed to a weaker combination of 4(a,b,d). In (8) we see an example of a total relation that satisfies all of 4(a,b,d) and that has no reflexive point. On the other hand, relation R in (2) satisfies Conditions 4(a,b,d) but does not satisfy 4(c), while it has a unique reflexive point and thus a least one. So Conditions 4(a,b,c,d) do not cover all possible situations. This illustrates the need for other conditions, such as those presented in Section 2.

3.4 Reflexive Points of Partial Relations

We suppose here that R is not total, i.e., there is an $s \in S$ such that $sR = \emptyset$. As we will see, this introduces a strong constraint, because $\bigsqcap sR = \bigsqcap \emptyset = \top$ and $\bigsqcup sR = \bigsqcup \emptyset = \bot$. We exclude the trivial case where the lattice S contains only one element, since in this case the only partial relation is $\emptyset$, and it has no reflexive point. So, assume $\bot \neq \top$.

We could show that the conjunction of Conditions 4(a,b,c,d) is sufficient to ensure the existence of a least reflexive point. However, this is a bit too strong. We start by exhibiting combinations of Conditions 4(a,b,c,d) that do not ensure the existence of a least reflexive point. This will help pinpointing the essential conditions.

- The relation R in (6) has three reflexive points, but no least one. This relation satisfies 4(b,c,d), but not 4(a). This shows that 4(a) is essential.
- The relation R in (7) has no reflexive point, hence no least one. This relation satisfies 4(a,c,d), but not 4(b). This shows that 4(b) is essential.

Thus, 4(a,b) are essential. However, they are not sufficient. The relation

$$R := \{(\bot,\mathsf{a}),(\bot,\mathsf{b}),(\mathsf{a},\mathsf{a}),(\mathsf{a},\mathsf{b}),(\mathsf{b},\mathsf{a}),(\mathsf{b},\mathsf{b})\} \tag{14}$$

on the lattice $\{\bot,\mathsf{a},\mathsf{b},\top\}$, with ordering $\bot < \mathsf{a} < \top$ and $\bot < \mathsf{b} < \top$, satisfies 4(a,b) (it satisfies none of 4(c,d)) and has no least reflexive point. See Appendix A(f).

We will show that each of the combinations 4(a,b,c) and 4(a,b,d) is sufficient. Before dealing with the first one, we derive a consequence of Conditions 4(a,b).

Lemma 2 *Let R be a relation satisfying* 4(a,b) *and let s be such that $sR = \emptyset$. Then*
$$\forall(y : s < y : yR = \{\top\}).$$

Note that Condition 4(b) implies that any two elements that are not in the domain of R are not related by $\leq$.

Proposition 4 *Let R be a partial relation satisfying* 4(a,b,c). *Then R has a least reflexive point.*

In (9), we have already given an example of a partial relation that satisfies 4(a,b,c) but not 4(d). Another example is the lattice $\{\bot,\mathsf{a},\mathsf{b},\mathsf{c},\mathsf{d},\mathsf{e},\mathsf{f},\mathsf{g},\top\}$, with ordering
$$\bot < \mathsf{a} < \top,\ \bot < \mathsf{b} < \mathsf{d} < \mathsf{e} < \mathsf{g} < \top \ \text{ and } \ \bot < \mathsf{c} < \mathsf{d} < \mathsf{f} < \mathsf{g},$$
and relation

$$\{(\bot,\mathsf{b}),(\bot,\mathsf{c}),(\mathsf{b},\mathsf{d}),(\mathsf{c},\mathsf{d}),(\mathsf{d},\mathsf{d}),(\mathsf{e},\mathsf{e}),(\mathsf{f},\mathsf{f}),(\mathsf{g},\mathsf{g}),(\top,\top)\}. \tag{15}$$

See Appendix A(g). Note that the least reflexive point is d, which means that the constraints 4(a,b,c) may lead to a somewhat more interesting situation than the next case that we analyze, where the least reflexive point is always one of $\bot$ or $\top$.

Now we tackle combination 4(a,b,d) and state a lemma similar to Lemma 2:

Lemma 3 *Let R be a relation satisfying* 4(b,d) *and let s be such that $sR = \emptyset$. Then*
$$\forall(x : x < s : xR = \{\bot\}).$$

Now we obtain

Proposition 5 *Let R be a partial relation satisfying* 4(a,b,d)*. Then R has a least reflexive point, which is either $\bot$ or $\top$.*

Proof: Let s be such that $sR = \emptyset$. On the basis of Lemma 2 and Lemma 3, we distinguish three cases according to the value of s.

1. $s = \bot$: then $\bot R = \emptyset$ and $tR = \{\top\}$ for every $t \neq \bot$. There is a unique reflexive point, namely $\top$, which is thus the least and greatest reflexive point.
2. $s = \top$: then $\top R = \emptyset$ and $tR = \{\bot\}$ for every $t \neq \top$. There is a unique reflexive point, namely $\bot$, which is thus the least and greatest reflexive point.
3. $s \neq \bot$ and $s \neq \top$: then $\bot R = \{\bot\}$ and $\top R = \{\top\}$. There is thus a least reflexive point, $\bot$, and a greatest reflexive one, $\top$.

□

In case 3 of the proof of the previous proposition, there might be other reflexive points. As an example, take S to be the lattice $\{\bot, \mathsf{a}, \mathsf{b}, \top\}$ with ordering $\bot < \mathsf{a} < \top$ and $\bot < \mathsf{b} < \top$, and the partial function

$$R := \{(\bot, \bot), (\mathsf{b}, \mathsf{b}), (\top, \top)\}. \tag{16}$$

Here, $s = \mathsf{a}$. See Appendix A(h).

And now an example of a partial relation that satisfies 4(a,b,d) but does not satisfy 4(c): the lattice is $\{\bot, \mathsf{a}, \mathsf{b}, \mathsf{c}, \mathsf{d}, \mathsf{e}, \top\}$, with ordering $\bot < \mathsf{a} < \top$, $\bot < \mathsf{b} < \mathsf{c} < \mathsf{e} < \top$ and $\mathsf{b} < \mathsf{d} < \mathsf{e}$, and the relation is

$$\{(\bot, \bot), (\mathsf{b}, \mathsf{c}), (\mathsf{b}, \mathsf{d}), (\mathsf{c}, \mathsf{b}), (\mathsf{c}, \mathsf{e}), (\mathsf{d}, \mathsf{b}), (\mathsf{d}, \mathsf{e}), (\mathsf{e}, \mathsf{c}), (\mathsf{e}, \mathsf{d}), (\top, \top)\}. \tag{17}$$

See Appendix A(i). Note that the subrelation on elements $\{\mathsf{b}, \mathsf{c}, \mathsf{d}, \mathsf{e}\}$ is the same (modulo renaming) as the one in (8), which was used as a total relation illustrating that 4(a,b,d) does not imply 4(c).

4 Insights on the Structure of the Set of Reflexive Points

4.1 The Roles of $m_\sqcap$ and $m_\sqcup$

The next proposition shows that every reflexive point is above $m_\sqcap$.

Proposition 6 $\forall(x : xRx : m_\sqcap \leq x)$.

Based on the previous proposition, the next proposition shows that the reflexive points between $m_\sqcap$ and $m_\sqcup$ are linearly ordered; this only requires Condition 4(c). The relation R need not be total.

Proposition 7 *Assume Condition* 4(c)*. Then*

$$\forall(x : xRx \text{ and } yRy : x \leq y \text{ or } y \leq x \text{ or } (m_\sqcup \leq x \text{ and } m_\sqcup \leq y)).$$

There can be an infinite number of reflexive points between $m_\sqcap$ and $m_\sqcup$. Let S be the lattice $\mathbf{N} \cup \{\infty\}$, with the usual ordering, where $\mathbf{N}$ is the set of natural numbers. Take

$$R := \{(m, n) \mid m = n \text{ or } (m \in \mathbf{N} \text{ and } n = m + 1)\}. \tag{18}$$

One can check that R satisfies 4(c), $m_\sqcap = 0, m_\sqcup = \infty$, and that every element of S is related to itself, and thus 0 is the least reflexive point. See Appendix A(j).

One may also have $m_\sqcap = m_\sqcup$. This is the case, for instance, of the relation in (15), see Appendix A(g). Note that for this relation, all the reflexive points are above $m_\sqcup$ $(= m_\sqcap = \bot)$ and that they are not linearly ordered.

It is even possible to have $m_\sqcup < m_\sqcap$. Consider for instance the relation $\{(\top, \top)\}$ on the lattice $\{\bot, \top\}$, for which $m_\sqcap = \top$ and $m_\sqcup = \bot$.

However, for total R we have $\{x \mid \bigsqcup xR \leq x\} \subseteq \{x \mid \bigsqcap xR \leq x\}$ and hence $m_\sqcap \leq m_\sqcup$.

4.2 Lattice Structure of the Reflexive Points

We now come to the main result of this section, viz. the generalization of Tarski's result [10] on the fixed points of a monotonic *total function* to the relational case.

Theorem 3 *Let R be a total relation and assume Condition 4(c). Then the set of reflexive points of R is a complete lattice.*

Moreover, we have

Proposition 8 *For a total relation R satisfying Condition 4(c) define $l := \sqcap\{x \mid xRx\}$. Then $m_\sqcap \leq l \leq m_\sqcup$.*

Finally, combining our results for the partial case with Theorem 3, we obtain

Proposition 9 *Let R be a partial relation that satisfies 4(a,b,c) and $\top R \neq \emptyset$. Then the set of reflexive points is a complete lattice.*

The condition $\top R \neq \emptyset$ is mandatory: the dual of Appendix A(a) satisfies 4(a,b,c) but has no greatest reflexive point, so that the set of reflexive points is not a complete lattice.

Unfortunately, the combination 4(a,b,d) does not guarantee a complete lattice of reflexive points, not even when $\top R \neq \emptyset$. This is shown in the example of Appendix A(n): the lattice is $\{\bot, \mathsf{a}, \mathsf{b}, \mathsf{c}, \mathsf{d}, \mathsf{e}, \mathsf{f}, \mathsf{g}, \mathsf{h}, \mathsf{i}, \top\}$, with ordering $\bot < \mathsf{a} < \top$, $\bot < \mathsf{b} < \mathsf{c} < \mathsf{e} < \mathsf{f} < \mathsf{g} < \mathsf{i} < \top$ and $\mathsf{b} < \mathsf{d} < \mathsf{e} < \mathsf{f} < \mathsf{h} < \mathsf{i}$, and the relation is

$$\begin{aligned}\{&(\bot,\bot), (\mathsf{b},\mathsf{c}), (\mathsf{b},\mathsf{d}), (\mathsf{c},\mathsf{c}), (\mathsf{c},\mathsf{d}), (\mathsf{d},\mathsf{c}), (\mathsf{d},\mathsf{d}), (\mathsf{e},\mathsf{c}), (\mathsf{e},\mathsf{d}),\\ &(\mathsf{f},\mathsf{g}), (\mathsf{f},\mathsf{h}), (\mathsf{g},\mathsf{g}), (\mathsf{g},\mathsf{h}), (\mathsf{h},\mathsf{g}), (\mathsf{h},\mathsf{h}), (\mathsf{i},\mathsf{g}), (\mathsf{i},\mathsf{h}), (\top,\top)\}.\end{aligned} \tag{19}$$

The set $\{\bot, \mathsf{c}, \mathsf{d}, \mathsf{g}, \mathsf{h}, \top\}$ of reflexive points is not a lattice, since, e.g., the subset $\{\mathsf{c}, \mathsf{d}\}$ has two minimal upper bounds, namely g and h. Note how the total relation of Appendix A(l), which also satisfies 4(a,b,d), is used twice as a sublattice of the partial relation of Appendix A(n).

5 Greatest Reflexive Points

We can obtain results for greatest reflexive points using properties of the least reflexive points in the dual of the given lattice. It suffices to replace $\leq, \sqcap, \sqcup$ and "least" by $\geq, \sqcup, \sqcap$ and "greatest", respectively. Doing so reveals that Conditions 4(a), 4(b), 4(c) and 4(d) are dual to 4(d), 4(b), 4(c) and 4(a), respectively. Properties of total or partial relations then easily follow from the previous text.

By Proposition 4, the set of Properties 4(a,b,c) guarantees the existence of a least reflexive point. Its dual, 4(b,c,d), guarantees the existence of a greatest reflexive point. The question arises whether it is possible to have a relation satisfying 4(b,c,d), and thus having a greatest reflexive point, that does not have a least reflexive point. The answer is yes and is illustrated by the relation given in (6) and Appendix A(a).

6 Can Monotonicity be Characterized Another Way?

Since we are working in a relational setting, an obvious question is whether monotonicity can be characterized in a purely algebraic, point-free style.

6.1 Candidate Conditions and Their Interrelation

For the case of total functions it is well known how to do this; when R is a total function, the following four conditions 20(a',b',c',d') all are equivalent to the usual pointwise definition of monotonicity. But for the same reasons as in discussed in Section 3.2, we will work with Conditions 20(a,b,c,d) (see also Section 6.2):

$$\begin{array}{llll}
\text{(a)} & <;R \subseteq R;\leq, & \text{(a')} & \leq;R \subseteq R;\leq,\\
\text{(b)} & < \subseteq R;\leq;R^{\smile}, & \text{(b')} & \leq \subseteq R;\leq;R^{\smile},\\
\text{(c)} & R^{\smile};<;R \subseteq \leq, & \text{(c')} & R^{\smile};\leq;R \subseteq \leq,\\
\text{(d)} & R^{\smile};< \subseteq \leq;R^{\smile}, & \text{(d')} & R^{\smile};\leq \subseteq \leq;R^{\smile}.
\end{array} \tag{20}$$

However, in the case of general relations they are *not* equivalent. There, we have that Conditions 20(a',b',c',d') are equivalent to the following pointwise expressions.

$$\begin{array}{ll} 20(\mathrm{a'}) & \forall(x,y : x \le y : \forall(v : yRv : \exists(u : xRu : u \le v))) \\ 20(\mathrm{b'}) & \forall(x,y : x \le y : \exists(u,v : xRu \text{ and } yRv : u \le v)) \\ 20(\mathrm{c'}) & \forall(x,y : x \le y : \forall(u,v : xRu \text{ and } yRv : u \le v)) \\ 20(\mathrm{d'}) & \forall(x,y : x \le y : \forall(u : xRu : \exists(v : yRv : u \le v))) \end{array}$$

On the basis of the structure of these expressions, one could say that Conditions 20(a',b',c',d') respectively define *upward existential monotonicity, existential monotonicity, universal monotonicity* and *downward existential monotonicity.*

A first consequence of these properties is stated in

Proposition 10 20(a,d) $\Rightarrow$ $R = \emptyset$ or *R total.*

6.2 Interdependence of the Conditions

As in the case of our earlier Conditions 4, we state the connections between the unprimed and primed versions

$$\begin{array}{llll} 20(\mathrm{a'}) \Leftrightarrow 20(\mathrm{a}), & & 20(\mathrm{c'}) \Leftrightarrow 20(\mathrm{c}) \text{ and } R \text{ functional}, \\ 20(\mathrm{b'}) \Leftrightarrow 20(\mathrm{b}) \text{ and } R \text{ total}, & & 20(\mathrm{d'}) \Leftrightarrow 20(\mathrm{d}). \end{array} \tag{21}$$

Hence, the situation is almost identical to that with Conditions 4 (see (12)), except for the equivalence of 20(b') and 20(b). Rather than using all eight Conditions 20, we will simply use Conditions 20(a,b,c,d) together with totality or functionality, like we did for Conditions 4. One can show (see [5,6]) that the same implications hold between Conditions 20(a,b,c,d) as between 4(a,b,c,d) (see Equation (5)):

$$\begin{array}{ccc} & 20(\mathrm{c}) \Rightarrow 20(\mathrm{a}) & \\ R \text{ total:} & \Downarrow \qquad\quad \Downarrow & \\ & 20(\mathrm{d}) \Rightarrow 20(\mathrm{b}) & \end{array} \qquad \begin{array}{ccc} & 20(\mathrm{c}) \Leftarrow 20(\mathrm{a}) & \\ R \text{ function:} & \Uparrow \qquad\quad \Uparrow & \\ & 20(\mathrm{d}) \Leftarrow 20(\mathrm{b}) & \end{array} \tag{22}$$

Here too, one can deduce from these diagrams that if R is a total function, all four Conditions 20(a,b,c,d) are equivalent, as was mentioned in the introductory part of Section 6.1.

In view of these striking similarities, one might expect that corresponding properties in 4(a,b,c,d) and 20(a,b,c,d) are equivalent. However, this is not the case, as the following theorem shows.

Theorem 4 *The following relationships hold between Conditions* 4 *and* 20. *All implications are strict (i.e., equivalence does not hold):*

$$\begin{array}{ll} 20(\mathrm{a}) \Rightarrow 4(\mathrm{a}) & 20(\mathrm{b}) \Rightarrow 4(\mathrm{b}) \\ 20(\mathrm{c}) \Leftrightarrow 4(\mathrm{c}) & 20(\mathrm{d}) \Rightarrow 4(\mathrm{d}) \end{array}$$

A single example (see Appendix A(k)) can be used to show that the three implications are strict: Take S to be the lattice $\{\bot, \mathsf{a}, \mathsf{b}, \mathsf{c}, \mathsf{d}, \top\}$ with ordering $\bot < \mathsf{a} < \mathsf{c} < \top$ and $\bot < \mathsf{b} < \mathsf{d} < \top$, and

$$R := \{(\bot,\mathsf{c}),(\bot,\mathsf{d}),(\mathsf{a},\mathsf{c}),(\mathsf{a},\mathsf{d}),(\mathsf{b},\mathsf{c}),(\mathsf{b},\mathsf{d}),(\mathsf{c},\mathsf{a}),(\mathsf{c},\mathsf{b}),(\mathsf{d},\mathsf{a}),(\mathsf{c},\mathsf{b}),(\top,\mathsf{a}),(\top,\mathsf{b})\}. \tag{23}$$

This total relation satisfies all of 4(a,b,d) and none of 20(a,b,d). The partial relation given in (17) is another example showing the strictness of the implications.

Hence, the question whether these conditions are independent cannot be reduced to our independence results in Section 3. A separate investigation shows the following:

1. Condition 20(b) does not follow from 20(a,c,d). This is shown by the relation given in (7) (Appendix A(b)).
2. Condition 20(c) does not follow from 20(a,b,d). This is shown by the lattice $\{\bot, \mathsf{a}, \top\}$ with ordering $\bot < \mathsf{a} < \top$ and the relation (see Appendix A(m))
$$R := \{(\bot,\mathsf{a}),(\bot,\top),(\mathsf{a},\mathsf{a}),(\mathsf{a},\top),(\top,\top)\}. \tag{24}$$
3. Conditions 20(a,d) both follow from 20(b,c).
4. Conditions 20(a,d) imply 20(b) or 20(c).

Thus the independence properties of Conditions 20 are different from those of Conditions 4 and one cannot have combinations like 20(a,b,c) and 20(b,c,d) without having all of 20(a,b,c,d). However, all the remaining combinations of our properties can be exemplified with lattices and relations, see Appendix B.

6.3 On the Existence of Least Reflexive Elements

Do the algebraic Conditions 20 ensure the existence of least reflexive elements of *partial* relations in the same manner as Conditions 4? Unfortunately, this is *not* the case, as can be seen as follows. The example in (7) (Appendix A(b)) satisfies 20(a,c,d), but has no reflexive point, so that 20(b) is essential for the existence.

On the other hand, it can be shown that 20(b) implies totality of R. In sum, all this means that Conditions 20 are not useful for studying the reflexive points of *partial* relations.

For *total* relations, we have concluded the corresponding Section 3.3 with the remark that Condition 4(c) in Theorem 2 cannot be relaxed to a weaker combination of some of 4(a,b,d). Because Conditions 20(a,b,d) are stronger than 4(a,b,d) (see Theorem 4), one might conjecture that 4(c) (equivalently, 20(c)) could be weakened to a combination of some of 20(a,b,d). However, this is not the case. The lattice $\{\bot, \mathsf{a}, \mathsf{b}, \top\}$, with ordering $\bot < \mathsf{a} < \top$ and $\bot < \mathsf{b} < \top$, and the relation

$$R := \{(\bot, \mathsf{a}), (\bot, \mathsf{b}), (\mathsf{a}, \mathsf{a}), (\mathsf{a}, \mathsf{b}), (\mathsf{b}, \mathsf{a}), (\mathsf{b}, \mathsf{b}), (\top, \mathsf{a}), (\top, \mathsf{b})\} \tag{25}$$

(see Appendix A(l)) provide an example of a total relation that satisfies all Conditions 20(a,b,d) and that has no least reflexive point.

6.4 Other Conditions Ensuring Reflexive Elements

In this paper we have solely treated the case of complete lattices. However, it is well known that under suitable assumptions Tarski's original fixed point theorem carries over to weaker structures such as chain-complete partial orders (cpo's).

In this connection we want to mention the paper [7] by Fujimoto. It studies *total* relations, represented as set-valued functions, on a cpo with a least element. The monotonicity condition used is 20(d'). Fujimoto shows that under the additional assumption that all image sets under R are inductively ordered, i.e., contain for each subchain an upper bound, there is an R-reflexive element, although not necessarily a least one.

It will be interesting to investigate further sufficient conditions with weaker assumptions.

7 Conclusion

This paper provides a first survey on the structure of the set of reflexive points of relations on complete lattices. We have exhibited suitable adaptations of the notion of monotonicity of a total function to the relational case. It may come as a certain surprise that, unless other conditions are added (as in Section 6.4), the direct relational formulations of monotonicity of a total function are of no use in this setting and have to be replaced by new conditions. With the help of these we have shown an analogue of Tarski's classical result [10] on existence and lattice structure of the reflexive points. Another advantage of the new conditions is that they are checked manually much more easily than the relational ones. What is still missing is a suitable generalization of the notion of continuity and, following that, an investigation when least reflexive elements can be obtained by iteration as in Kleene's Theorem [8] (the process being also already mentioned in [10]).

Another open question in connection with iteration (see Section 2) is whether there are (not too strong) conditions guaranteeing the existence of a least reflexive point when the strict ordering is progressively finite and the relation is inflationary. Moreover, which additional conditions ensure that any path by R from $\bot$ leads to this least reflexive point? We conclude this paper with a last counterexample showing that Conditions 4(a,b,d) are not sufficient to guarantee this. Consider again the relation R from (24) (see Appendix A(m)). That relation is inflationary and satisfies 4(a,b,d), the strict ordering is progressively finite, there is a least and a greatest reflexive point, but there is a path from $\bot$ to $\top$ that does not go through the least reflexive point, which is a.

Acknowledgments

We are grateful to the referees of paper [6] for their helpful remarks, in particular to Oege de Moor for his hint to look at Condition 20(a). Valuable comments were also provided by Thorsten Ehm and Georg Struth.

A Examples and Counterexamples

This section contains diagrams of some of the lattices and relations presented in the paper. Please see overleaf for further explanation.

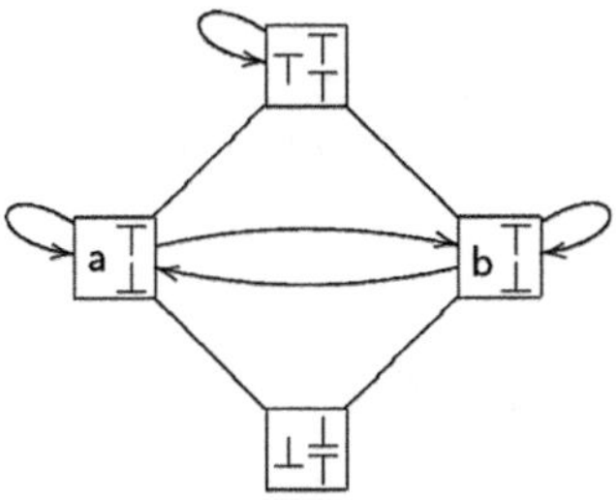

a) 4(b,c,d)
20(c,d)
Equation 6

b) 4(a,c,d)
20(a,c,d)
Equation 7

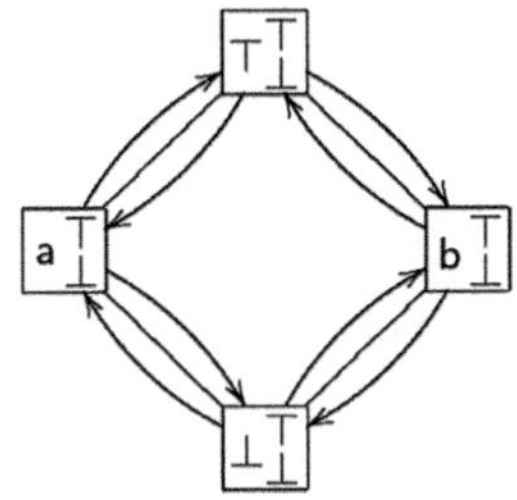

c) 4(a,b,d)
20(b)
Equation 8

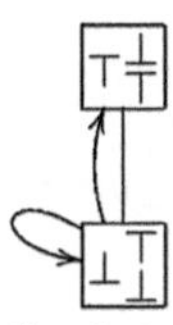

d) 4(a,b,c)
20(a,c)
Equation 9

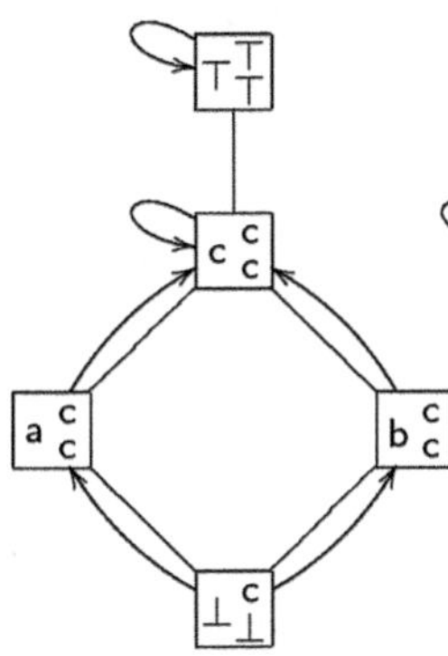

e) 4(a,b,c,d)
20(a,b,c,d)
Inflationary
Equation 13

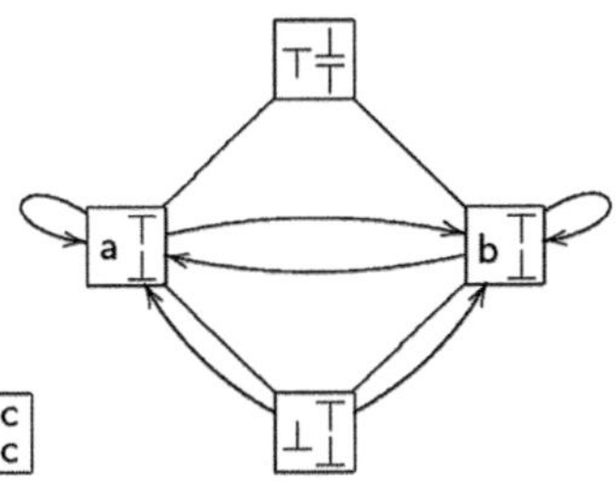

f) 4(a,b)
20(a)
Equation 14

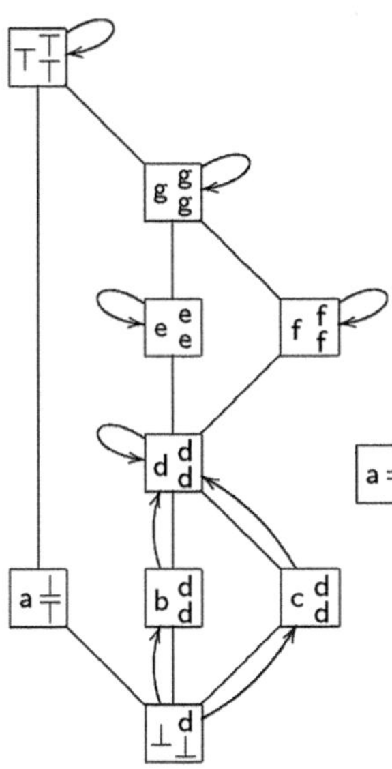

g) 4(a,b,c)
20(c)
Equation 15

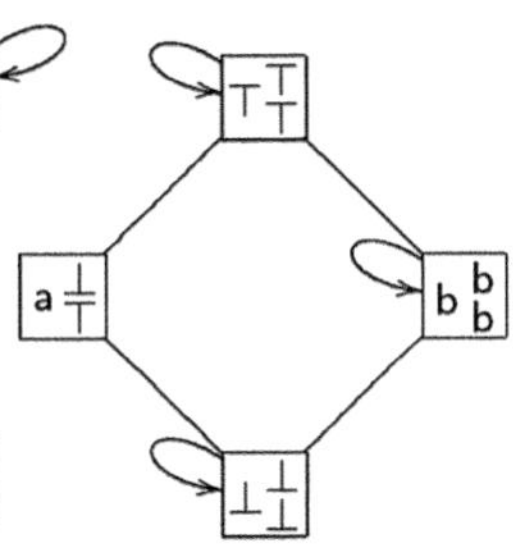

h) 4(a,b,c,d)
20(c)
Equation 16

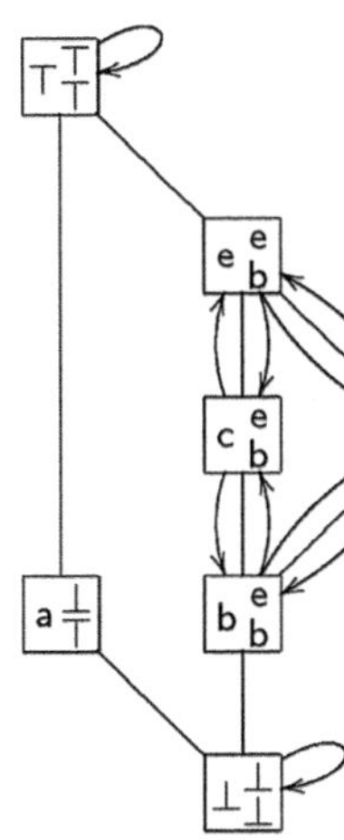

i) 4(a,b,d)
Equation 17

Each lattice $(S, \sqcap, \sqcup, \bot, \top, \leq)$ is described by the Hasse diagram of its corresponding partial order, with boxes representing vertices (elements of S) and straight lines representing edges. The relation R on the lattice is represented by arrows linking the boxes. Each box contains three pieces of information: (i) on the left is the element $s \in S$, (ii) on the bottom right is $\bigsqcap sR$, and (iii) on the top right is $\bigsqcup sR$. The following information is given under each diagram: (i) the list of properties among 4(a,b,c,d) that hold for this diagram, (ii) the list of properties among 20(a,b,c,d) that hold for this diagram, (iii) the word "Inflationary", if the relation is inflationary, and (iv) the equation where the relation is defined.

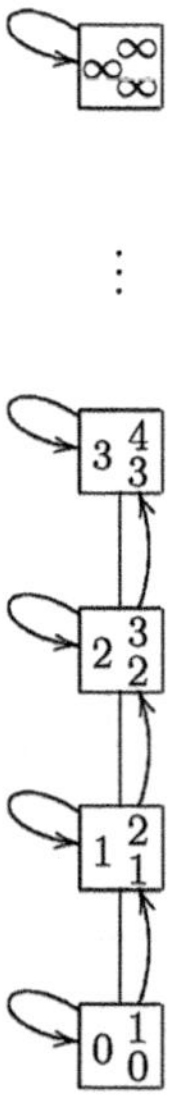

j) 4(a,b,c,d)
20(a,b,c,d)
Inflationary
Equation 18

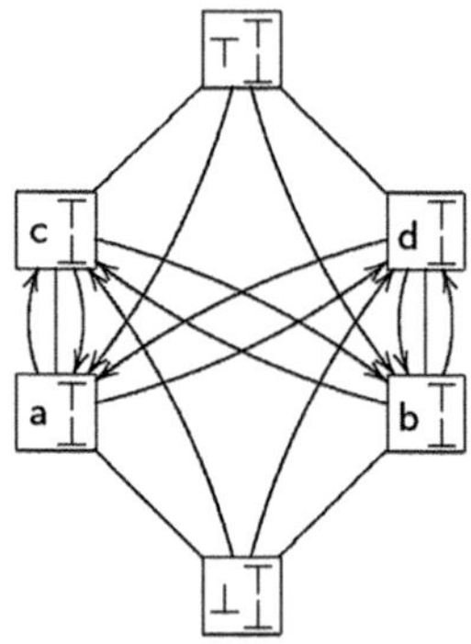

k) 4(a,b,d)
Equation 23

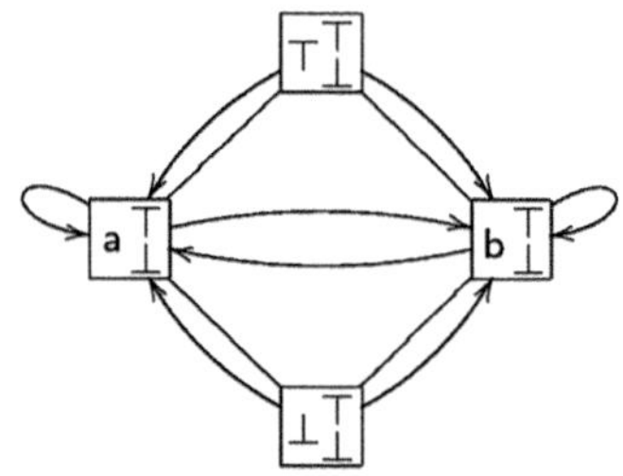

l) 4(a,b,d)
20(a,b,d)
Equation 25

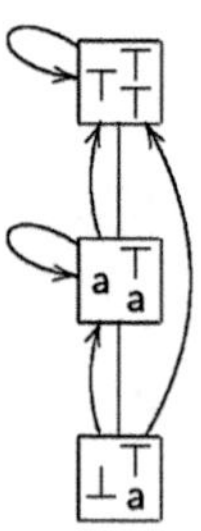

m) 4(a,b,d)
20(a,b,d)
Inflationary
Equation 24

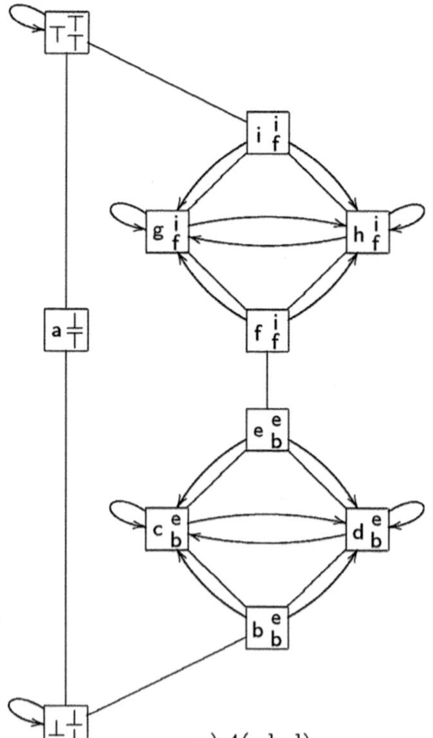

n) 4(a,b,d)
Equation 19

B Combinations of Properties

Here are examples of lattices and relations showing all possible combinations of Conditions 4 and all possible combinations of Conditions 20. For all but one example, there is no need of sophisticated lattices: two linear orders suffice! Each example uses a minimal lattice, except possibly relation (p) in the tables below; also, the trivial lattice with $\top = \bot$ with the empty or the universal relation could be used instead of (j) to illustrate 4(a,b,c,d) and 20(a,b,c,d).

The labels of the following relations correspond to those of the diagrams that follow the presentation of the relations.

1. Lattice $\{\bot, \top\}$ with ordering $\bot < \top$.

	Relation	Properties
(a)	$\{(\bot,\top),(\top,\bot)\}$	None
(b)	$\{(\bot,\bot),(\bot,\top),(\top,\bot)\}$	4(a,b), 20(a,b)
(c)	$\{(\bot,\top)\}$	4(a,c), 20(a,c)
(d)	$\{(\bot,\top),(\top,\bot),(\top,\top)\}$	4(b,d), 20(b,d)
(e)	$\{(\top,\bot)\}$	4(c,d), 20(c,d)
(f)	$\{(\bot,\bot),(\bot,\top)\}$	4(a,b,c), 20(a,c)
(g)	$\{(\bot,\bot),(\bot,\top),(\top,\bot),(\top,\top)\}$	4(a,b,d), 20(a,b,d)
(h)	$\emptyset$	4(a,c,d), 20(a,c,d)
(i)	$\{(\top,\bot),(\top,\top)\}$	4(b,c,d), 20(c,d)
(j)	$\{(\bot,\bot),(\top,\top)\}$	4(a,b,c,d), 20(a,b,c,d)

2. Lattice $\{\bot, \mathsf{a}, \top\}$ with ordering $\bot < \mathsf{a} < \top$.

	Relation	Properties
(k)	$\{(\bot,\bot),(\bot,\top),(\mathsf{a},\mathsf{a})\}$	4(a), 20(a)
(l)	$\{(\bot,\mathsf{a}),(\bot,\top),(\mathsf{a},\bot),(\mathsf{a},\mathsf{a}),(\top,\mathsf{a})\}$	4(b), 20(b)
(m)	$\{(\mathsf{a},\bot),(\mathsf{a},\mathsf{a})\}$	4(c), 20(c)
(n)	$\{(\mathsf{a},\mathsf{a}),(\top,\bot),(\top,\top)\}$	4(d), 20(d)
(o)	$\{(\bot,\bot),(\bot,\mathsf{a}),(\top,\mathsf{a}),(\top,\top)\}$	4(b,c), 20(c)

3. Lattice $\{\bot, \mathsf{a}, \mathsf{b}, \mathsf{c}, \mathsf{d}, \top\}$ with ordering $\bot < \mathsf{a} < \mathsf{c} < \top$ and $\bot < \mathsf{b} < \mathsf{d} < \top$.

	Relation	Properties
(p)	$\{(\bot,\bot),(\mathsf{a},\mathsf{a}),(\mathsf{a},\mathsf{b}),(\mathsf{c},\mathsf{c}),(\mathsf{c},\mathsf{d}),(\top,\top)\}$	4(a,d)

a) None

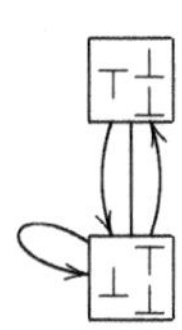
b) 4(a,b)
20(a,b)

c) 4(a,c)
20(a,c)

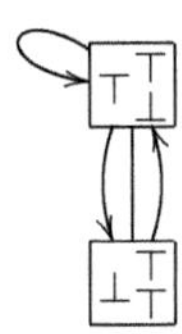
d) 4(b,d)
20(b,d)

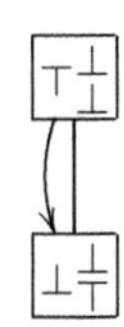
e) 4(c,d)
20(c,d)

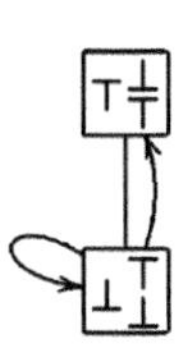
f) 4(a,b,c)
20(a,c)

g) 4(a,b,d)
20(a,b,d)

h) 4(a,c,d)
20(a,c,d)

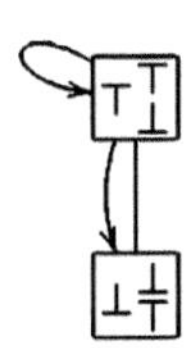
i) 4(b,c,d)
20(c,d)

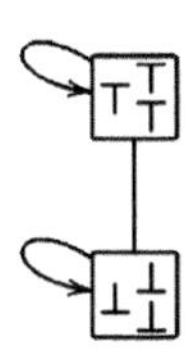
j) 4(a,b,c,d)
20(a,b,c,d)

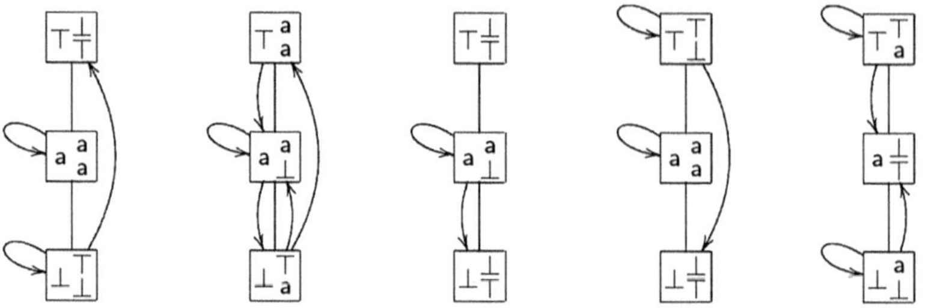

k) 4(a) 20(a) l) 4(b) 20(b) m) 4(c) 20(c) n) 4(d) 20(d) o) 4(b,c) 20(c)

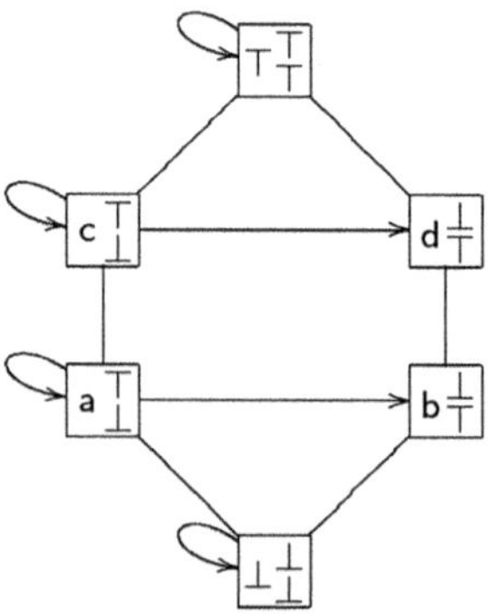

p) 4(a,d)

References

1. R. Backhouse et al.: Fixed point calculus. *Information Processing Letters* 53(3), 131–136, 1995.
2. J. Cai and R. Paige: Program derivation by fixed point computation. *Science of Computer Programming* 11(3), 197–261, 1989.
3. J. Cai and R. Paige: Languages polynomial in the input plus output. In: *Second International Conference on Algebraic Methodology and Software Technology* (AMAST '91), London, Springer, 287–300, 1992.
4. B. A. Davey and H. A. Priestley: *Introduction to Lattices and Order.* Cambridge Mathematical Textbooks. Cambridge University Press, Cambridge, 1990.
5. J. Desharnais and B. Möller: Least reflexive points of relations. Research Report, Institut für Informatik, Universität Augsburg, Germany, 2002.
6. J. Desharnais and B. Möller: Least reflexive points of relations. *Higher-Order and Symbolic Computation.* Special Issue in memory of Robert Paige. 18(1/2), 51–77, 2005.
7. T. Fujimoto: An extension of Tarski's fixed point theorem and its application to isotone complementarity problems. *Mathematical Programming* 28, 116–118, 1984.
8. S. C. Kleene: *Introduction to Metamathematics.* Van Nostrand, New York, 1952.
9. G. Schmidt and T. Ströhlein: *Relations and Graphs.* EATCS Monographs in Computer Science. Springer, Berlin, 1993.
10. A. Tarski: A lattice-theoretical fixpoint theorem and its applications. *Pacific Journal of Mathematics* 5, 285–309, 1955.

Efficient Type Matching

Somesh Jha[1], Jens Palsberg[2], Tian Zhao[3], and Fritz Henglein[4]

[1]Computer Science Department, University of Wisconsin, Madison, WI 53706, USA
jha@cs.wisc.edu
[2]UCLA Computer Science Department, University of California, Los Angeles, CA 90095, USA
palsberg@ucla.edu
[3]Department of Electrical Engineering and Computer Science, University of Wisconsin, Milwaukee, WI 53211, USA. tzhao@cs.uwm.edu
[4]Department of Computer Science (DIKU), University of Copenhagen, DK-2100 Copenhagen, Denmark. henglein@diku.dk

Summary. Palsberg and Zhao [25] presented an $O(n^2)$ time algorithm for matching two recursive types; that is, deciding type isomorphism with associative-commutative product type constructors. In this paper, we present an $O(n \log n)$-time algorithm for matching recursive types and an $O(n)$-time algorithm for matching nonrecursive types. The linear-time algorithm for nonrecursive types works without hashing or pointer arithmetic, by employing multiset discrimination techniques due to Paige et al. [9,10,21–24]. The $O(n \log n)$ algorithm for recursive types works by reducing the type matching problem to the problem of finding a size-stable partition of a graph, which has $O(n \log n)$ algorithms due to Cardon/Crochemore and Paige/Tarjan. The key to these algorithms is the use of a "modify-the-smaller-half" approach pioneered by Hopcroft and Ullman for DFA minimization.

Our results may help improve systems, such as Polyspin and Mockingbird, that are designed to facilitate interoperability of software components. We also discuss possible applications of our algorithm to Java. Issues related to subtyping of recursive types are also discussed.

Keywords: graph algorithms, size-stable partitions, recursive types.

1 Introduction

Interoperability is a fundamental problem in software engineering. Interoperability issues arise in various contexts, such as software reuse, distributed programming, use of legacy components, and integration of software components developed by different organizations. Interoperability of software components has to address two fundamental problems: *matching* and *bridging*. Matching deals with determining whether two components A and B are compatible, and bridging allows one to use component B using the interface defined for component A.

Matching: A common technique for facilitating matching is to associate signatures with components. These signatures can then be used as keys to retrieve relevant components from an existing library of components. Use of finite types as signatures was first proposed by Rittri [27]. Zaremski and Wing [32, 33] used a similar approach for retrieving components from an ML-like functional library. Moreover, they also emphasized flexibility and support for user-defined types. Aponte and Cosmo [2] had also studied a notion of type isomorphism for equating module signatures in functional languages.

Bridging: In a multilingual context, *bridge code* for "gluing" components written in different languages (such as C, C++, and Java) has to be developed. CORBA [20], PolySpin [5], and Mockingbird [3, 4] allow composing components implemented in different languages. Software components are considered to be of two kinds: *objects*, which provide public interfaces, and *clients*, which invoke the methods of the objects and thereby use the services provided by the objects.

O. Danvy et al. (eds.), Automatic Program Development – A Tribute to Robert Paige, 229–246.

The Problem: Assume that we use types as signatures for components. Thus, the type matching problem reduces to the problem of determining whether two types are equivalent. Much previous work on type matching focuses on nonrecursive types [2,8,14,19,26–29,32]. In this paper, we consider equivalence of recursive types. Equality and subtyping of recursive types have been studied in the 1990s by Amadio and Cardelli [1]; Kozen, Palsberg, and Schwartzbach [18]; Brandt and Henglein [7]; Jim and Palsberg [17]; and others. These papers concentrate on the case where two types are considered equal if their infinite unfoldings are identical. In this case, type equivalence can be decided in $O(n\alpha(n))$ time. If we allow a product-type constructor to be associative and commutative, then two recursive types may be considered equal *without* their infinite unfoldings being identical. Alternatively, think of a product type as a multiset, by which associativity and commutativity are obtained for free. Such flexibility has been advocated by Auerbach, Barton, and Raghavachari [3]. Palsberg and Zhao [25] presented a definition of type equivalence that supports this idea. They also presented an $O(n^2)$ time algorithm for deciding their notion of type equivalence.

A notion of subtyping defined by Amadio and Cardelli [1] can be decided in $O(n^2)$ time [18]. We also briefly discuss subtyping of recursive types with associative-commutative products in this paper.

Our results: We present $O(n \log n)$ and $O(n)$ time algorithms for deciding type equivalence with and without type recursion, respectively. The $O(n \log n)$ algorithm improves upon the $O(n^2)$ algorithm of Palsberg and Zhao [25]. It works by reducing the type matching problem to the well-understood problem of finding a size-stable partition of a graph [12,23]. The $O(n)$ algorithm employs multiset discrimination due to Paige et al.

Our algorithms and their bounds extend to type matching for an arbitrary number of types and apply to shared type definitions (type abbreviations); e.g., all components (methods, functions) in a library and an application can be partitioned in $O(n \log n)$, respectively $O(n)$, such that each resulting partition contains pairwise matching components, where n is the size of the library and application combined. Furthermore, with such partitioning as a preprocessing step, all pairwise type matching queries for components in the combined set can be answered in constant time.

The organization of the paper is as follows. A small example is described in Section 2. This example will be used throughout the paper for illustrative purposes. In Section 3 we recall the notions of terms and term automata [1, 11, 16, 18], and we state the definitions of types and type equivalence from the paper by Palsberg and Zhao [25]. In Section 4 we present our $O(n)$ algorithm for finite (nonrecursive) types, and in Section 5 we present our $O(n \log n)$ algorithm for recursive types. An implementation of our $O(n \log n)$ algorithm is discussed in Section 6. Subtyping of recursive types is discussed in Section 7. For simplicity, the technical development is for pairwise type matching. In the Conclusion, Section 8, we point out the extensibility of the algorithms to partitioning, compare our work to recent related work and discuss possibilities for applications in practical type matching.

2 Example

In this section we provide a small example which will be used throughout the paper. It is straightforward to map a `Java` type to a recursive type of the form considered in this paper. A collection of method signatures can be mapped to a product type, a single method signature can be mapped to a function type, and in case a method has more than one argument, the list of arguments can be mapped to a product type. Recursion, direct or indirect, is expressed with the μ operator. This section provides an example of `Java` interfaces and provides an illustration of our algorithm.

Suppose we are given the four Java interfaces shown in Figure 1. We would like to find out whether interface I_1 is "structurally equal" to or "matches" with interface J_2. We want a notion of equality for which interface names and method names do not matter, and for which the order of the methods in an interface and the order of the arguments of a method do not matter.

```
interface I1 {              interface I2 {              interface J1 {              interface J2 {
    float m1(I1 a);             I1 m3(float a);             J1 n1(float a);             int n3(J1 a);
    int m2(I2 a);               I2 m4(float a);             J2 n2(float a);             float n4(J2 a);
}                           }                           }                           }
```

Fig. 1. Interfaces I_1, I_2, J_1, J_2.

Notice that interface I_1 is recursively defined. The method m_1 takes an argument of type I_1 and returns a floating point number. In the following, we use names of interfaces and methods to stand for their type structures. The type of method m_1 can be expressed as $I_1 \to \mathit{float}$. The symbol $\to$ stands for the function type constructor. Similarly, the type of m_2 is $I_2 \to \mathit{int}$. We can then capture the structure of I_1 with conventional μ-notation for recursive types:

$$I_1 = \mu\alpha.(\alpha \to \mathit{float}) \times (I_2 \to \mathit{int})$$

The symbol α is the type variable bound to the type I_1 by the symbol μ. The interface type I_1 is a product type with the symbol $\times$ as the type constructor. Since we think of the methods of interface I_1 as unordered, we could also write the structures of I_1 and I_2 as

$$I_1 = \mu\alpha.(I_2 \to \mathit{int}) \times (\alpha \to \mathit{float}) \qquad I_2 = \mu\delta.(\mathit{float} \to I_1) \times (\mathit{float} \to \delta)$$

In the same way, the structures of the interfaces J_1 and J_2 are:

$$J_1 = \mu\beta.(\mathit{float} \to \beta) \times (\mathit{float} \to J_2) \qquad J_2 = \mu\eta.(J_1 \to \mathit{int}) \times (\eta \to \mathit{float}).$$

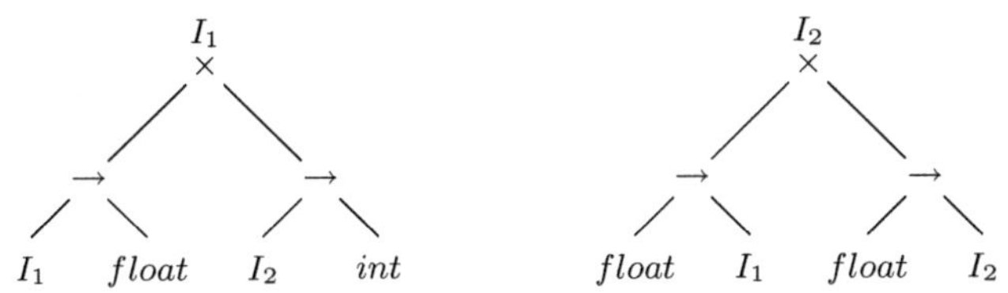

Fig. 2. Trees for interfaces I_1 and I_2.

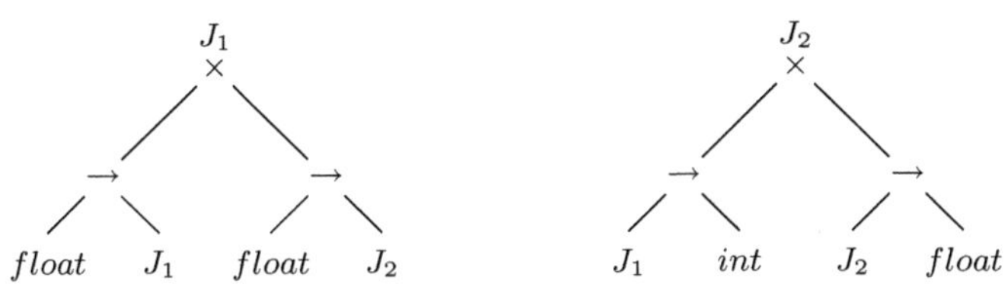

Fig. 3. Trees for interfaces J_1 and J_2.

Trees corresponding to the two types are shown in Figures 2 and 3. The interface types I_1, J_2 are equivalent iff there exists a one-to-one mapping or a bijection from the methods in I_1 to the methods in J_2 such that each pair of methods in the bijection relation have the same type. The types of two methods are equal iff the types of the arguments and the return types are equal.

The equality of the interface types I_1 and J_2 can be determined by trying out all possible orderings of the methods in each interface and comparing the two types in the form of finite automata. In this case, there are only few possible orderings. However, if the number of methods is large and/or some methods take many arguments, the above approach becomes time consuming because the number of possible orderings grows exponentially. An efficient algorithm for determining equality of recursive types will be given later in the paper.

3 Definitions

A (uniform) recursive type is a type described by a set of equations involving the μ operator. An example of a recursive type was provided in Section 2. This section provides representation of recursive types as terms and term automata.

Term automata and representation of types are described in Subsection 3.1. A definition of type equivalence for recursive types in terms of bisimulation is given in Subsection 3.2. An efficient algorithm for determining whether two types are equivalent is given in Section 5.

3.1 Terms and Term Automata

Here we give a general definition of (possibly infinite) terms over an arbitrary finite ranked alphabet Σ. Such terms are essentially labeled trees, which we model as partial functions labeling strings over ω (the natural numbers) with elements of Σ.

Let Σ_n denote the set of elements of Σ of arity n. Let ω denote the set of natural numbers and let ω^* denote the set of finite-length strings over the alphabet ω.

A *term* over Σ is a partial function $t : \omega^* \to \Sigma$ satisfying the following properties:
(i) the domain of t is nonempty and prefix-closed, and
(ii) if $t(\alpha) \in \Sigma_n$, then we have that $\{\, i \mid \alpha i \in$ the domain of $t \,\} = \{0, 1, \ldots, n-1\}$.

Let t be a term and $\alpha \in \omega^*$. Define the partial function $t{\downarrow}\alpha : \omega^* \to \Sigma$ by $t{\downarrow}\alpha(\beta) = t(\alpha\beta)$. If $t{\downarrow}\alpha$ has nonempty domain, then it is a term, and is called the *subterm of t at position α*.

A term t is said to be *regular* if it has only finitely many distinct subterms; that is, if $\{t{\downarrow}\alpha \mid \alpha \in \omega^*\}$ is a finite set.

Every regular term over a finite ranked alphabet Σ has a finite representation in terms of a special type of automaton called a *term automaton*. A term automaton over Σ is a tuple $A = (Q,\ \Sigma,\ q_0,\ \delta,\ \ell)$ where: (i) Q is a finite set of *states*, (ii) $q_0 \in Q$ is the *start state*, (iii) $\delta : Q \times \omega \to Q$ is a partial function called the *transition function*, and (iv) $\ell : Q \to \Sigma$ is a (total) *labeling function*, such that for any state $q \in Q$, if $\ell(q) \in \Sigma_n$ then we have that: $\{i \mid \delta(q,i) \text{ is defined}\} = \{0, 1, \ldots, n-1\}$.

The partial function δ extends naturally to an inductively defined partial function $\widehat{\delta} : Q \times \omega^* \to Q$ such that: (i) $\widehat{\delta}(q, \epsilon) = q$, and (ii) $\widehat{\delta}(q, \alpha\, i) = \delta(\widehat{\delta}(q, \alpha), i)$.

For any $q \in Q$, the domain of the partial function $\lambda\alpha.\widehat{\delta}(q, \alpha)$ is nonempty (it always contains ϵ) and prefix-closed. Moreover, because of the condition on the existence of i-successors in the definition of term automata, the partial function $\lambda\alpha.\ell(\widehat{\delta}(q, \alpha))$ is a term.

Let A be a term automaton. The term *represented by* A is the term $t_A = \lambda\alpha.\ell(\widehat{\delta}(q_0, \alpha))$. A term t is said to be *representable* if $t = t_A$ for some A.

Intuitively, $t_A(\alpha)$ is determined by starting in the start state q_0 and scanning the input α, following transitions of A as far as possible. If it is not possible to scan all of α because some i-transition along the way does not exist, then $t_A(\alpha)$ is undefined. If on the other hand A scans the entire input α and ends up in state q, then $t_A(\alpha) = \ell(q)$.

It is straightforward to show that a term t is regular if and only if it is representable. Moreover, a term t is regular if and only if it can be described by a finite set of equations involving the μ operator [18].

3.2 Equivalence of Recursive Types

A recursive type is a regular term over the ranked alphabet

$$\Sigma = \Gamma \ \cup\ \{\to\}\ \cup\ \{\textstyle\prod^n \mid n \geq 2\},$$

where Γ is a set of base types, $\to$ is binary, and $\prod^n$ is of arity n. Given a type σ, if $\sigma(\epsilon) = \to$, $\sigma(0) = \sigma_1$, and $\sigma(1) = \sigma_2$, then we write the type as $\sigma_1 \to \sigma_2$. If $\sigma(\epsilon) = \prod^n$ and $\sigma(i) = \sigma_i\ ,\ \ \forall i \in \{0, 1, \ldots, n-1\}$, then we write the type σ as $\prod_{i=0}^{n-1} \sigma_i$.

A syntactic presentation of a recursive type can be generated by the grammar:

$$t ::= \Gamma \mid t \rightarrow t \mid \prod_{i=0}^{n-1} t_i \mid \mu\alpha.t \mid \alpha$$

where α ranges over type variables. There are standard algorithms for translating syntactic presentations of types into term automata, and vice versa, see [18] for a summary.

For example, a recursive type t generated by the above grammar can be reduced to a term automaton $(Q, \Sigma, q_0, \delta, \ell)$, where

- q_0 corresponds to the term t;
- for each subterm σ of t that is either a base type, a function type, or a product type, there is a distinct state $q \in Q$ such that $\ell(q) = \sigma(\epsilon)$ and $\delta(q, i) = q_i$, where state q_i corresponds to $\sigma(i)$;
- for each subterm $\mu\alpha.t'$ of t, where state q' corresponds to t', if $\exists\sigma$ in t' such that $\sigma(i) = \alpha$, and state q corresponds to σ, then let $\delta(q, i) = q'$.

Palsberg and Zhao [25] presented three equivalent definitions of type equivalence. Here we will work with the one which is based on the idea of bisimilarity. A relation R on types is called a *bisimulation* if it satisfies the following three conditions:

- if $(\sigma, \tau) \in R$, then $\sigma(\epsilon) = \tau(\epsilon)$
- if $(\sigma_1 \rightarrow \sigma_2, \tau_1 \rightarrow \tau_2) \in R$, then $(\sigma_1, \tau_1) \in R$ and $(\sigma_2, \tau_2) \in R$
- if $(\prod_{i=0}^{n-1} \sigma_i, \prod_{i=0}^{n-1} \tau_i) \in R$, then there exists a bijection $b : \{0 \ldots n-1\} \rightarrow \{0 \ldots n-1\}$ such that $\forall i \in \{0 \ldots n-1\}$, $(\sigma_i, \tau_{b(i)}) \in R$.

Bisimulations are closed under union, therefore, there exists a largest bisimulation $\mathcal{R} = \bigcup \{ \; R \mid R \text{ is a bisimulation } \}$. It is straightforward to show that $\mathcal{R}$ is an equivalence relation. Two types τ_1 and τ_2 are said to be *equivalent* (denoted by $\tau_1 \cong \tau_2$) iff $(\tau_1, \tau_2) \in \mathcal{R}$.

4 Linear-Time Type Equivalence for Nonrecursive Types

Assume that we are given two nonrecursive types that are represented as two term automata. In this section we shall demonstrate that type matching of nonrecursive types can be decided in $O(n)$ worst-case time, based on a result of Paige and Yang.

Multiset discrimination is a simple, highly efficient technique for finding and eliminating duplicate values in a structured collection of values, even in the presence of associative (A), associative-commutative (AC), and associative-commutative-idempotent (ACI) operators; that is, list, multiset and set comprehensions. It was pioneered by Paige, Tarjan, and Bonic [21, 23] for improved DFA minimization and lexicographic sorting; it has been applied to an array of language processing problems demonstrating that hashing and pointer (array index) arithmetic can be avoided, with improved worst-case complexity [9, 10]; it has been extended to provide complete dagification of forests and acyclic dags even in the presence of list, bag and set operators [22, 24].

In the terminology of this paper, complete dagification is the transformation of a term automaton $A = (Q,\ \Sigma,\ q_0,\ \delta,\ \ell)$ to a term automaton $A^d = (Q^d,\ \Sigma,\ q_0^d,\ \delta^d,\ \ell^d)$ such that the following properties hold.

1. There is a surjective function $f : Q \longrightarrow Q^d$ such that q and $f(q)$ represent equivalent terms (are bisimilar) for all $q \in Q$.
2. For all $q, q' \in Q$, q and q' represent equivalent terms (are bisimilar) if and only if $f(q) = f(q')$.

Intuitively, complete dagification collapses all equivalent states in the original term automaton into a single node. We can associate the state $f(q)$ with q by representing q as a record with a field containing its dagified state $f(q)$. After complete dagification of A we can answer any equivalence query: "Given $q, q' \in A$, are q and q' bisimilar (do they represent equivalent types)?" in constant time: Look up $f(q)$ and $f(q')$ in the records for q and q', respectively, and check if they are equal. If so, q and q' are bisimilar; if not, q and q' are not bisimilar.

Definition 1 (Acyclic Term Automaton) *A term automaton is* acyclic *if there exists a total order $<$ on Q such that $\delta(q, i) > q$ for all q, i for which $\delta(q, i)$ is defined.*

In other words, a term automaton is acyclic if it contains no cycle of transitions. Acyclic term automata represent all and only nonrecursive types (types formed without the use of the fixpoint operator μ). A term automaton is a forest (set of trees), if every state has at most one predecessor; that is, for each $q \in Q$ there is at most one $q' \in Q$ such that $\delta(q', i) = q$ for some i.

Theorem 1 [Paige and Yang [22,24]] *Let $A = (Q,\ \Sigma,\ q_0,\ \delta,\ \ell)$ be an acyclic term automaton and $q_1, q_2 \in Q$. We can decide in time $O(n+m)$ whether q_1 and q_2 represent equivalent terms; that is, whether $(q_1, q_2) \in \mathcal{R}$. Here, $n = |Q|$, the size of Q, and $m = |\delta|$, the number of transitions in δ.*

Proof: If A is a pair of trees rooted at q_1 and q_2, respectively, then this theorem is a direct consequence of Paige and Yang [24, Theorem 2]. The complete dagification of the term automaton corresponds to Stage 2 of Paige and Yang's transformation of an input string in their *external language* to an *abstract syntax dag* (*ASD*). The term $\tau \rightarrow \tau'$ here corresponds to the the tuple (list) $[\rightarrow, \tau, \tau']$ in their external language, and $\tau_1 \times \ldots \times \tau_n$ corresponds to $[\times, \langle \tau_1, \ldots, \tau_n \rangle]$. Note that the angled brackets in $\langle \tau_1, \ldots, \tau_n \rangle$ signal a multiset expression, which captures the associativity and commutativity of the $\prod^n$-operator in our term language.

Paige and Yang's ASD construction also works in time $O(m+n)$ if A is already partially dagified. This follows from Lemmas 1 and 2 in their paper. □

Let us define the size of a term automaton $A = (Q, \Sigma, q_0, \delta, l)$ to be $|Q| + |\delta|$; i.e., the sum of the number of states and transitions in the automaton.

Corollary 2 *For nonrecursive types τ_1, τ_2 represented by acyclic term automata A_1, A_2, each of size at most n, we can decide $(\tau_1, \tau_2) \in \mathcal{R}$ in $O(n)$ time.*

Proof: The two automata A_1, A_2 can be turned into a single automaton of size at most $2n$, by taking the disjoint union of their states and transitions. The result then follows from Theorem 1. □

5 $O(n \log n)$-Time Type Equivalence for Recursive Types

Assume that we are given two recursive types τ_1 and τ_2 that are represented as two term automata A_1 and A_2. Lemma 1 proves that $\tau_1 \cong \tau_2$ (or $(\tau_1, \tau_2) \in \mathcal{R}$) if and only if there is a reflexive bisimulation C between A_1 and A_2 such that the initial states of the term automata A_1 and A_2 are related by C. Lemma 3 essentially reduces the problem of finding a reflexive bisimulation C between A_1 and A_2 to finding a size-stable coarsest partition [12,23]. Theorem 3 uses the algorithm of Paige and Tarjan to determine in $O(n \log n)$ time (n is the sum of the sizes of the two term automata) whether there exists a reflexive bisimulation C between A_1 and A_2.

Throughout this section, we will use A_1, A_2 to denote two term automata over the alphabet Σ:

$$A_1 = (Q_1, \Sigma, q_{01}, \delta_1, \ell_1) \qquad A_2 = (Q_2, \Sigma, q_{02}, \delta_2, \ell_2).$$

We assume that $Q_1 \cap Q_2 = \emptyset$. Define $Q = Q_1 \cup Q_2$, $\delta : Q \times \omega \rightarrow Q$ where $\delta = \delta_1 \oplus \delta_2$, and $\ell : Q \rightarrow \Sigma$, where $\ell = \ell_1 \oplus \ell_2$, where $\oplus$ denotes disjoint union of two functions. We say that A_1, A_2 are *bisimilar* if and only if there exists a relation $C \subseteq Q \times Q$, called a bisimulation between A_1 and A_2, such that:

- if $(q, q') \in C$, then $\ell(q) = \ell(q')$
- if $(q, q') \in C$ and $\ell(q) = \rightarrow$,
  then $(\delta(q, 0), \delta(q', 0)) \in C$ and $(\delta(q, 1), \delta(q', 1)) \in C$
- if $(q, q') \in C$ and $\ell(q) = \prod^n$, then there exists a bijection $b : \{0 \ldots n-1\} \rightarrow \{0 \ldots n-1\}$ such that for all $i \in \{0 \ldots n-1\}$ we have that: $(\delta(q, i), \delta(q', b(i))) \in C$.

Notice that the bisimulations between A_1 and A_2 are closed under union, therefore, there exists a largest bisimulation between A_1 and A_2. It is straightforward to show that the identity relation on Q is a bisimulation, and that any reflexive bisimulation is an equivalence relation. Hence, the largest bisimulation is an equivalence relation.

Lemma 1 *For types τ_1, τ_2 that are represented by the term automata A_1, A_2, respectively, we have $(\tau_1, \tau_2) \in \mathcal{R}$ if and only if there is a reflexive bisimulation C between A_1 and A_2 such that $(q_{01}, q_{02}) \in C$.*

Proof: Suppose $(\tau_1, \tau_2) \in \mathcal{R}$. Define: $C = \{ (q, q') \in Q \times Q \mid (\lambda\alpha.\ell(\widehat{\delta}(q,\alpha)), \lambda\alpha.\ell(\widehat{\delta}(q',\alpha))) \in \mathcal{R} \}$. It is straightforward to show that C is a bisimulation between A_1 and A_2, and that $(q_{01}, q_{02}) \in C$; we omit the details.

Conversely, let C be a reflexive bisimulation between A_1 and A_2 such that $(q_{01}, q_{02}) \in C$. Define: $R = \{ (\sigma_1, \sigma_2) \mid (q, q') \in C \wedge \sigma_1 = \lambda\alpha.\ell(\widehat{\delta}(q,\alpha)) \wedge \sigma_2 = \lambda\alpha.\ell(\widehat{\delta}(q',\alpha)) \}$. From $(q_{01}, q_{02}) \in C$, we have $(\tau_1, \tau_2) \in R$. It is straightforward to prove that R is a bisimulation; again, we omit the details. From $(\tau_1, \tau_2) \in R$ and R being a bisimulation, we conclude that $(\tau_1, \tau_2) \in \mathcal{R}$. □

A *partitioned graph* is a 3-tuple (U, E, P), where U is a set of nodes, $E \subseteq U \times U$ is an edge relation, and P is a *partition* of U. A partition P of U is a set of pairwise disjoint subsets of U whose union is all of U. The elements of P are called its *blocks*. If P and S are partitions of U, then S is a *refinement* of P if and only if every block of S is contained in a block of P.

A partition S of a set U can be characterized by an equivalence relation K on U such that each block of S is an equivalence class of K. If U is a set and K is an equivalence relation on U, then we use U/K to denote the partition of U into equivalence classes for K.

A partition S is *size-stable* with respect to E if and only if for all blocks $B_1, B_2 \in S$, and for all $x, y \in B_1$, we have $|E(x) \cap B_2| = |E(y) \cap B_2|$, where $E(x)$ is the set of neighbors $\{y \mid (x, y) \in E\}$. If E is clear from the context, we will simply use size-stable. We will repeatedly use the following characterization of size-stable partitions.

Lemma 2 *For an equivalence relation K, we have that U/K is size-stable if and only if for all $(u, u') \in K$, there exists a bijection $\pi : E(u) \to E(u')$ such that for all $u_1 \in E(u)$, we have $(u_1, \pi(u_1)) \in K$.*

Proof: Suppose that U/K is size-stable. Let $(u, u') \in K$. Let B_1 be the block of U/K which contains u and u'. For each block B_2 of U/K, we have that $|E(u) \cap B_2| = |E(u') \cap B_2|$. So, for each block B_2 of U/K, we can construct a bijection from $E(u) \cap B_2$ to $E(u') \cap B_2$, such that for all $u_1 \in E(u) \cap B_2$, we have $(u_1, \pi(u_1)) \in K$. These bijections can then be merged to a single bijection $\pi : E(u) \to E(u')$ with the desired property.

Conversely, suppose that for all $(u, u') \in K$, there exists a bijection $\pi : E(u) \to E(u')$ such that for all $u_1 \in E(u)$, we have $(u_1, \pi(u_1)) \in K$. Let $B_1, B_2 \in U/K$, and let $x, y \in B_1$. We have that $(x, y) \in K$, so there exists a bijection $\pi : E(x) \to E(y)$ such that for all $u_1 \in E(x)$, we have $(u_1, \pi(u_1)) \in K$. Each element of $E(x) \cap B_2$ is mapped by π to an element of $E(y) \cap B_2$. Moreover, each element of $E(y) \cap B_2$ must be the image under π of an element of $E(x) \cap B_2$. We conclude that π restricted to $E(x) \cap B_2$ is a bijection to $E(y) \cap B_2$, so $|E(x) \cap B_2| = |E(y) \cap B_2|$. □

Given two term automata A_1, A_2, we define a partitioned graph (U, E, P):

$$
\begin{aligned}
U &= Q \cup \{ \langle q, i \rangle \mid q \in Q \wedge \delta(q,i) \text{ is defined } \} \\
E &= \{ (q, \langle q, i \rangle) \mid \delta(q,i) \text{ is defined } \} \cup \{ (\langle q, i \rangle, \delta(q,i)) \mid \delta(q,i) \text{ is defined } \} \\
L &= \{ (q, q') \in Q \times Q \mid \ell(q) = \ell(q') \} \\
&\quad \cup \{ (\langle q, i \rangle, \langle q', i' \rangle) \mid \ell(q) = \ell(q') \text{ and if } \ell(q) = \rightarrow, \text{ then } i = i' \} \\
P &= U/L.
\end{aligned}
$$

The graph contains one node for each state and transition in A_1, A_2. Each transition in A_1, A_2 is mapped to two edges in the graph. This construction ensures that if a node in the graph corresponds to a state labeled $\prod^n$, then that node will have n distinct successors in the graph. This is convenient when establishing a bijection between the successors of two nodes labeled $\prod^n$.

The equivalence relation L creates a distinction between the two successors of a node that corresponds to a state labeled $\rightarrow$. This is done by ensuring that if $(\langle q,i\rangle,\langle q,i'\rangle)\in L$ and $\ell(q)=\rightarrow$, then $i=i'$. This is convenient when establishing a bijection between the successors of two nodes labeled $\rightarrow$.

Lemma 3 *There exists a reflexive bisimulation C between A_1 and A_2 such that $(q_{01},q_{02})\in C$ if and only if there exists a size-stable refinement S of P such that q_{01} and q_{02} belong to the same block of S.*

Proof: Let $C\subseteq Q\times Q$ be a reflexive bisimulation between A_1 and A_2 such that $(q_{01},q_{02})\in C$. Define an equivalence relation $K\subseteq U\times U$ as follows:

$$\begin{aligned}K = C &\cup \{\, (\langle q,i\rangle,\langle q',i\rangle) \mid (q,q')\in C \wedge \ell(q)=\ell(q')=\rightarrow \}\\ &\cup \{\, (\langle q,i\rangle,\langle q',i'\rangle) \mid (q,q')\in C \wedge (\delta(q,i),\delta(q',i'))\in C \wedge \ell(q)=\ell(q')\\ &\qquad \wedge \ell(q)\neq\rightarrow \}\\ S = U/K&.\end{aligned}$$

From $(q_{01},q_{02})\in C$, we have $(q_{01},q_{02})\in K$, so q_{01} and q_{02} belong to the same block of S. We will now show that S is a size-stable refinement of P.

Let $(u,u')\in K$. From Lemma 2 we have that it is sufficient to show that there exists a bijection $\pi: E(u)\rightarrow E(u')$, such that for all $u_1\in E(u)$, we have $(u_1,\pi(u_1))\in K$. There are three cases. First, suppose $(u,u')\in C$. We have

$$E(u)=\{\, \langle u,i\rangle \mid \delta(u,i) \text{ is defined }\} \quad\text{and}\quad E(u')=\{\, \langle u',i'\rangle \mid \delta(u',i') \text{ is defined }\}.$$

Let us consider each of the possible cases of u and u'. If $\ell(u)=\ell(u')\in\Gamma$, then $E(u)=E(u')=\emptyset$, and the desired bijection exists trivially. Next, if $\ell(u)=\ell(u')=\rightarrow$, then

$$E(u)=\{\, \langle u,0\rangle,\langle u,1\rangle \,\} \quad\text{and}\quad E(u')=\{\, \langle u',0\rangle,\langle u',1\rangle \,\}$$

so the desired bijection is $\pi: E(u)\rightarrow E(u')$, where $\pi(\langle u,0\rangle)=\langle u',0\rangle$ and $\pi(\langle u,1\rangle)=\langle u',1\rangle$, because $(\langle u,0\rangle,\langle u',0\rangle)\in K$ and $(\langle u,1\rangle,\langle u',1\rangle)\in K$. Finally, if $\ell(u)=\ell(u')=\prod^n$, then

$$E(u)=\{\, \langle u,i\rangle \mid \delta(u,i) \text{ is defined}\} \quad\text{and}\quad E(u')=\{\, \langle u',i'\rangle \mid \delta(u',i') \text{ is defined}\}.$$

From $(u,u')\in C$, we have a bijection $b:\{0\ldots n-1\}\rightarrow\{0\ldots n-1\}$ such that $\forall i\in\{0\ldots n-1\}: (\delta(u,i),\delta(u',b(i)))\in C$. From that, the desired bijection can be constructed.

Second, suppose $u=\langle q,i\rangle$ and $u'=\langle q',i\rangle$, where $(q,q')\in C$, and $\ell(q)=\ell(q')=\rightarrow$. We have:

$$E(u)=\{\, \delta(q,i) \,\} \quad\text{and}\quad E(u')=\{\, \delta(q',i) \,\},$$

and from $(q,q')\in C$ we have $(\delta(q,i),\delta(q',i))\in C\subseteq K$, so the desired bijection exists.

Third, suppose $u=\langle q,i\rangle$ and $u'=\langle q',i'\rangle$, where $(q,q')\in C$, $(\delta(q,i),\delta(q',i'))\in C$, $\ell(q)=\ell(q')$, and $\ell(q)\neq\rightarrow$. We have

$$E(u)=\{\, \delta(q,i) \,\} \quad\text{and}\quad E(u')=\{\, \delta(q',i') \,\},$$

and $(\delta(q,i),\delta(q',i'))\in C\subseteq K$, so the desired bijection exists.

Conversely, let S be a size-stable refinement of P such that q_{01} and q_{02} belong to the same block of S. Define:

$$\begin{aligned}K&=\{\, (u,u')\in U\times U \mid u,u' \text{ belong to the same block of } S \,\}\\ C&=K\cap(Q\times Q).\end{aligned}$$

Notice that $(q_{01},q_{02})\in C$ and that C is reflexive. We will now show that C is a bisimulation between A and A'.

First, suppose $(q,q')\in C$. From S being a refinement of P we have $(q,q')\in L$, so $\ell(q)=\ell(q')$.

Second, suppose $(q,q')\in C$ and $\ell(q)=\rightarrow$. From the definition of E we have

$$E(q)=\{\, \langle q,0\rangle,\langle q,1\rangle \,\} \quad\text{and}\quad E(q')=\{\, \langle q',0\rangle,\langle q',1\rangle \,\}.$$

From S being size-stable, $(q, q') \in C \subseteq K$, and Lemma 2 we have that there exists a bijection $\pi : E(q) \to E(q')$ such that for all $u \in E(q)$ we have that $(u, \pi(u)) \in K$. From $K \subseteq L$ and $\ell(q) = \to$ we have that there is only one possible bijection π:

$$\pi(\langle q, 0\rangle) = \langle q', 0\rangle \text{ and } pi(\langle q, 1\rangle) = \langle q', 1\rangle,$$

so $(\langle q, 0\rangle, \langle q', 0\rangle) \in K$ and $(\langle q, 1\rangle, \langle q', 1\rangle) \in K$. From the definition of E we have, for $i \in \{0, 1\}$,

$$E(\langle q, i\rangle) = \delta(q, i) \text{ and } E(\langle q', i\rangle) = \delta(q', i),$$

and since S is size-stable, we have, for $i \in \{0, 1\}$, $(\delta(q, i), \delta(q', i)) \in K$. Moreover, for $i \in \{0, 1\}$, we have $(\delta(q, i), \delta(q', i)) \in Q \times Q$, and so we can conclude $(\delta(q, i), \delta(q', i)) \in C$.

Third, suppose $(q, q') \in C$ and $\ell(q) = \prod^n$. From the definition of E we have

$$E(q) = \{ \langle q, i\rangle \mid \delta(q, i) \text{ is defined } \} \text{ and } E(q') = \{ \langle q', i\rangle \mid \delta(q', i) \text{ is defined } \}.$$

Notice that $|E(q)| = |E(q')| = n$. From S being size-stable, $(q, q') \in C \subseteq K$, and Lemma 2, we have that there exists a bijection $\pi : E(q) \to E(q')$ such that for all $u \in E(q)$ we have that $(u, \pi(u)) \in K$. From π we can derive a bijection $b : \{0 \ldots n-1\} \to \{0 \ldots n-1\}$ such that $\forall i \in \{0 \ldots n-1\}$: $(\langle q, i\rangle, \langle q', b(i)\rangle) \in K$. From the definitions of E and E' we have that for $i \in \{0 \ldots n-1\}$,

$$E(\langle q, i\rangle) = \{ \delta(q, i) \} \text{ and } E(\langle q', i\rangle) = \{ \delta(q', i) \},$$

and since S is size-stable, and, for all $i \in \{0 \ldots n-1\}$, $(\langle q, i\rangle, \langle q', b(i)\rangle) \in K$, we have $(\delta(q, i), \delta(q', b(i))) \in K$. Moreover, we have $(\delta(q, i), \delta(q', b(i))) \in Q \times Q$, and so we can conclude $(\delta(q, i), \delta(q', b(i))) \in C$. □

Recall that the size of a term automaton $A = (Q, \Sigma, q_0, \delta, l)$ is $|Q| + |\delta|$; i.e., the sum of the number of states and transitions in the automaton.

Theorem 3 *For types τ_1, τ_2 represented by term automata A_1, A_2 of size at most n, we can decide $(\tau_1, \tau_2) \in \mathcal{R}$ in $O(n \log n)$ time.*

Proof: From Lemma 1 we have that $(\tau_1, \tau_2) \in \mathcal{R}$ if and only if there is a reflexive bisimulation C between A_1 and A_2 such that $(q_{01}, q_{02}) \in C$. From Lemma 3 we have that there exists a reflexive bisimulation C between A_1 and A_2 such that $(q_{01}, q_{02}) \in C$ if and only if there exists a size-stable refinement S of P such that q_{01} and q_{02} belong to the same block of S.

Paige and Tarjan [23] give an $O(m \log p)$ algorithm to find the coarsest size-stable refinement of P, where m is the size of E and p is the size of the universe U.

Our algorithm first constructs (U, E, P) from A_1 and A_2, then runs the Paige-Tarjan algorithm to find the coarsest size-stable refinement S of P, and finally checks whether q_{01} and q_{02} belong to the same block of S. If A_1 and A_2 are of size at most n, then the size of E is at most $2n$, and the size of U is at most $2n$, so the total running time of our algorithm is $O(2n \log(2n)) = O(n \log n)$. □

Next, we illustrate how our algorithm determines that equivalence between the types. Details of the algorithm can be found in [23]. Consider two types I_1 and J_1 defined in Section 2. The set of types corresponding to the two interfaces are:

$$\{I_1, I_2, m_1, m_2, m_3, m_4, \mathit{int}, \mathit{float}\} \text{ and } \{J_1, J_2, n_1, n_2, n_3, n_4, \mathit{int}, \mathit{float}\}$$

Note that we abuse notation and use m_1, m_2, etc, to denote the *types* of the methods with those names. Figure 4 shows various steps of our algorithm. For simplicity, the figure only shows the blocks of actual types, but not the blocks of the extra nodes of the form $\langle q, i\rangle$. The blocks in the first row are based on labels, e.g., states labeled with $\times$ are in the same block. In the next step, the block containing the methods are split based on the type of the result of the method, e.g., methods m_1 and n_4 both return *float*, so they are in the same block. In the next step (corresponding to the third row) the block $\{I_1, I_2, J_1, J_2\}$ are split. The final partition, where block $\{m_3, m_4, n_1, n_2\}$ is split, is shown in the fourth row.

Our algorithm can be tuned to take specific user needs into account. This is done simply by modifying the definition of the equivalence relation L. For example, suppose a user cares

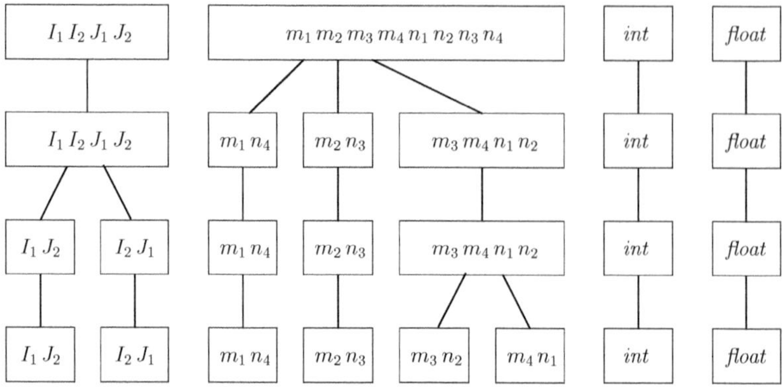

Fig. 4. Blocks of types.

about the order of the arguments to a method. This means that the components of the product type that models the argument list should not be allowed to be shuffled during type matching. We can prevent shuffling by employing the same technique that the current definition of L uses for function types. The idea is to insist that two component types may only be matched when they have the same component index.

Another example of the tunability of our algorithm involves the modifiers in `Java`. Suppose a programmer is developing a product that is multi-threaded. In this case the programmer may only want to match `synchronized` methods with other `synchronized` methods. This can be handled easily in our framework by changing L such that two method types may only be matched when they are both synchronized. On the other hand if the user is working on a single-threaded product, the keyword `synchronized` can be ignored. The same observation applies to other modifiers such as `static`. See the discussion, Section 8.4, for other variations of type matching that can be handled by our algorithm.

6 Our Implementation

We have implemented our algorithm in `Java` and the current version is based on the code written by Wanjun Wang. The implementation and documentation are freely available at `http://www.cs.purdue.edu/homes/tzhao/matching/matching.htm` .

The current version has a graphical user interface so that users may input type definitions written in a file and also may specify restrictions on type isomorphism.

Suppose we are given the following file with four `Java` interfaces.

```
interface I1 {                      interface I2 {
   float m1 (I1 a, int b);             J2 m3 (float a);
     int m2 (I2 a);                    I1 m4 (float a);
}                                   }
interface J1 {                      interface J2 {
   I1 n1 (float a);                      int n3 (J1 a);
   J2 n2 (float a);                    float n4 (int a, J2 b);
}                                   }
```

The implementation, as illustrated in the Figure 5, will read and parse the input file and then transform the type definitions into partitions of numbers with each type definition and dummy type assigned a unique number. The partitions will be refined by the Paige–Tarjan algorithm until it is *size-stable* as defined in this paper. Finally, we will be able to read the results from the final partitions. Two types are isomorphic if the numbers assigned to them are in the same partition.

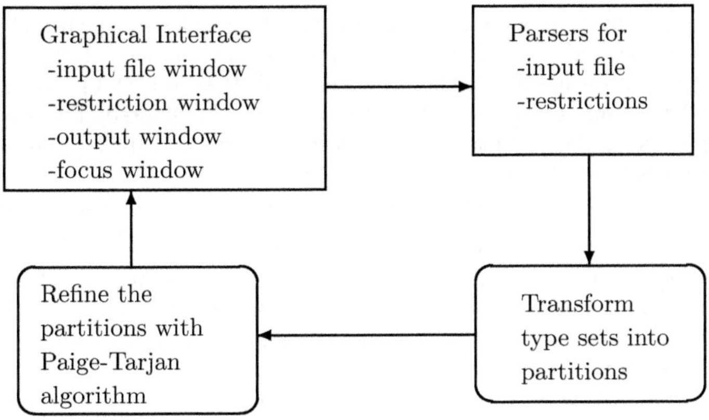

Fig. 5. Schematic diagram for the implementation.

The implementation will give the following output:

$I_1 = J_2, \quad I_2 = J_1$

$I_1.m_1 = J_2.n_4, \quad I_1.m_2 = J_2.n_3, \quad \text{and} \quad I_2.m_3 = I_2.m_4 = J_1.n_1 = J_1.n_2$.

We can see that the types of interfaces I_2 and J_1 are isomorphic and moreover, all method types of I_2, J_1 match. Suppose that we have additional information about the method types such that only method m_3 and n_1 should have isomorphic types. We can restrict the type matching by adding $I_2.m_3 = J_1.n_1$ to the *restrictions* window of the user interface. The new matching result is:

$I_1 = J_2, \quad I_2 = J_1$

$I_1.m_1 = J_2.n_4, \quad I_1.m_2 = J_2.n_3, \quad I_2.m_3 = J_1.n_1, \quad \text{and} \quad I_2.m_4 = J_1.n_2$.

We can focus on the matching of two interface types such as I_2, J_1 in the *focus* windows of the user interface, which will match their methods one to one.

7 Subtyping of Recursive Types

In this section we discuss subtyping and formalize it using a simulation relation. We also discuss reasons why the algorithm given in Section 5 is not applicable to subtyping of recursive types. Consider the interfaces I_1 and I_2 shown in Figure 6, and suppose a user is looking for I_2. The interfaces I_1 and I_2 can be mapped to the following recursive types:

$$\tau_1 = \mu\alpha.((\mathit{float} \times \mathit{boolean}) \to \alpha) \times (\alpha \to \mathit{boolean})$$
$$\tau_2 = \mu\beta.(\mathit{int} \times \mathit{boolean}) \to \beta)$$

```
interface I1 {
        I1 m (float a, boolean b);
    boolean p (I1 j);
}
```

```
interface I2 {
    I2 m (int i, boolean b);
}
```

Fig. 6. Interfaces I_1 and I_2.

Assuming that *int* is a subtype of *float* (we can always coerce integers into floats) we have that τ_1 is a subtype of τ_2. Therefore, the user can use the interface I_1. There are several points to notice from this example. In the context of subtyping, we need two kinds of products:

one that models a collection of methods and another that models sequence of parameters. In our example, the user only specified a type corresponding to method m. Therefore, during the subtyping algorithm method p should be ignored. However, the parameters of method m are also modeled using products and none of these can be ignored. Therefore, we consider two types of product type constructors in our type systems and the subtyping rule for these two types of products are different.

As stated before, a type is a regular term, in this case over the ranked alphabet

$$\Sigma = \Gamma \cup \{\rightarrow\} \cup \{\prod^n, n \geq 2\} \cup \{\times^n, n \geq 2\}.$$

Roughly speaking, $\prod^n$ and $\times^n$ will model collection of parameters and methods respectively. Also assume that we are given a subtyping relation on the base types Γ. If τ_1 is a subtype of τ_2, we will write it as $\tau_1 \preceq \tau_2$. A relation S is called a *simulation* on types if it satisfies the following conditions:

- if $(\sigma, \tau) \in S$ and $\sigma(\epsilon) \in \Gamma$, then $\tau(\epsilon) \in \Gamma$ and $\sigma(\epsilon) \preceq \tau(\epsilon)$.
- if $(\sigma, \tau) \in S$ and $\sigma(\epsilon) \in (\{\rightarrow\} \cup \{\prod^n, n \geq 2\})$, then $\sigma(\epsilon) = \tau(\epsilon)$.
- if $(\sigma_1 \rightarrow \sigma_2, \tau_1 \rightarrow \tau_2) \in S$, then $(\tau_1, \sigma_1) \in S$ and $(\sigma_2, \tau_2) \in S$.
- if $(\prod_{i=0}^{n-1} \sigma_i, \prod_{i=0}^{n-1} \tau_i) \in S$, then there exists a bijection $b : \{0 \dots n-1\} \rightarrow \{0 \dots n-1\}$ such that for all $i \in \{0 \dots n-1\}$, we have $(\sigma_i, \tau_{b(i)}) \in S$.
- Suppose $(\sigma, \tau) \in S$, $\sigma(\epsilon) = \times^n$, and $\sigma = \times_{i=0}^{n-1} \sigma_i$. If $\tau(\epsilon) \notin \{\times^m, m \geq 2\}$, then there exists a $j \in \{0 \dots n-1\}$ such that $(\sigma_j, \tau) \in S$. Otherwise, assume that $\tau(\epsilon) = \times^m$, where $m \leq n$ and $\tau = \times_{i=0}^{m-1} \tau_i$. In this case, then there exists an injective function $c : \{0 \dots m-1\} \rightarrow \{0 \dots n-1\}$ such that for all $i \in \{0 \dots m-1\}$, we have $(\sigma_{c(i)}, \tau_i) \in S$. Notice that this rule allows ignoring certain components of σ.

As is the case with bisimulations, simulations are closed under union, therefore there exists a largest simulation (denoted by $\mathcal{S}$).

Let A_1, A_2 denote two term automata over Σ:

$$A_1 = (Q_1, \Sigma, q_{01}, \delta_1, \ell_1) \text{ and } A_2 = (Q_2, \Sigma, q_{02}, \delta_2, \ell_2).$$

We assume that $Q_1 \cap Q_2 = \emptyset$. Define $Q = Q_1 \cup Q_2$, $\delta : Q \times \omega \rightarrow Q$ where $\delta = \delta_1 \oplus \delta_2$, and $\ell : Q \rightarrow \Sigma$, where $\ell = \ell_1 \oplus \ell_2$, where $\oplus$ denotes disjoint union of two functions. We say that A_2 *simulates* A_1 (denoted by $A_1 \preceq A_2$) if and only if there exists a relation $D \subseteq Q \times Q$, called a *simulation relation* between A_1 and A_2, such that:

- if $(q, q') \in D$ and $\ell(q) \in \Gamma$, then $\ell(q') \in \Gamma$ and $\ell(q) \preceq \ell(q'))$.
- if $(q, q') \in D$ and $\ell(q) \in (\{\rightarrow\} \cup \{\prod^n, n \geq 2\})$, then $\ell(q) = \ell(q')$.
- if $(q, q') \in D$ and $\ell(q) = \rightarrow$, then $(\delta(q', 0), \delta(q, 0)) \in D$ and $(\delta(q, 1), \delta(q', 1) \in D$.
- if $(q, q'), \in D$ and $\ell(q) = \prod^n$, then there exists a bijection $b : \{0 \dots n-1\} \rightarrow \{0 \dots n-1\}$ such that for all $i \in \{0 \dots n-1\}$, we have: $(\delta(q, i), \delta(q' i)) \in D$.
- Suppose $(q, q') \in D$ and $\ell(q) = \times^n$. If $\ell(q') \notin \{\times^m, m \geq 2\}$, then there exists a $j \in \{0 \dots n-1\}$ such that $(\delta(q, j), q') \in D$. Otherwise, assume that $\ell(q') = \times^m$. Then $m \leq n$ and there exists an injective function $c : \{0 \dots m-1\} \rightarrow \{0 \dots n-1\}$ such that for all $i \in \{0 \dots m-1\}$, we have $(\delta(q, c(i)), \delta(q', i)) \in D$.

Notice that the simulations between A_1 and A_2 are closed under union, therefore there exists a largest simulation between A_1 and A_2. The proof of Lemma 4 is similar to the proof of Lemma 1 and is omitted.

Lemma 4 *For types τ_1, τ_2 that are represented by the term automata A_1, A_2, respectively, we have $(\tau_1, \tau_2) \in \mathcal{S}$ if and if only there is a reflexive simulation D between A_1 and A_2 such that $(q_{01}, q_{02}) \in D$.*

The largest simulation between the term automata A_1 and A_2 is given by the following greatest fixed point $\nu D.\forall q, q'.sim(q, q', D)$, where $D \subseteq Q_1 \times Q_2$ and the predicate $sim(q, q', D)$ is the conjunction of the five conditions which appear in the definition of the simulation relation between two automata. Let n and m be the size of the term automata A_1 and A_2, respectively. Since nm is a bound on the size of D, the number of iterations in computing the greatest fixed point is bounded by nm. In general, the relation D (or for that matter the simulation relation) is not symmetric. On the other hand, the bisimulation relation was an equivalence relation, and so could be represented as a partition on the set $Q_1 \cup Q_2$, or in other words, partitions give us a representation of an equivalence relation that is linear in the sum of the sizes of the set of states Q_1 and Q_2. The Paige–Tarjan algorithm uses the partition representation of the equivalence relation. Since D is not symmetric (and thus not an equivalence relation), it cannot be represented by a partition. This is the crucial reason why our previous algorithm cannot be applied to subtyping.

8 Conclusion

In this paper we addressed the problem of matching recursive and nonrecursive types. We presented algorithms with $O(n \log n)$ and $O(n)$ time complexities that decide whether two types are equivalent. To our knowledge, these are the most efficient algorithms for type matching with and without type recursion, respectively. Our results are applicable to the problem of matching signatures of software components. Applications to `Java` were also discussed. Issues related to subtyping of recursive types were also addressed.

We conclude by discussing type matching of sets of types; related work; conceivable applications; and future work.

8.1 Multiple Type Matching

Recall Theorem 3. The algorithm employed actually works on more than 2 types and preserves its complexity even if types are represented by graphs with nodes shared amongst multiple types (corresponding to type abbreviations). Without explicit proof we state that Theorem 3 can be generalized as follows:

Theorem 4 *For types $\tau_1, \tau_2, \ldots, \tau_m$ represented by nodes $Q_0 = q_1, q_2, \ldots, q_m$ in term automaton A of size at most n, we can preprocess A in time $O(n \log n)$ such that:*

- *Q_0 can be partitioned into blocks of pairwise matching types in time $O(m)$, and the blocks can be output in the same time.*
- *For any $1 \leq i, j \leq m$ it can be decided in $O(1)$ time whether (the types denoted by) q_i and q_j match or not.*
- *For any $1 \leq i \leq m$ the set $[q_i] \subseteq Q_0$ of (nodes denoting) types matching q_i can be output in time $O(|[q_i]|)$.*

The corresponding strengthening of Theorem 1, with $O(n)$ instead of $O(n \log n)$, holds for acyclic A. These are strengthenings due to the particular algorithms used; they are not a property of the problem. To wit, type matching *without* commutativity of our k-ary products can be done in time $O(n\alpha(n, n)$ for a pair of nodes q_1, q_2 using unification closure, but the best known algorithm for doing it for m nodes $q_1, \ldots, q_m$ is our algorithm (Theorem 4), which requires time $\Theta(n \log n)$. (Note that solving this problem using unification closure on all pairs q_i, q_j takes time $O(nm^2)$.) Furthermore, the strengthening *does not* hold for binary associative-commutative operators, as in Zibin, Gil and Considine [34]. This is due to an exponential blow-up in the preprocessing of binary associativity in a setting with shared data to a version with lists (n-ary products in our setting); see the related work discussion below.

8.2 Related Work

Word problem with Associative-Commutative Operators
Zibin, Gil, and Considine [34] present a linear-time type equivalence algorithm for nonrecursive types and an $O(n \log^2 n)$ algorithm for a richer equivalence with distributivity. Their

linear-time algorithm is incomparable to ours, as it handles an equivalence around a binary associative-commutative product operator, but requires a nonshared input data representation to achieve linearity. The equivalence solved by their linear-time algorithm is defined by:

(1)	$A \times 1 = A$	
(2)	$A \to 1 = 1$	
(3)	$1 \to A = A$	
(4)	$A \times B = B \times A$	(commutativity)
(5)	$A \times (B \times C) = A \times (B \times C)$	(associativity)
(6)	$A \to (B \to C) = (A \times B) \to C$	(Currying)

where 1 is the unit type and A, B, C range over simple types with function type and product constructors and a nonempty set of constant types. Their solution consists of first preprocessing the input, a pair of type terms, by rewriting them according to the axioms, excepting commutativity, in left-to-right direction. The preprocessed terms are then solved using basic multiset discrimination techniques. The combined time of preprocessing and multiset discrimination is linear if the inputs are represented as trees; that is, without sharing of subterms. We observe that with shared representations, preprocessing becomes exponential in time and space; this is due to associativity rewriting. (Note that Currying does not make it worse, and the neutrality axioms are harmless.) To wit, consider $A_{i+1} = A_i \times A_i$ for $i \geq 0$, where A_0 denotes some constant type. Represented as a term automaton, A_n requires $O(n)$ space, but associativity rewriting is applicable $O(2^n)$ times, resulting in a term type representation $(\ldots(A_0 \times A_0)\ldots \times A_0)$ with 2^n occurrences of A_0 and a space requirement of $\Theta(2^n)$, with or without sharing.

Zibin, Gil, and Considine rediscover some of the basic multiset discrimination techniques for integers. (They are apparently unaware of Paige et al.'s previous work, since no reference to any of it is given in their article.) They require, however, a constant-time allocated integer array of size U for multiset discrimination of integers in the interval $[1 \ldots U]$. This is impractical since 32-bit unsigned integers require an array with 4 billion elements, and 64-bit unsigned integers would require approximately 10^{20} bytes. Furthermore, their computational model assumes that arrays of arbitrary size (including arrays whose length is exponential in the size of the input) can be allocated in constant time. Using a hash table implementation would be a practical alternative, but destroy the worst-case linear asymptotic bound of linear type isomorphism. Instead, as observed by Paige previously, we propose using tuple multiset discrimination on large numbers: each nonnegative integer v can be represented by its number representation $a_k \ldots a_0$ with $0 \leq a_i < r$ for $0 \leq i \leq k$ for some radix r such that $v = \Sigma_{i=0}^{k} a_i r^i$. Multiset discrimination of such representations requires only one auxiliary array with r elements and can be performed in linear time and space. (Input size is counted as the number of bits. No word level parallelism is exploited here.) For example, 64-bit integers can be discriminated realistically by treating each number as an 8-tuple of bytes ($r = 2^8 = 256$) using a single global array of 256 elements or as a 4-tuple of 16-bit values ($r = 2^{16} = 65536$) using an array with 65536 elements.

Subtyping with Commutative Products

In a continuation of our work, Di Cosmo, Pottier, and Rémy [15] present an algorithm for deciding the subtyping problem discussed in Section 7. Their algorithm follows the same algorithmic strategy: a bipartite matching algorithm is iterated a quadratic number of times.

8.3 Potential Applications

Next we discuss potential applications of our algorithms.

CORBA: The CORBA approach utilizes a separate definition language called IDL. Objects are associated with language-independent interfaces defined in IDL. These interfaces are then translated into the language being used by the client. The translated interface then enables the clients to call the objects. Since the IDL interfaces have to be translated into several languages, their type system is very restrictive. Therefore, IDL interfaces lack expressive power because, intuitively speaking, the type system used in IDL has to be the intersection of the type systems of the languages language it supports. The drawbacks of CORBA-style approaches to interoperability are well articulated in [4, 5].

Polyspin and Mockingbird: The Polyspin and Mockingbird approaches do not require a common interface language, such as IDL. In both these approaches, clients and objects are written in languages with separate type systems, and an operation that crosses the language boundary is supported by bridge code that is automatically generated. Therefore, systems such as Polyspin and Mockingbird support seamless interoperability since the programmer is not burdened with writing interfaces in a special interface language such as IDL in CORBA. Polyspin supports only finite types. Mockingbird on the other hand supports recursive types, including records, linked lists, and arrays. The type system used in Mockingbird is called the *Mockingbird Signature Language* or *MockSL*. The problem of deciding type equivalence for MockSL remains open [3]. In this paper we considered a type system which is related to the one used in Mockingbird. However, we are investigating a translation from *MockSL* to recursive types.

Megaprogramming: Techniques suitable for very large software systems have been a major goal of software engineering. The term *megaprogramming* was introduced by DARPA to motivate this goal [6]. Roughly speaking, in megaprogramming, *megamodules* provide a higher level of abstraction than modules or components. For example, a megamodule can encapsulate the entire logistics of ground transportation in a major city. Megaprogramming is explained in detail in [31]. Interoperability issues arise when megaprograms are constructed using megamodules, see [31, Section 4.2]. We believe that the framework presented in this paper can be used to address mismatch between interfaces of megamodules.

8.4 Discussion and Future Work

Possible future work includes investigating type inference for programs in the presence of implicit type matching (type isomorhism) and subtyping as studied in this paper. The recent paper of Coppo [13] on type inference with recursive type equations may contain applicable techniques.

Brandt and Henglein [7] show how to derive semantically unique coercions (bridge code) interpreting Amadio-Cardelli style (read: no commutativity) subtyping and type isompor-phism. It should be noted, however, that completely automatic generation of adapter code in the presence of type matching *with commutativity* is risky since it is semantically ambiguous: Any method with two parameters of equal type matches another method in at least two semantically different ways.

Rittri [26] motivated type matching based on type isomorphisms by the problem of searching existing function libraries. Dating back to Thatte's work on synthesizing interface adapters [30] and more recently emphasized by Di Cosmo, Pottier, and Rémy [15], it has been argued that record *subtyping* is important for type-based component matching in an object-oriented language setting since it models "ignoring" methods in an implementation that are not required for an application. Apart from subtyping, other notions of matching may be of practical interest, especially in a multilanguage setting such as Mockingbird; e.g., including functions operating on arrays that require an explicit size parameter in a search for functions that operate on arrays (only).

Type matching has been formulated as a "one-on-one" problem: Does this desired function type signature match one particular (library) function? A practically more natural formulation, however, is to find the *set* of functions a desired type signature matches in a given library and, more generally yet, to do so for multiple desired type signatures at a time. In other words, a natural application setting is where a m desired interfaces are matched against a potentially large library of n components. This corresponds to performing type matching for a potentially large set of types. If only a type matching function of two type arguments (one-on-one matching) is available this requires mn applications of such a function, which by itself results in at least quadratic time requirements. As discussed in Section 8.1, our algorithms generalize to processing an arbitrary number of arguments: they partition the desired interfaces and all components in one go in linear time without recursive types and, with a logarithmic factor, for recursive types. Since the algorithms appear to be implementable with interactive response times for even large libraries, it appears feasible to use them in an interactive environment where type signatures are interactively changed for matching purposes (only). Each iteration produces a partitioning, which may then be used as a basis for refinement in further iterations. For example, in a first pass all primitive types might be treated as equivalent. Or some types might be treated as 1, the neutral element for product types. Doing so provides some of the benefits of record subtyping, but also allows treatment beyond that: e.g., treating type *int* as 1 will match a method invocation $f(\mathit{float}[]$ with a C-library function $g(\mathit{float}[], \mathit{int})$ that requires an explicit array length parameter; and *vice versa*. Note that the presently known best algorithms for type matching with record subtyping work in a one-on-one fashion and have high complexity, which makes them unlikely candidates for use in such an iterative and interactive fashion.

Acknowledgments

A preliminary version of this paper, excluding the results of Section 4, and authored by Jha, Palsberg, and Zhao was presented at FOSSACS 2002. Palsberg was supported by an NSF CAREER award, CCR–9734265, and by IBM. Henglein would like to acknowledge support by the Danish Natural Sciences Research Council under Project PLI.

References

1. R. M. Amadio and L. Cardelli: Subtyping recursive types. *ACM Transactions on Programming Languages and Systems*, 15(4):575–631, 1993. Also in Proceedings POPL '91.
2. M.-V. Aponte and R. Di Cosmo: Type isomorphisms for module signatures. In *Proceedings of PLILP '96*, 334–346. Springer (LNCS 1140), 1996.
3. J. Auerbach, C. Barton, and M. Raghavachari: Type isomorphisms with recursive types. Research Report RC 21247, IBM Research Division, T. J. Watson Research Center, Yorktown Heights, NY, August 1998.
4. J. Auerbach and M. C. Chu-Carroll: The mockingbird system: A compiler-based approach to maximally interoperable distributed programming. Research Report RC 20718, IBM Research Division, T. J. Watson Research Center, Yorktown Heights, NY, February 1997.
5. D. J. Barrett, A. Kaplan, and J. C. Wileden: Automated support for seamless interoperability in polylingual software systems. In *ACM FSE '96, Fourth Symposium on the Foundations of Software Engineering*, San Francisco, California, October 1996.
6. B. Boehm and B. Scherlis: Megaprogramming. In *Proceedings of DARPA Software Technology Conference*, April 28–30, Meridien Corporation, Arlington, VA 1992.
7. M. Brandt and F. Henglein: Coinductive axiomatization of recursive type equality and subtyping. *Fundamenta Informaticae*, 33:309–338, 1998. Invited submission to special issue featuring a selection of contributions to the 3rd Int'l Conf. on Typed Lambda Calculi and Applications (TLCA), 1997.
8. K. B. Bruce, R. Di Cosmo, and G. Longo: Provable isomorphisms of types. *Mathematical Structures in Computer Science*, 2(2):231–247, 1992.
9. J. Cai and R. Paige: Look ma, no hashing, and no arrays neither. In *Proc. 18th Annual ACM Symp. on Principles of Programming Languages, Orlando, Florida*, 143–154, 1991.

10. J. Cai and R. Paige: Using multiset discrimination to solve language processing problems without hashing. *Theoretical Computer Science*, 145(1–2)(1–2):189–228, 1995.
11. L. Cardelli and P. Wegner: On understanding types, data abstraction, and polymorphism. *ACM Computing Surveys*, 17(4):471–522, December 1985.
12. A. Cardon and M. Crochemore: Partitioning a graph in $O(|A| \log_2 |V|)$. *Theoretical Computer Science*, 19:85–98, 1982.
13. M. Coppo: Type inference with recursive type equations. In *Proceedings of FOSSACS '01, Foundations of Software Science and Computation Structures*, 184–198. Springer-Verlag (LNCS 2030), 2001.
14. R. Di Cosmo: *Isomorphisms of Types: from λ-calculus to information retrieval and language design*. Birkhäuser, 1995.
15. R. Di Cosmo, F. Pottier, and D. Rémy: Subtyping recursive types modulo associative commutative products. In *Proceedings of TLCA '05, 7th International Conference on Typed Lambda Calculus and Applications*, 179–193. Springer (LNCS 3461), 2005.
16. B. Courcelle: Fundamental properties of infinite trees. *Theoretical Computer Science*, 25(1):95–169, 1983.
17. T. Jim and J. Palsberg: Type inference in systems of recursive types with subtyping. Manuscript, 1997.
18. D. Kozen, J. Palsberg, and M. I. Schwartzbach: Efficient recursive subtyping. *Mathematical Structures in Computer Science*, 5(1):113–125, 1995. Preliminary version in Proceedings of POPL '93, Twentieth Annual Sigplan–Sigact Symposium on Principles of Programming Languages, 419–428, Charleston, South Carolina, January 1993.
19. P. Narendran, F. Pfenning, and R. Statman: On the unification problem for Cartesian closed categories. In *Proceedings, Eighth Annual IEEE Symposium on Logic in Computer Science*, 57–63. IEEE Computer Society Press, 1993.
20. OMG: The common object request broker: Architecture and specification. Technical Report, Object Management Group, Version 2.3.1, 1999.
21. R. Paige, R. Tarjan, and R. Bonic: A linear time solution to the single function coarsest partition problem. *Theoretical Computer Science*, 40:67–84, 1985.
22. R. Paige: Efficient translation of external input in a dynamically typed language. In B. Pehrson and I. Simon, editors, *Proc. 13th World Computer Congress, Vol. 1*. Elsevier Science B.V. (North-Holland), 1994.
23. R. Paige and R. Tarjan: Three partition refinement algorithms. *SIAM Journal on Computing*, 16(6):973–989, December 1987.
24. R. Paige and Z. Yang: High level reading and data structure compilation. In *Proc. 24th ACM Sigplan-Sigact Symp. on Principles of Programming Languages, Paris, France*, 456–469, ACM Press. January 1997. http://www.acm.org
25. J. Palsberg and T. Zhao: Efficient and flexible matching of recursive types. *Information and Computation*, 171:364–387, 2001. Preliminary version in Proceedings of LICS '00, Fifteenth Annual IEEE Symposium on Logic in Computer Science, 388–398, Santa Barbara, California, June 2000.
26. M. Rittri: Retrieving library identifiers via equational matching of types. In M. E. Stickel, editor, *Proceedings of the 10th International Conference on Automated Deduction*, volume 449 of *LNAI*, 603–617, Kaiserslautern, FRG. Springer Verlag. July 1990.
27. M. Rittri: Using types as search keys in function libraries. *Journal of Functional Programming*, 1(1):71–89, 1991.
28. M. Rittri: Retrieving library functions by unifying types modulo linear isomorphism. *RAIRO Theoretical Informatics and Applications*, 27(6):523–540, 1993.
29. S. V. Soloviev: The category of finite sets and cartesian closed categories. *Journal of Soviet Mathematics*, 22:1387–1400, 1983.
30. S. Thatte: Automated synthesis of interface adapters for reusable classes. In *Proceedings of POPL '94, 21st Annual Sigplan–Sigact Symposium on Principles of Programming Languages*, 174–187, 1994.

31. G. Wiederhold, P. Wegner, and S. Ceri: Towards Megaprogramming: A paradigm for component-based programming. *Communications of the ACM*, 35(11):89–99, November 1992.
32. A. M. Zaremski and J. M. Wing: Signature matching: a tool for using software libraries. *ACM Transactions on Software Engineering Methodology*, 4(2):146–170, April 1995.
33. A. M. Zaremski and J. M. Wing: Specification matching of software components. In *Proceedings of 3rd ACM SIGSOFT Symposium on the Foundation of Software Engineering*, 6–17, 1995.
34. Y. Zibin, Y. Gil, and J. Considine: Efficient algorithms for isomorphisms of simple types. In *Proceedings of POPL '03, Sigplan-Sigact Symposium on Principles of Programming Languages*, 160–171, 2003.

Aspects as Invariants

Douglas R. Smith

Kestrel Institute, 3260 Hillview Avenue, Palo Alto, California 94304, USA. smith@kestrel.edu

Summary. Aspect-Oriented Programming (AOP) offers new insights and tools for the modular development of systems with crosscutting features. Current tool support for AOP is provided mainly in the form of code-level constructs. This paper presents a way to express crosscutting features as logical invariants and then to generate the kind of code that is usually produced from manually written aspects. In order to state invariants that express crosscutting features, we often need to reify certain extra-computational values such as history or the runtime call stack. The invariant approach is illustrated by a variety of examples.

Keywords: invariants, aspect-oriented programming.

1 Introduction

Aspect-Oriented Programming (AOP) contributes to the broad goal of modular programming, with a particular focus on crosscutting concerns [2, 5]. A concern is crosscutting if its manifestation cuts across the dominant hierarchical structure of a program. A simple example is an error logging policy — the requirement to log all errors in a system in a standard format. Error logging necessitates the addition of code that is distributed throughout the system code, even though the concept is easy to state in itself. Crosscutting concerns explain a significant fraction of the code volume and interdependencies of a system. The interdependencies complicate the understanding, development, and evolution of the system.

In this paper we focus on aspects as expressed in AspectJ [9] and recent extensions of it. AspectJ aspects can be thought of as providing a kind of "whenever" construct: whenever an event of type e occurs during execution, perform action a. For example, whenever an exception is thrown, perform a logging action. The runtime events are called *join points* and descriptions of join points are called *pointcuts*. One can think of pointcuts as defining a type whose elements are joinpoints. The method-like actions to apply at joinpoints are called *advice* and the process of inserting advice at code locations in the base code that satisfy a pointcut is called *weaving*. An *aspect* is a modular treatment of a crosscutting concern that is composed of pointcuts, advice, and other Java code. See [11] for an introduction and many practical examples.

Our goal is to explore how aspects can be specified more abstractly than in current languages. We focus primarily on AspectJ, the most widely used implementation of AOP, for which there is flurry of activity to extend its expressiveness and range of applicability. This paper explores the proposition that aspects can be specified as invariants, and that weaving is invariant maintenance.

AspectJ has attracted a a wide user community partly because it is well integrated with Java: aspects are written in a classlike syntax and have a semantics that closely adheres to Java semantics. Yet despite its attractiveness to programmers, some issues arise due to the operational nature of aspects:

O. Danvy et al. (eds.), Automatic Program Development – A Tribute to Robert Paige, 247–263.

1. *Intent* — What is the intent of an aspect? The program-like nature of an aspect often obscures it's intention. It would be desirable to have a more semantic characterization of aspects, at least as an alternate description. What is the specification for which the aspect is an implementation?
2. *Pointcut Completeness* — Does a pointcut exactly characterize the intended runtime events? AspectJ pointcuts depend on the base program conforming to certain naming conventions, rather than more semantic considerations. It may sometimes be difficult to catch all relevant joinpoints in the pointcut. For security aspects in particular, it is important not to overlook a potential joinpoint.
3. *Advice Correctness* — Does an aspect's advice correctly realize its intent?

In this paper we present a generative approach to AspectJ that addresses these issues. The key idea is that an invariant captures the intent of an aspect. Aspect weaving is then the process of maintaining the invariant by generating and inserting code fragments at appropriate locations in the base code. In this approach, a pointcut specification is derived from the invariant and it characterizes the set of code points that might disrupt the invariant. For each such disruption point in the code, a specification for maintenance code is derived from the invariant. Code that is generated from the maintenance specification corresponds to statically woven advice, and could be expressed either directly in AspectJ, or by direct generation and insertion of code into the system. By expressing an aspect as an invariant we more clearly separate its intent from its implementation/realization.

The generative techniques in this paper derive from transformational work on incremental computation, in particular Bob Paige's pioneering work on Finite Differencing [15]. Finite Differencing is intended to optimize programs by replacing expensive expressions in loops by new data structures and incremental computation. It achieves this by maintaining invariants of the form $c = f(x)$ where c is a fresh variable, x is a vector of program variables, and $f(x)$ is an expensive expression (usually in a loop). Code to maintain the invariant is automatically generated and inserted at points where the dependent variables change.

In addition to addressing the three issues listed above, the invariant maintenance approach also provides novel insights and approaches to the problems of (1) context-specialization of advice, (2) aspect interference, and (3) evolution in response to base code changes. After introducing some notation, we work through a variety of examples. We conclude by revisiting the issues listed above and examining how the invariant approach provides answers and fresh insight to them.

2 Preliminaries

For purposes of this paper, a behavior of a program can be represented graphically as alternating states and actions

$$state_0 \xrightarrow{act_0} state_1 \xrightarrow{act_1} state_2 \xrightarrow{act_2} state_3 \cdots$$

or more formally as a sequence of triples of the form $\langle state_i, act_i, state_{i+1}\rangle$, where states are a mapping from variables to values, and actions are state-changing operations (i.e., program statements). The details of representing an action are not important here, although some form of concrete or abstract syntax suffices. The representation is a system-, language- and application-specific decision. The operators *nil*, written $[]$, and $append(S, a)$, written $S :: a$ for sequence S and element a, construct sequences, including behaviors. The selectors on behaviors are

$$\begin{aligned}
&preState(\langle state_0, act, state_1\rangle) = state_0\\
&action(\langle state_0, act, state_1\rangle) = act\\
&postState(\langle state_0, act, state_1\rangle) = state_1
\end{aligned}$$

If x is a state variable and s a state, then $s.x$ denotes the value of x in s. Further, in the context of the action triple $\langle state_0, act, state_1\rangle$, x will refer to the value of x in the preState, $state_0.x$, and x' refers to the value in the postState, $state_1.x$.

Two higher-order operators will be useful:
image: Written f^*S, computes the image of f over a sequence S:

$$f^*nil = nil$$
$$f^*(S :: a) = (f^*S) :: f(a)$$

filter: Written $p \triangleright S$, computes the subsequence of S comprised of elements that satisfy p:

$$p \triangleright nil = nil$$
$$p \triangleright (S :: a) = \mathit{if}\ p(a)\ \mathit{then}\ (p \triangleright S) :: a\ \mathit{else}\ p \triangleright S$$

We specify actions in a pre- and postcondition style. For example, the specification

assume: $x \geq 0$
achieve: $x' * x' = x \ \wedge\ x' \geq 0$

is satisfied by the action $x := \sqrt{x}$.

This paper presents its results in a generic imperative language framework, even though most AOP approaches target object-oriented languages and even though some of the details of static analysis and code generation are necessarily language-specific. The specifications that we work with are sufficiently abstract that we believe it will not be difficult to generate code in most current programming and modeling languages.

3 An Example

A simple example serves to introduce the technique: maintaining an error log for a system. More precisely, whenever an exception handler is invoked, we require that an entry be made in an error log.

The overall approach is to specify an invariant that gives a declarative semantical definition of our requirement, and then to generate aspectual code from it. First, what does the error log mean as a data structure? Informally, the idea is that at any point in time t, the error log records a list of all exceptions that have been raised by the program up to time t. In order to formalize this we need some way to discuss the history of the program at any point in time.

Maintaining a History Variable
The execution history of the program can be reified into the state by means of a *virtual* variable (also called a shadow or ghost variable). That is, imagine that with each action taken by the program there is a concurrent action to update a variable called *hist* that records the history up until the current state.

$$s_0 \xrightarrow[hist := hist::\langle s_0, act_0, s_1\rangle]{act_0} s_1 \xrightarrow[hist := hist::\langle s_1, act_1, s_2\rangle]{act_1} s_2 \xrightarrow[hist := hist::\langle s_2, act_2, s_3\rangle]{act_2} s_3 \cdots$$

Obviously this would be an expensive variable, but it is only needed for specification purposes, and usually only a residue of it will appear in the executable code.

Invariant
Given the history variable, $action^*hist$ represents the sequence of actions so far in the execution history. To express the invariant, we need a test for whether an action represents an error; i.e. whether it represents the invocation of an exception handler. Let $error?(act)$ be true when act is an exception, so $error? \triangleright action^*hist$ is the sequence of error actions so far in the execution history.

We can now represent the semantics of the error log:

$$\text{Invariant:}\ \ errlog \ = \ error? \triangleright action^*hist$$

i.e., in any state, the value of the variable *errlog* is the sequence of error actions that have occurred previously. The idea is that the programmer asserts this formula as a requirement on the code. It is a crosscutting requirement since exceptions can be raised anywhere in the code, regardless of its structure.

Establishing the Invariant

In order to correctly realize the invariant in the target code, we proceed by induction. The first step is to generate code to establish the invariant initially, by satisfying the following specification:

$$\begin{array}{l}\textbf{assume: } hist = [\,]\\ \textbf{achieve: } errlog \;=\; error? \triangleright action^{*} hist\end{array}$$

The postcondition can be simplified as follows:

$$\begin{array}{ll} & errlog \;=\; error? \triangleright action^{*} hist \\ \equiv & \quad \{\text{using the definition of } hist\} \\ & errlog \;=\; error? \triangleright action^{*}[\,] \\ \equiv & \quad \{\text{simplifying}\} \\ & errlog \;=\; [\,] \end{array}$$

which is satisfied by the initialization code: $errlog := [\,]$.

More generally, when the invariant contains reified variables, the following scheme specifies code for establishing the invariant $I(x)$:

$$\begin{array}{l}\textbf{assume}: hist = [\,] \;\wedge\; \ldots \text{ initial values of other reified variables } \ldots \\ \qquad\qquad \wedge \ldots \text{ base code preconditions } \ldots \\ \textbf{achieve}: I(x)\end{array}$$

In Section 4.3, we give an example that does not mention reified variables. It uses a slightly different scheme for specifying the establishment of the invariant.

Specifying Disruptive Code and Deriving the Pointcut

To proceed with the inductive argument, we must maintain the invariant for all actions of the target code. Since most actions of the target code have no effect on the invariant, it is useful to focus on those actions that might disrupt the invariant. We will then generate code for maintaining the invariant in parallel with the disruptive action.

The set of all code points that might disrupt the invariant corresponds to the AspectJ concept of code points that satisfy a pointcut. The maintenance code that we generate for each such disruptive code point corresponds to a point-specific instance of the advice of an aspect. An exact characterization of the disruption points is given by

$$I(x) \neq I(x'). \tag{1}$$

That is, any action that satisfies (1) as a postcondition is a disruption point. More generally, any action that satisfies a necessary condition on (1) is a potential disruption point. In our example, we set up the following inference task:

$$\begin{array}{l}\textbf{assume}: \; errlog \;=\; error? \triangleright action^{*} hist \wedge hist' = hist :: \langle \_, act, \_\rangle \;\wedge\; errlog' = errlog \\ \textbf{simplify}: (errlog = error? \triangleright action^{*} hist) \;\neq\; (errlog' = error? \triangleright action^{*} hist')\end{array}$$

In words, we assume that the invariant holds before an arbitrary action *act*, and that the *hist* variable is updated in parallel with *act*. Moreover, we add in a frame axiom that asserts that *act* does not change *errlog* since it is a fresh variable introduced by the invariant.

We calculate a pointcut specification as follows:

$$\begin{array}{ll} & (errlog \;=\; error? \triangleright action^{*} hist) \;\neq\; (errlog' \;=\; error? \triangleright action^{*} hist') \\ \equiv & \quad \{\text{ using the frame axiom and simplifying }\} \\ & error? \triangleright action^{*} hist \;\neq\; error? \triangleright action^{*} hist' \\ \equiv & \quad \{\text{ using the definition of } hist' \,\} \\ & error? \triangleright action^{*} hist \;\neq\; error? \triangleright action^{*}(hist :: \langle \_, act, \_\rangle) \\ \equiv & \quad \{\text{ distributing } action^{*} \text{ over } :: \,\} \end{array}$$

$$\begin{aligned}
& \mathit{error?} \rhd \mathit{action}^* \mathit{hist} \ \neq \ \mathit{error?} \rhd ((\mathit{action}\ ^* \mathit{hist}) :: \mathit{act}) \\
\equiv \quad & \{ \text{ distributing } \mathit{error?} \rhd \text{ over } :: \} \\
& \mathit{error?} \rhd \mathit{action}^* \mathit{hist} \ \neq \ (\mathit{if}\ \neg \mathit{error?}(\mathit{act}) \ \ \mathit{then} \ \ \mathit{error?} \rhd \mathit{action}^* \mathit{hist} \\
& \qquad\qquad\qquad\qquad\qquad\quad \mathit{else}\ (\mathit{error?} \rhd \mathit{action}\ ^* \mathit{hist}) :: \mathit{act}) \\
\equiv \quad & \{ \text{ distributing the conditional outward } \} \\
& \quad \mathit{if}\ \neg \mathit{error?}(\mathit{act}) \ \ \mathit{then} \ \ \mathit{error?} \rhd \mathit{action}^* \mathit{hist} \ \neq \ \mathit{error?} \rhd \mathit{action}^* \mathit{hist} \\
& \quad \mathit{else}\ \mathit{error?} \rhd \mathit{action}^* \mathit{hist} \ \neq \ (\mathit{error?} \rhd \mathit{action}\ ^* \mathit{hist}) :: \mathit{act} \\
\equiv \quad & \{ \text{ simplifying } \} \\
& \mathit{if}\ \neg \mathit{error?}(\mathit{act})\ \mathit{then}\ \mathit{false}\ \mathit{else}\ \mathit{true} \\
\equiv \quad & \{ \text{ simplifying } \} \\
& \mathit{error?}(\mathit{act}).
\end{aligned}$$

A static analyzer would scan the code (i.e., the abstract syntax representation of the code) looking for all actions that satisfy this derived pointcut. More generally, the task to infer a pointcut specification is given by an instance of the following scheme:

$$\begin{aligned}
& \textbf{assume} : I(x) \ \ \wedge \ \ \mathit{hist}' = \mathit{hist} :: \langle _, \mathit{act}, _ \rangle \\
& \qquad\qquad \wedge \ldots \text{ updates of other reified variables } \ldots \ \wedge \ \ldots \text{ relevant frame conditions } \ldots \\
& \textbf{simplify} : I(x) \ \neq \ I(x')
\end{aligned}$$

The simplified result will typically contain a mixture of terms, some of which can be evaluated statically (i.e. on the abstract syntax of the source code) and some of which must be evaluated dynamically (i.e. on the runtime data). Since we only need a necessary condition on (1), we can weaken the derived pointcut specification by discarding those subformulas that can only be evaluated dynamically. This weakening process means that the pointcut specification may allow false positives, but, as will be seen in later examples, the generated maintenance code incorporates the relevant semantics of the discarded dynamic tests.

Specification and Derivation of Maintenance Code

To complete the induction, we must find each potentially disruptive action (using the derived pointcut specification) and then generate maintenance code to reestablish the invariant in parallel with it. Suppose that *act* is an action such that *error?*(*act*). In order to preserve the invariant, we need to perform a maintenance action that satisfies

$$\begin{aligned}
& \textbf{assume} : \mathit{errlog} \ = \ \mathit{error?} \rhd \mathit{action}^* \mathit{hist} \ \wedge \ \mathit{error?}(\mathit{act}) \ \wedge \ \mathit{hist}' = \mathit{hist} :: \langle _, \mathit{act}, _ \rangle \\
& \textbf{achieve} : \mathit{errlog}' \ = \ \mathit{error?} \rhd \mathit{action}^* \mathit{hist}'
\end{aligned}$$

The postcondition can be simplified as follows:

$$\begin{aligned}
& \mathit{errlog}' \ = \ \mathit{error?} \rhd \mathit{action}^* \mathit{hist}' \\
\equiv \quad & \{ \text{ using the definition of } \mathit{hist} \} \\
& \mathit{errlog}' \ = \ \mathit{error?} \rhd \mathit{action}\ ^*(\mathit{hist} :: \langle _, \mathit{act}, _ \rangle) \\
\equiv \quad & \{ \text{ distributing } \mathit{action}^* \text{ over } :: \} \\
& \mathit{errlog}' \ = \ \mathit{error?} \rhd ((\mathit{action}\ ^* \mathit{hist}) :: \mathit{act}) \\
\equiv \quad & \{ \text{ distributing } \mathit{error?} \rhd \text{ over } ::, \text{ using assumption that } \mathit{error?}(\mathit{act}) \ \} \\
& \mathit{errlog}' \ = \ (\mathit{error?} \rhd \mathit{action}^* \mathit{hist}) :: \mathit{act} \\
\equiv \quad & \{ \text{ using the precondition/invariant inductively } \} \\
& \mathit{errlog}' \ = \ \mathit{errlog} :: \mathit{act}
\end{aligned}$$

which is satisfied by the simple update $\mathit{errlog} \ := \ \mathit{errlog} :: \mathit{act}$. This maintenance action may be performed in parallel with *act*. More generally, suppose that static analysis has identified an action *act* as potentially disruptive of invariant $I(x)$. If *act* satisfies the specification

$$\begin{aligned}
& \textbf{assume} : P(x) \\
& \textbf{achieve} : Q(x, x')
\end{aligned}$$

then the maintenance code *maint* can be specified as

$$\begin{aligned}
& \textbf{assume} : P(x) \ \wedge \ I(x) \ \wedge \ \mathit{hist}' = \mathit{hist} :: \langle s_0, \mathit{act}, s_1 \rangle \ \wedge \ldots \text{ updates to other reified vars } \ldots \\
& \textbf{achieve} : Q(x, x') \ \wedge \ I(x')
\end{aligned}$$

In this schematic specification we compose the aspect with the base code by means of a conjunction. Note that this specification preserves the effect of *act* while additionally reestablishing the invariant I. If it is inconsistent to achieve both, then the specification is unrealizable.

The generated code for *maint* may take the form of a parallel composition $act||update$ of the actions *act* and *update*, or it may take a sequential form. Besides conceptual clarity, an advantage to treating the maintenance action as parallel to the disruptive action is that the invariant is always observed to hold in all states. Most work on programming with invariants (e.g. [3, 14]), as well as AspectJ, sequentializes the maintenance action. If the maintenance action is sequentialized, say for purposes of optimization, the generator needs to take care that no external process that depends on the invariant could observe the state between the two actions and notice that the invariant is (temporarily) violated. One technique for assuring that no observation of the intermittent violation can be made is to lock the relevant variables while the maintenance is being performed.

4 More Examples

4.1 Procedure Calls and Dynamic Context

This exercise treats procedure calls and the reification of dynamic procedure call context.

• PROBLEM: Maintain a global that flags when a Sort procedure is executing.

• REIFICATION: This problem requires that we reify and maintain the call stack, analogously to the way that history is maintained in *hist*. To reify the call stack, it is necessary to elaborate the model of behavior presented in Section 2. A call to procedure P, $s_0 \xrightarrow{x:=P(x)} s_1$, can be elaborated to a subbehavior

$$s_0 \xrightarrow{eval\ args} s_{00} \xrightarrow[parms := argvals]{enter\ P} s_{01} \xrightarrow{execute\ P} s_{02} \xrightarrow[x := result]{exit\ P} s_1$$

With this elaboration, it is straightforward to maintain a call stack variable cs with operators *InitStack*, *push*, and *pop*:

$$s_0 \xrightarrow{eval\ args} s_{00} \xrightarrow[cs := push(cs, \langle P, argvals\rangle)]{enter\ P} s_{01} \xrightarrow{execute\ P} s_{02} \xrightarrow[cs := pop(cs)]{exit\ P} s_1$$

Procedural languages abstract away these details so a static analyzer must take this finer-grain model into account when appropriate.

• DOMAIN THEORY: The boolean variable *sorting*? is to be true exactly when a call to *Sort* is on the call stack cs. In the invariant, we use a predicate $pcall?(act, f)$ that is true exactly when action *act* is a procedure call to f.

• INVARIANT: $sorting? = \exists(call)(call \in cs \ \wedge \ pcall?(call, Sort))$
Incrementally maintaining a boolean value is difficult, and a standard technique is to transform a quantified expression into an equivalent set-theoretic form that is easier to maintain [15]:

$$sorting? = size(\{call \mid call \in cs \ \wedge \ pcall?(call, Sort)\}) > 0$$

and introduce a second invariant:

$$sortcnt = size(\{call \mid call \in cs \ \wedge \ pcall?(call, Sort)\})$$

By maintaining *sortcnt*, we can replace *sorting*? by $sortcnt > 0$ everywhere it occurs.

• ESTABLISHING THE INVARIANT: The code to establish the *sortcnt* invariant is specified as

$$\begin{aligned} &\textbf{assume}: hist = [\,] \ \wedge \ cs = InitStack() \\ &\textbf{achieve}: sortcnt \ = \ size(\{call \mid call \in cs \ \wedge \ pcall?(call, Sort)\}) \end{aligned}$$

The postcondition can be simplified as follows:

$$\begin{aligned} & sortcnt \ = \ size(\{call \mid call \in cs \ \wedge \ pcall?(call, Sort)\}) \\ \equiv \quad & \{\text{using the assumption about } cs\} \end{aligned}$$

$$
\begin{array}{ll}
& sortcnt \;=\; size(\{call \mid call \in InitStack() \;\wedge\; pcall?(call, Sort)\}) \\
\equiv & \quad \{\text{simplifying}\} \\
& sortcnt \;=\; 0
\end{array}
$$

which is satisfied by the initialization code: $sortcnt \;:=\; 0.$

• Disruptive Actions: The following task to infer a pointcut specification assumes a frame axiom that characterizes the effect of an arbitrary base code action on the call stack variable – it either effects a *push*, a *pop*, or has no effect.

$$
\begin{array}{l}
\textbf{assume: } sortcnt \;=\; size(\{call \mid call \in cs \wedge pcall?(call, Sort)\}) \\
\qquad \wedge\; hist' = hist :: \langle \_, act, \_\rangle \;\wedge\; (cs' = push(cs, \langle P, argvals\rangle) \\
\qquad\qquad\qquad\qquad \vee\; (cs' = pop(cs) \;\wedge\; top(cs) = \langle P, argvals\rangle) \\
\qquad\qquad\qquad\qquad \vee\; cs' = cs) \\
\qquad \wedge\; sortcnt' = sortcnt \\
\textbf{simplify: } (sortcnt \;=\; size(\{call \mid call \in cs \;\wedge\; pcall?(call, Sort)\})) \\
\qquad\qquad \neq\; (sortcnt' \;=\; size(\{call \mid call \in cs' \;\wedge\; pcall?(call, Sort)\}))
\end{array}
$$

We calculate

$$
\begin{array}{ll}
& (sortcnt \;=\; size(\{call \mid call \in cs \;\wedge\; pcall?(call, Sort)\})) \\
& \neq\; (sortcnt' \;=\; size(\{call \mid call \in cs' \;\wedge\; pcall?(call, Sort)\})) \\
\equiv & \quad \{\text{using the frame axiom for } sortcnt \text{ and simplifying}\} \\
& size(\{call \mid call \in cs \;\wedge\; pcall?(call, Sort)\}) \\
& \neq\; size(\{call \mid call \in cs' \;\wedge\; pcall?(call, Sort)\})
\end{array}
$$

Using the disjunctive frame axiom on the call stack variable, we can proceed by cases:

$$
\begin{array}{ll}
\equiv & \quad \{\text{ Case 1: assume } cs' = push(cs, \langle P, argvals\rangle) \;\} \\
& size(\{call \mid call \in cs \;\wedge\; pcall?(call, Sort)\}) \\
& \neq\; size(\{call \mid call \in push(cs, \langle P, argvals\rangle) \;\wedge\; pcall?(call, Sort)\}) \\
\equiv & \quad \{\text{ eliding the lhs and distributing call stack membership over } push \;\} \\
& \ldots \neq\; size(\{call \mid (call \in cs \;\vee\; call = \langle P, argvals\rangle) \;\wedge\; pcall?(call, Sort)\}) \\
\equiv & \quad \{\text{ distributing }\} \\
& \ldots \neq\; size(\{call \mid call \in cs \;\wedge\; pcall?(call, Sort)\}) \\
& \qquad + size(\{call \mid call = \langle P, argvals\rangle \;\wedge\; pcall?(call, Sort)\}) \\
\equiv & \quad \{\text{ simplifying }\} \\
& \ldots \neq\; size(\{call \mid call \in cs \;\wedge\; pcall?(call, Sort)\}) \\
& \qquad + \mathit{if}\; pcall?(\langle P, argvals\rangle, Sort)\; \mathit{then}\; 1\; \mathit{else}\; 0 \\
\equiv & \quad \{\text{ distributing the conditional outwards }\} \\
& \mathit{if}\; pcall?(\langle P, argvals\rangle, Sort)\; \mathit{then}\; \mathit{true}\; \mathit{else}\; \mathit{false} \\
\equiv & \quad \{\text{ simplifying }\} \\
& pcall?(\langle P, argvals\rangle, Sort).
\end{array}
$$

The derived pointcut specification in this case is

$$cs' = push(cs, \langle P, argvals\rangle) \;\wedge\; pcall?(\langle P, argvals\rangle, Sort).$$

Continuing with the case analysis, we calculate

$$
\begin{array}{ll}
\equiv & \quad \{\text{ Case 2: assume } (cs' = pop(cs) \;\wedge\; top(cs) = \langle P, argvals\rangle) \\
& \qquad\qquad \text{which implies } (cs = push(cs', \langle P, argvals\rangle) \;\} \\
& size(\{call \mid call \in push(cs', \langle P, argvals\rangle) \;\wedge\; pcall?(call, Sort)\}) \\
& \neq\; size(\{call \mid call \in cs' \;\wedge\; pcall?(call, Sort)\}) \\
\equiv & \quad \{\text{ using similar reasoning to previous case}\} \\
& pcall?(\langle P, argvals\rangle, Sort).
\end{array}
$$

The derived semantic pointcut in this case is

$$cs' = pop(cs) \;\wedge\; top(cs) = \langle P, argvals\rangle \;\wedge\; pcall?(\langle P, argvals\rangle, Sort).$$

And the final step in the case analysis is

$\equiv$ { Case 3: assume $cs' = cs$ }
$size(\{call \mid call \in cs \ \wedge \ pcall?(call, Sort)\})$
$\neq \ size(\{call \mid call \in cs' \ \wedge \ pcall?(call, Sort)\})$
$\equiv$ { using the assumption, and simplifying}
false

Combining the case assumptions with their derived pointcut specifications, we obtain

$$\begin{aligned} & cs' = push(cs, \langle P, argvals\rangle) \ \wedge \ pcall?(\langle P, argvals\rangle, Sort) \\ \vee \ & cs' = pop(cs) \ \wedge \ top(cs) = \langle P, argvals\rangle \ \wedge \ pcall?(\langle P, argvals\rangle, Sort) \\ \vee \ & cs' = cs \ \wedge \ false \end{aligned}$$

or simply

$$\begin{aligned} & cs' = push(cs, \langle P, argvals\rangle) \ \wedge \ pcall?(\langle P, argvals\rangle, Sort) \\ \vee \ & cs' = pop(cs) \ \wedge \ top(cs) = \langle P, argvals\rangle \ \wedge \ pcall?(\langle P, argvals\rangle, Sort). \end{aligned}$$

which specifies entrances and exits of calls to *Sort* respectively.

• SPECIFICATION AND DERIVATION OF MAINTENANCE CODE: The pointcut specification gives rise to two cases: *push* and *pop* operations. For a push operation of the form

$cs := push(cs, \langle Sort, \_\rangle)$

the maintenance specification is

$$\begin{aligned} \textbf{assume}: \ & sortcnt \ = \ size(\{call \mid call \in cs \ \wedge \ pcall?(call, Sort)\}) \\ \textbf{achieve}: \ & cs' = push(cs, \langle Sort, argvals\rangle) \\ & \wedge \ sortcnt' \ = \ size(\{call \mid call \in cs' \ \wedge \ pcall?(call, Sort)\}) \end{aligned}$$

which an easy calculation shows to be satisfied by the concurrent assignment

$$cs \ := \ push(cs, \langle Sort, argvals\rangle) \ || \ sortcnt \ := \ sortcnt + 1$$

on entrance to procedure *Sort*. For a pop operation of the form $cs \ := \ pop(cs)$ where $top(cs) = \langle Sort, \_\rangle$, the maintenance specification is

$$\begin{aligned} \textbf{assume}: \ & cs \neq initStack() \ \wedge \ \ top(cs) = \langle Sort, \_\rangle \\ & \wedge \ sortcnt \ = \ size(\{call \mid call \in cs \ \wedge \ pcall?(call, Sort)\}) \\ \textbf{achieve}: \ & cs' = pop(cs) \ \wedge \ \ sortcnt' \ = \ size(\{call \mid call \in cs' \ \wedge \ pcall?(call, Sort)\}) \end{aligned}$$

which is satisfied by the concurrent assignment

$$cs \ := \ pop(cs) \ || \ sortcnt \ := \ sortcnt - 1$$

The concurrent formulation of the maintenance code can be implemented by sequentializing the *sortcnt* updates into the body of the procedure, just after entry and just before return.

4.2 Counting Swaps in a Sort Routine

This problem builds on the previous problem and illustrates the execution of advice within dynamic contexts, a key feature of AspectJ.

• PROBLEM: Count the number of calls to a *swap* procedure that are invoked during the execution of a sort procedure *Sort*.

• DOMAIN THEORY: As in the previous problem, let *cs* be the reified call stack, with operators *InitStack*, *push*, and *pop*.

• INVARIANT: The invariant uses a sequence comprehension notation, so that *swpcnt* is the length of a sequence of actions satisfying various properties. Also, recall that the notation $s_0.cs$ refers to the value of variable *cs* in state s_0.

$$\begin{aligned} swpcnt = length([\ & act_0 \mid \langle st_0, act_0, st_1\rangle \in hist \ \wedge \ pcall?(act_0, swap) \\ & \wedge \ \exists(pc)(pc \in st_0.cs \ \wedge \ pcall?(pc, Sort))]) \end{aligned}$$

or, more simply, using the invariant from the previous example

$$swpcnt = length([\ act_0 \mid \langle st_0, act_0, st_1\rangle \in hist \ \wedge \ pcall?(act_0, swap) \ \wedge \ st_0.sortcnt > 0 \ ])$$

- Establishing the invariant: The code to establish the *sortcnt* invariant is specified as

$$\textbf{assume} : hist = [\,] \ \wedge \ cs = initStack()$$
$$\textbf{achieve} : swpcnt = length([\ act_0 \mid \langle st_0, act_0, st_1\rangle \in hist \ \wedge pcall?(act_0, swap) \ \wedge \ st_0.sortcnt > 0\ ]).$$

The postcondition can be simplified as follows:

$$\begin{array}{ll} & swpcnt \ = \ length([\ act_0 \mid \langle st_0, act_0, st_1\rangle \in hist \ \wedge \ \ldots]) \\ \equiv & \{ \text{ using the assumption about } hist \ \} \\ & swpcnt \ = \ length([\ act_0 \mid \langle st_0, act_0, st_1\rangle \in [\,] \ \wedge \ \ldots]) \\ \equiv & \{ \text{ simplifying } \} \\ & swpcnt \ = \ 0 \end{array}$$

which is satisfied by the initialization code $swpcnt \ := \ 0$.

- Disruptive Actions: The inference task to infer a pointcut is specified as follows

$$\begin{array}{l} \textbf{assume} : swpcnt = length([\ act \mid \langle st_0, act_0, st_1\rangle \in hist \ \wedge \ pcall?(act_0, swap) \\ \qquad\qquad\qquad\qquad\qquad\qquad \wedge \ st_0.sortcnt > 0]) \\ \qquad\quad \wedge \ hist' = hist :: \langle s_0, act, s_1\rangle \ \wedge \ swpcnt' = swpcnt \\ \textbf{simplify} : (swpcnt = length([\ act_0 \mid \langle st_0, act_0, st_1\rangle \in hist \ \wedge \ pcall?(act_0, swap) \\ \qquad\qquad\qquad\qquad\qquad\qquad \wedge \ st_0.sortcnt > 0])) \\ \qquad\quad \neq \ (swpcnt' = length([\ act_0 \mid \langle st_0, act_0, st_1\rangle \in hist' \ \wedge \ pcall?(act_0, swap) \\ \qquad\qquad\qquad\qquad\qquad\qquad \wedge \ st_0.sortcnt > 0])) \end{array}$$

We calculate a pointcut specification as follows:

$$\begin{array}{ll} & (swpcnt = length([\ act_0 \mid \langle st_0, act_0, st_1\rangle \in hist \ \wedge \ pcall?(act_0, swap) \\ & \qquad\qquad \wedge \ st_0.sortcnt > 0])) \\ & \neq \ (swpcnt' = length([\ act_0 \mid \langle st_0, act_0, st_1\rangle \in hist' \ \wedge \ pcall?(act_0, swap) \\ & \qquad\qquad \wedge \ st_0.sortcnt > 0])) \\ \equiv & \{ \text{ using the frame axiom, and simplifying } \} \\ & \quad length([\ act_0 \mid \langle st_0, act_0, st_1\rangle \in hist \ \wedge \ pcall?(act_0, swap) \ \wedge \ st_0.sortcnt > 0]) \\ & \neq \ length([\ act_0 \mid \langle st_0, act_0, st_1\rangle \in hist' \ \wedge \ pcall?(act_0, swap) \ \wedge \ st_0.sortcnt > 0]) \\ \equiv & \{ \text{ eliding the left-hand side and using the assumption about } hist' \ \} \\ & \ldots \neq \ length([\ act_0 \mid \langle st_0, act_0, st_1\rangle \in hist :: \langle s_0, act, s_1\rangle \ \wedge \ pcall?(act_0, swap) \\ & \qquad\qquad \wedge \ st_0.sortcnt > 0]) \\ \equiv & \{ \text{ distributing } \} \\ & \ldots \neq \ length([\ act_0 \mid (\langle st_0, act_0, st_1\rangle \in hist \ \vee \ \langle st_0, act_0, st_1\rangle = \langle s_0, act, s_1\rangle) \\ & \qquad\qquad \wedge \ pcall?(act_0, swap) \ \wedge \ st_0.sortcnt > 0]) \\ \equiv & \{ \text{ distributing the disjunction outwards } \} \\ & \ldots \neq \ length([\ act_0 \mid \langle st_0, act_0, st_1\rangle \in hist \ \wedge \ pcall?(act_0, swap) \ \wedge \ st_0.sortcnt > 0]) \\ & \quad + length([\ act_0 \mid \langle st_0, act_0, st_1\rangle = \langle s_0, act, s_1\rangle \ \wedge \ pcall?(act_0, swap) \ \wedge \ st_0.sortcnt > 0]) \\ \equiv & \{ \text{ distributing the equality in the second addend } \} \\ & \ldots \neq \ length([\ act_0 \mid \langle st_0, act_0, st_1\rangle \in hist \ \wedge \ pcall?(act_0, swap) \ \wedge \ st_0.sortcnt > 0]) \\ & \quad + \ length([\ act \mid pcall?(act, swap) \ \wedge \ s_0.sortcnt > 0]) \\ \equiv & \{ \text{ simplifying } \} \\ & \ldots \neq \ length([\ act_0 \mid \langle st_0, act_0, st_1\rangle \in hist \ \wedge \ pcall?(act_0, swap) \ \wedge \ st_0.sortcnt > 0]) \\ & \quad + \ if \ pcall?(act, swap) \ \wedge \ s_0.sortcnt > 0 \ then \ 1 \ else \ 0 \\ \equiv & \{ \text{ distributing the conditional outwards and simplifying } \} \\ & if \ pcall?(act, swap) \ \wedge \ s_0.sortcnt > 0 \ then \ true \ else \ false \\ \equiv & \{ \text{ simplifying } \} \\ & pcall?(act, swap) \ \wedge \ s_0.sortcnt > 0. \end{array}$$

The derived pointcut specification is $pcall?(act, swap) \ \wedge \ sortcnt > 0$.

Note that the second conjunct is not statically determinable in general, so we weaken the pointcut to *pcall?*(*act*, *swap*) (recall that we only need a necessary condition on the disruption of the invariant). When we derive the maintenance code below, the extra condition $sortcnt > 0$ will show up as a runtime test.

• SPECIFICATION AND DERIVATION OF MAINTENANCE CODE: Any call to *swap* is a potentially disruptive action. The following specification jointly achieves the effect of *act* and maintains the invariant:

$$\begin{aligned}
\textbf{assume}\text{: } & \mathit{precondition}(\mathit{act}) \wedge \mathit{swpcnt} = \mathit{length}([\, \mathit{act}_0 \mid \langle \mathit{st}_0, \mathit{act}_0, \mathit{st}_1\rangle \in \mathit{hist} \\
& \qquad\qquad \wedge\, \mathit{pcall?}(\mathit{act}_0, \mathit{swap}) \wedge \mathit{st}_0.\mathit{sortcnt} > 0]) \\
& \wedge\, \mathit{hist}' = \mathit{hist} :: \langle s_0, \mathit{act}, s_1\rangle \wedge \mathit{pcall?}(\mathit{act}, \mathit{swap}) \\
\textbf{achieve}\text{: } & \mathit{postcondition}(\mathit{act}) \wedge \mathit{swpcnt}' = \mathit{length}([\, \mathit{act}_0 \mid \langle \mathit{st}_0, \mathit{act}_0, \mathit{st}_1\rangle \in \mathit{hist}' \\
& \qquad\qquad \wedge\, \mathit{pcall?}(\mathit{act}_0, \mathit{swap}) \wedge s_0.\mathit{sortcnt} > 0])
\end{aligned}$$

The second conjunct of the postcondition can be simplified as follows:

$$\begin{aligned}
& \mathit{swpcnt}' = \mathit{length}([\, \mathit{act}_0 \mid \langle \mathit{st}_0, \mathit{act}_0, \mathit{st}_1\rangle \in \mathit{hist}' \\
& \qquad\qquad \wedge\, \mathit{pcall?}(\mathit{act}_0, \mathit{swap}) \wedge \mathit{st}_0.\mathit{sortcnt} > 0]) \\
\equiv\; & \{ \text{ using the assumption about } \mathit{hist}' \} \\
& \mathit{swpcnt}' = \mathit{length}([\, \mathit{act}_0 \mid \langle \mathit{st}_0, \mathit{act}_0, \mathit{st}_1\rangle \in \mathit{hist} :: \langle s_0, \mathit{act}, s_1\rangle \\
& \qquad\qquad \wedge\, \mathit{pcall?}(\mathit{act}_0, \mathit{swap}) \wedge \mathit{st}_0.\mathit{sortcnt} > 0]) \\
\equiv\; & \{ \text{ distributing } \in \text{ over } :: \} \\
& \mathit{swpcnt}' = \mathit{length}([\, \mathit{act}_0 \mid \mathit{pcall?}(\mathit{act}_0, \mathit{swap}) \wedge \mathit{st}_0.\mathit{sortcnt} > 0 \\
& \qquad \wedge\, (\langle \mathit{st}_0, \mathit{act}_0, \mathit{st}_1\rangle \in \mathit{hist} \vee \langle \mathit{st}_0, \mathit{act}_0, \mathit{st}_1\rangle = \langle s_0, \mathit{act}, s_1\rangle)]) \\
\equiv\; & \{ \text{ driving } \vee \text{ outward through } \wedge \text{, sequence-former, and } \mathit{length} \} \\
& \mathit{swpcnt}' = \mathit{length}([\, \mathit{act}_0 \mid \langle \mathit{st}_0, \mathit{act}_0, \mathit{st}_1\rangle \in \mathit{hist} \\
& \qquad\qquad \wedge\, \mathit{pcall?}(\mathit{act}_0, \mathit{swap}) \wedge \mathit{ts}_0.\mathit{sortcnt} > 0\,]) \\
& \qquad + \mathit{length}([\, \mathit{act}_0 \mid \langle \mathit{st}_0, \mathit{act}_0, \mathit{st}_1\rangle = \langle s_0, \mathit{act}, s_1\rangle \\
& \qquad\qquad \wedge\, \mathit{pcall?}(\mathit{act}_0, \mathit{swap}) \wedge \mathit{st}_0.\mathit{sortcnt} > 0\,]) \\
\equiv\; & \{ \text{ using assumption about } \mathit{swpcnt} \text{, and distributing equality in sequence-former } \} \\
& \mathit{swpcnt}' = \mathit{swpcnt} + \mathit{length}([\, \mathit{act} \mid \mathit{pcall?}(\mathit{act}, \mathit{swap}) \wedge s_0.\mathit{sortcnt} > 0\,]) \\
\equiv\; & \{ \text{ using assumption about } \mathit{act} \text{, and simplifying} \} \\
& \mathit{swpcnt}' = \mathit{swpcnt} + \mathit{length}([\, \mathit{act} \mid s_0.\mathit{sortcnt} > 0\,]) \\
\equiv\; & \{ \text{ using independence of } \mathit{act} \text{ from the sequence-former predicate } \} \\
& \mathit{swpcnt}' = \mathit{swpcnt} + (\mathit{if}\ s_0.\mathit{sortcnt} > 0\ \mathit{then}\ \mathit{length}([\, \mathit{act} \mid \mathit{true}]) \\
& \qquad\qquad\qquad\qquad \mathit{else}\ \mathit{length}([\, \mathit{act} \mid \mathit{false}]) \\
\equiv\; & \{ \text{ simplifying } \} \\
& \mathit{swpcnt}' = \mathit{swpcnt} + (\mathit{if}\ \mathit{st}_0.\mathit{sortcnt} > 0\ \mathit{then}\ 1\ \mathit{else}\ 0) \\
\equiv\; & \{ \text{ pulling the conditional outward and simplifying } \} \\
& \mathit{if}\ \mathit{st}_0.\mathit{sortcnt} > 0\ \mathit{then}\ \mathit{swpcnt}' = \mathit{swpcnt} + 1\ \mathit{else}\ \mathit{swpcnt}' = \mathit{swpcnt}.
\end{aligned}$$

Consequently, the maintenance specification is satisfied by the parallel statement

$$\mathit{act} \,||\, \mathit{if}\ \mathit{sortcnt} > 0\ \mathit{then}\ \mathit{swpcnt} := \mathit{swpcnt} + 1.$$

Note that a residue of the invariant appears in the maintenance code. The test $\mathit{sortcnt} > 0$ could not be decided statically, so it falls through as a runtime test.

4.3 Maintaining the Length of a List

This example does not require the reification of an extra-computational entity. It is presented as an example of our technique that cannot currently be treated in AspectJ because it is handled at the assignment level, rather than at the method-call level.

• PROBLEM: Maintain the length of a list ℓ.

• DOMAIN THEORY: The list data type includes constructors (*nil*, *append*, *concat*), selectors (*first*, *rest*), *deleteElt*, as well as a length function and other operators.

• INVARIANT: $\mathit{llength} = \mathit{length}(\ell)$

• DISRUPTIVE ACTIONS: The derivation of the pointcut specification results in $\ell \neq \ell'$; i.e., any action that changes ℓ may disrupt the invariant. Static analysis looks for any action that changes ℓ, such as assignments to ℓ.

• ESTABLISHING THE INVARIANT: The invariants that were treated in previous examples all referred to reified variables, and consequently, the code to establish them belonged to the outermost program initialization phase. This example refers only to program variables (ℓ in particular), so its initialization code belongs to the scope of those variables. For this example, with one dependent variable, it is easy for static analysis to locate its initialization action, say *linit*. Then the code to establishment of the invariant initially is specified as

assume: $true$
achieve: $llength \ = \ length(\ell) \ \wedge \ linit$

That is, we wish to establish the invariant concurrently with the initialization of its dependent variable. To be concrete, if the static analyzer finds the initialization code $\ell \ := \ nil$ then the appropriate instance of the above scheme results in the concurrent code

$\ell \ := \ nil \,||\, llength \ := \ 0$

which establishes the invariant.

• SPECIFICATION AND DERIVATION OF MAINTENANCE CODE: For each potentially disruptive action *act*, we generate a specification for an action that jointly achieves the effect of *act* and maintains the invariant. For example, suppose that the pointcut specification matches an assignment $\ell \ := \ \ell :: elt$, an action *act* that appends an element onto ℓ results in the maintenance specification

assume: $llength \ = \ length(\ell)$
achieve: $\ell' = \ell :: elt \ \wedge \ llength' \ = \ length(\ell')$

from which one can easily calculate the satisfying concurrent assignment

$\ell \ := \ \ell :: elt \,||\, llength \ := \ llength + 1$

For other actions that change ℓ, we create the corresponding maintenance code specifications, and then generate code. Note that each change to ℓ can result in code that is completely different from other maintenance actions for ℓ.

4.4 Model-View Consistency Maintenance

The classic model-view problem is to maintain consistency between a data model and various graphical views when the program and/or user can change any of them. That is, whenever the program changes the data model, the graphical views should be updated to maintain consistency, and conversely, if the user changes one graphical view interactively, then the data model and the other views must be updated to reflect the change.

Note that this example has no newly introduced variables as in previous examples. The nature of the problem is to enforce a new constraint on existing variables.

• PROBLEM: Maintain consistency between a data model *md*:*Model* and a graphical view *vw*: *View*. Generalizing the example to multiple models with multiple views is discussed at the end of this section.

• DOMAIN THEORY: Assume that the data content of *md*:*Model* is given by an attribute $mValue : Model \rightarrow Value$ for some type *Value*, and, similarly, the data content of a view is given by $vValue : View \rightarrow Value$ for *View*. Although equality is used between these values to express consistency, in practical situations, a more complex predicate is needed.

• INVARIANT: $vw.vValue = md.mValue$

• DISRUPTIVE ACTIONS: We assume that the only changes that can be made to a model are via a call to its *update* method; similarly for views. Formally, the third assumption below is a frame axiom asserting that a system action either (i) leaves a model unchanged, or (ii) is a call to the model *update* method. The fourth assumption states the analogous constraint on views. The task to infer a pointcut is specified as follows:

assume: $s_0.vw.vValue = s_0.md.mValue \ \wedge \ hist' = hist :: \langle s_0, act, s_1 \rangle$
$\wedge \ (s_0.md = s_1.md \ \vee \ pcall?(act, md.update))$
$\wedge \ (s_0.vw = s_1.vw \ \vee \ pcall?(act, vw.update))$
simplify: $(s_0.vw.vValue = s_0.md.mValue) \ \neq \ (s_1.vw.vValue = s_1.md.mValue)$

We calculate a pointcut specification as follows:

$(s_0.vw.vValue = s_0.md.mValue) \neq (s_1.vw.vValue = s_1.md.mValue)$

$\equiv$ { using the assumption that the lhs holds and simplifying}

$s_1.vw.vValue \neq s_1.md.mValue.$

Using the disjunctive frame axiom on models, we can proceed by cases:

$\equiv$ { Case 1: assume $s_0.md = s_1.md$ }

$s_1.vw.vValue \neq s_0.md.mValue$

$\equiv$ { using the assumption that the invariant holds }

$s_1.vw.vValue \neq s_0.vw.mValue.$

Now, using the disjunctive frame axioms on views, we proceed by cases:

$\equiv$ { Case 1.1: assume $s_0.vw - s_1.vw$ }

false.

$\equiv$ { Case 1.2: assume $pcall?(act, vw.update)$ }

$s_1.vw.vValue \neq s_0.vw.mValue.$

Finally, popping up a level and proceeding with the case analysis:

$\equiv$ { Case 2: assume $pcall?(act, md.update)$ }

$s_1.vw.vValue \neq s_1.md.mValue.$

Combining the case assumptions with their derived pointcut specifications, we obtain

$$s_1.vw.vValue \neq s_1.md.mValue \wedge pcall?(act, md.update)$$
$$\vee\ s_1.vw.vValue \neq s_0.vw.mValue \wedge pcall?(act, vw.update)$$

which specifies calls to update either the model or the view.

• SPECIFICATION AND DERIVATION OF MAINTENANCE CODE: If action *act* has the form $md.update(newval)$, then we generate following the maintenance specification

assume: $s_0.vw.vValue = s_0.md.mValue$

achieve: $s_1.vw.vValue = s_1.md.mValue \wedge s_1.md.mValue = newval$

which is satisfied by the concurrent command $md.update(newval) \mid\mid vw.update(newval)$.

The same code is derived when action *act* has the form $vw.update(newval)$. In general, one would like to maintain consistency between a dynamic collection of data models and their corresponding views. To specify this would require quantifying over all currently allocated models and views, which in turn requires reifying the heap. It also requires taking into account the methods for creating, destroying, and associating models and views.

5 Remarks

This work may develop in a number of directions, some of which are discussed below.

1. *Implementation* — We anticipate implementing the techniques of this paper in our behavioral extension [16] of the Specware system [8]. The inference tasks in the examples are comparable in difficulty to those that were performed routinely and automatically in KIDS [19]. However, automated deduction requires the presence of an adequate inference-oriented theory of the language, data types, and application domain. As can be seen from the examples, most of the theorems needed are in the form of distributivity laws.

In general, the problem of synthesizing code from pre/postconditions is not decidable. However, two factors help to achieve tractability. First, note that the synthesis problem here is highly structured and incremental in nature. The goal is to reestablish an invariant that has just been perturbed by a given action. Second, synthesis can be made tractable by suitable restrictions on the language/logic employed. For example, in Paige's RAPT system [15], invariants and disruptive actions were restricted to finite-set-theoretic operations from the SETL language, and the corresponding maintenance code could be generated by table lookup.

2. *Granularity of Maintenance Code* — It may be convenient to treat a code block or a procedure/method as a single action for purposes of invariant maintenance. The main issue is that no (external) process that depends on the invariant could observe a state in which the invariant is violated. This notion suggests that static analysis could be used to check both (i) potential disruption points of the invariant, and (ii) the largest enclosing scope of dependent variables that is unobservable externally. An advantage of using a larger grain for maintenance is the performance advantage of bundling many changes at once, rather than eagerly updating at every dependent-variable-change. This is particularly advantageous when the update is relatively expensive.

3. *Constraint Maintenance: Maximization versus Invariance* — Sometimes a crosscutting feature may not have the form of an invariant for practical reasons. Consider, for example, the quality of service offered by a wireless communications substrate. Ideally, full capacity service is provided invariantly. However, physical devices are inherently more or less unreliable. There are at least two characterizations of constraint maintenance that make sense in this situation:

(i) *Maximize the uptime of the service* — That is, maximize the amount of time that a prescribed level of service is provided. Design-time maintenance might involve composing a fault-adaptive scheme to improve uptime.

(ii) *Maximize the provided bandwidth* — That is, continually make adjustments that provide maximal bandwidth given the circumstances.

4. *Enforcing Behavioral Policies* — This paper focuses on crosscutting concerns that can be specified as invariants. Behavioral invariants can be equivalently expressed as single-node automata with an axiom at the node. It is natural to consider crosscutting concerns that are specified by more complex automata and their corresponding temporal logic formulas. As mentioned earlier, some security policies disallow certain behavior patterns, as opposed to individual runtime events (see for example [6]). It is natural to consider generalizing the techniques of this paper to classes of policy automata. In recent work we developed a behavioral notion of pointcut, using automata to specify behavioral context for the application of advice [21]. Independently, several other research groups have been developing similar concepts [1,4,23,24].

5. *Maintaining Interacting Constraints* — Many application areas, including active databases with consistency constraints and combinatorial optimization problems with constraint propagation, have the characteristic that a single change ($x := e$) can stimulate extensive iteration until quiescence (a fixpoint) is reached. In terms of this paper, several invariants may have overlapping dependent variables and consequently their maintenance can interfere with each other's truth. That is, a change to maintain one constraint may cause the violation of another.

A sufficient condition that maintaining such a set of constraints leads to a fixpoint may be found in [17]. Constraints over a finite semilattice that are definite (a generalized Horn-clause form $x \sqsupseteq A(x)$ where x is a variable over the semilattice and A is monotone) can be solved in linear time. Using essentially the same theory, in [20, 26] we describe the process of automatically generating a customized constraint solver for definite constraints. The resulting solving process is an iterative refinement of variable values in the semilattice.

This context leads to a generalization of the formalism of this paper when (1) changes to certain variables can be treated as decreasing in a semilattice, and (2) constraints are definite. Then, a disruptive action ($x := e$) has postcondition ($x' \sqsupseteq e$) rather than the stronger ($x' = e$), and all constraint maintenance is downward in the semilattice, until a fixpoint is reached.

In a similar spirit, JMangler [10] implements a capability to iterate class transformations at class load-time when they have mutual dependencies. Under conditions on the transformations that satisfy the conditions above, the iteration converges to a fixpoint and the result guarantees that all transformations are fully applied.

6. *Comparison with AspectJ* — We conjecture that many aspects in AspectJ can be expressed as invariants, and that their effect can be achieved by means of the general process of this paper. However, the use of the around advice in AspectJ allows the replacement of a method call by arbitrary code, changing its semantics (e.g., consider the aspect that replaces every call to method m by advice that throws an exception). Our approach is restricted to maintenance that refines existing actions, so it is not complete with respect to AspectJ. On the other hand several of the examples in this paper cannot be carried out in AspectJ, so the two are expressively incomparable.

In the long run, it may be that only semantics-preserving aspects will be embraced in practice. If aspects are allowed to modify the behavior of code, then locality of program semantics is destroyed. This undercuts the potential for improved understandability of code due to the increased modularity that is the hallmark of aspects.

6 Recapitulation and Comparison with Related Work

We now summarize how the invariant maintenance approach treats the issues raised in the Introduction.

1. *What is the intention of an aspect?* — The intention of an aspect is expressed by an invariant property of state. The invariant can be thought of as a formal specification of an aspect.

2. *Is the pointcut complete?* — Does a pointcut express exactly the set of the intended run-time events? The invariant maintenance approach characterizes the joinpoints semantically, as those actions that could possibly disrupt the invariant. It thus does not require adherence to naming conventions. The pointcut is derived as a necessary condition on the violation of the invariant by a system action. Static analysis is then guaranteed to identify a superset of code locations that, at runtime, give rise to a violation of the invariant.

The derived pointcut is a specification of an AspectJ-style pointcut. To show the implementation relation between them, one would need to link the method calls and other joinpoints of the AspectJ pointcut to their semantic description (via pre/postconditions) to show that they satisfy the derived pointcut specification.

3. *Is the advice correct?* — Does an aspect correctly implement its intention? For each potentially disruptive action, our approach generates an action-specific specification, which, if realized by synthesis, maintains the invariant while still accomplishing the action. The overall argument that the invariant is enforced in the target system is essentially by induction. We establish the invariant initially, and then use static analysis to ensure that the invariant is maintained inductively. The invariant maintenance approach achieves correctness by construction. Other approaches to correctness of aspects and to woven code include verification through model checking [13], and runtime checking of contracts [12].

The invariant maintenance approach also provides novel contributions to the following issues:

1. *Context-Specialized Advice* — The advice of an AspectJ aspect may need to embody many case distinctions that cater for the various contexts that arise at the joinpoints, giving rise to inefficiency and complexity. It would be desirable if the aspect weaver could tailor the advice body to the specific context in which it will execute.

In AspectJ, and most extensions of it, the advice body is a code template that is instantiated with expressions from the context of a pointcut, and is parametric on runtime values. One can increase the range of context and parametricity, but the advice is still a code template. There have been a variety of extensions to AspectJ that aim to increase the range of context that can be captured at joinpoints. For example, LogicAJ [18] and Sally [7] use logical metavariables to supplement the pattern constructs of AspectJ. Both also allow patterns in pointcut and advice definitions where AspectJ only allows constants (e.g., metavariables that match types). This mechanism supports the binding of more pieces of context (e.g., types) than AspectJ allows and thus advice bodies can be more context-sensitive. The use of metavariables also adds consistency constraints in the matching process.

The invariant maintenance approach is fully context-sensitive in the sense that maintenance code/advice is unique to each code location satisfying the pointcut specification. It can be completely different in different contexts, not just different instances of a fixed template. The list example in Section 4.4 illustrates this. To obtain the same effect in AspectJ or LogicAJ would require either (1) a single disjunction pointcut together with a big context switch in the advice, or (2) multiple aspects, one for each context together with the appropriate advice template for that context. The invariant approach neatly specifies what to do in an unbounded number of contexts. Of course it only *specifies* what to do in each context — the prescription of what to do must be achieved by synthesis. The difference is generation of arbitrary code per disruptive action versus multiple instances of a single advice template.

The use of reification is a potentially unbounded technique for bringing context into play. In this paper we have mentioned reification of history, the call stack, and the heap. There are many other possibilities. For example, reifying instruction timing information would allow one to maintain real-time properties of the base code.

2. *Aspect Interference* — Aspects may interfere with one another — since the order in which they are applied makes a semantic difference, the burden is on the AspectJ programmer to order them and resolve interferences. Interference is a fundamental problem of aspect composition.

The invariant maintenance approach brings two extra degrees of freedom in treating interference relative to programmatic aspects: semantic abstraction and inherent concurrency. Since the maintenance specification is expressed in terms of pre/postconditions and the invariants to be maintained, the synthesizer has maximal freedom to design a mutually satisfactory behavior. The resulting generated code does not need to be the same as the default implementation of the aspects individually. Also, the maintenance code and the system code are conceptually concurrent. To render them into conventional programming languages, it is necessary to sequentialize them, which introduces the possibility for interference. Thus the interference detection/resolution tools that have been explored for dependency analysis and instruction ordering are needed.

On the other hand, in some situations the system code and the aspects simply conflict. In the invariant maintenance approach this is signaled by the inability to synthesize maintenance code at a pointcut location. Depending on the techniques used, a failed synthesis process may be able to return a counterexample that can be used by the developers to pinpoint the semantic discrepancy between their invariants and the system code.

Several projects have developed analysis tools for detecting aspect interference and inferring safe orderings; e.g., LogicAJ [18]. Reflex [22] provides programmatic mechanisms for detecting potential interference and prescribing how they should compose.

3. *Evolution* — Evolution of the base program may require extending the AspectJ pointcut and advice description to reference any new class members (which requires an understanding of the modified set of runtime events being targeted, and what code to execute in each specific context of occurrence). In our approach, if the invariant remains unchanged, then the derived pointcut specification doesn't change. Consequently, if the base code changes, then the static analyzer can simply run the pointcut specification over the new base code. Ideally, this can be done incrementally if the structure of the changes have been recorded. In other words, an invariant will tend to be more stable under base code changes than an aspect that implements it.

The generative techniques in this paper derive from transformational work on incremental computation, especially Paige's Finite Differencing transformation [14, 15]. Finite Differencing, as implemented in the RAPTS system, automatically maintains invariants of the form $c = f(x)$ where c is a fresh variable, x is a vector of program variables, and f is a composite of set-theoretic programming language operations. Maintenance code is generated by table lookup. In the KIDS system [19], we extended Finite Differencing by (1) allowing the maintenance of both language- and user-defined terms, and (2) using automatic simplifiers to calculate maintenance code at design-time. The functional language setting in KIDS naturally reveals the concurrency of disruptive code and maintenance updates.

As in Finite Differencing, other approaches to programming with invariants (e.g. [3]) work exclusively with program variables. This paper introduces the notion of reifying extra-computational information, enabling the expression of system-level crosscutting features as invariants.

A common use for AspectJ is to enforce preconditions, postconditions, and invariants (e.g. [11]). The intention is to insert code that dynamically *checks* those conditions and flags violations. In contrast, this paper focuses on adding code to *enforce* invariants when they would otherwise be violated. The resulting code is guaranteed to satisfy the invariant even though the base code may not.

7 Concluding Remarks

Aspect-Oriented Software Development aims to support a more modular approach to programming, with a special focus on crosscutting concerns. This paper explores techniques for specifying crosscutting concerns as invariants, and generating the code necessary to maintain them. The reification of extra-computational entities helps in expressing many crosscutting concerns.

Our invariants provide an abstract yet precise, semantic characterization of crosscutting concerns. The abstraction should aid in clarifying the intention of a concern and promote stability under evolution. The precise semantics means that the generation of maintenance code can be performed mechanically, with assurance that the result meets intentions.

The generally accepted semantics of AspectJ is based on call-stack reification [25], suggesting that AspectJ crosscutting concerns can be characterized as actions to take about method calls in a specified dynamic context. Our approach lifts to a more general perspective: what kinds of crosscutting concerns can be addressed when arbitrary extra-computational information is reified.

This work advocates a design process that focuses on generating a normal-case base program from high-level models or specifications, followed by the generation and insertion of extensions to implement various crosscutting concerns. Code structure simplifies to a clean natural decomposition of the basic business logic together with system-level invariants that specify crosscutting concerns. The improved modularity should help to lower the cost of development and evolution and provide increased assurance.

Acknowledgments

Thanks to Cordell Green, Gregor Kiczales, Peter Pepper, and Kevin Sullivan for discussions of this work. Thanks also to Lambert Meertens, Stephen Westfold, and the reviewers for useful comments on the text. This work was partially supported by the US Department of Defense and by the Office of Naval Research under Grant N00014-04-1-0727.

References

1. C. Allan, P. Avgustinov, A. S. Christensen, L. J. Hendren, S. Kuzins, O. Lhoták, O. de Moor, D. Sereni, G. Sittampalam, and J. Tibble: Adding trace matching with free variables to AspectJ. In: *Proceedings of OOPSLA*, 345–3640, 2005.
2. Aspect-Oriented Software Development website 2003, `http://www.aosd.net/`
3. X. Deng, M. Dwyer, J. Hatcliff, and M. Mizuno: Invariant-based specification, synthesis and verification of synchronization in concurrent programs. In: *Proceedings of the 24th International Conference on Software Engineering*, May 2002.

4. R. Douence, P. Fradet, and M. Suedholt: Composition, reuse and interaction analysis of stateful aspects. In: *Aspect-Oriented Software Development* (AOSD04), ACM Press, 141–150, 2004.
5. T. Elrad, R. Filman, and A. Bader (Eds.): *Special Issue on Aspect-Oriented Programming*. Vol. 44(10). Communications of the ACM, 2001.
6. U. Erlingsson and F. Schneider: SASI enforcement of security policies: A retrospective. In: *Proceedings of the New Security Paradigms Workshop*, Ontario, Canada, 1999.
7. S. Hanenberg and R. Unland: Parametric introductions. In: *Aspect-Oriented Software Development*, 2003.
8. Kestrel Institute, Specware System and documentation, 2003. Available at: `http://www.specware.org/`.
9. G. Kiczales et al.: An Overview of AspectJ. In: *Proc. ECOOP*, LNCS 2072, Springer, 327–353, 2001.
10. G. Kniesel, P. Costanza, and M. Austermann: JMangler – a powerful back-end for aspect-oriented programming. In R. Filman, T. Elrad, S. Clarke, and M. Aksit, (Eds.): *Aspect-Oriented Software Development*, Prentice-Hall, 2004.
11. R. Laddad: *AspectJ in Action*. Manning Publishing Co., 2003.
12. D. H. Lorenz and T. Skotiniotis: Extending design by contract for aspect-oriented programming. *Smithsonian/NASA Astrophysics Data System*. cs/0501070, 2005. `http://adsabs.harvard.edu/abs/2005cs`.
13. H. Li, S. Krishnamurthi, and K. Fischer: Modular verification of open features through three-valued model checking. *Automated Software Engineering Journal* 12, 349–382, 2005.
14. R. Paige: Programming with invariants. *IEEE Software* 3:11, 56–69, 1986.
15. R. Paige and S. Koenig: Finite differencing of computable expressions. *ACM Transactions on Programming Languages and Systems* 4:3, 402–454, 1982.
16. D. Pavlovic and D. R. Smith: Evolving specifications. Technical Report, Kestrel Institute, Palo Alto, CA, USA, 2004.
17. J. Rehof and T. Mogenson: Tractable constraints in finite semilattices. *Science of Computer Programming* 35, 191–221, 1999.
18. T. Rho and G. Kniesel: Uniform genericity for aspect languages. Technical Report IAI-TR-2004-4, Computer Science Department III, University of Bonn, Germany, 2004.
19. D. R. Smith: KIDS – a semi-automatic program development system. *IEEE Transactions on Software Engineering, Special Issue on Formal Methods in Software Engineering* 16, 1024–1043, 1990.
20. D. R. Smith, E. A. Parra, and S. J. Westfold: Synthesis of planning and scheduling software. In Tate, A. (Ed.): *Advanced Planning Technology*, AAAI Press, Menlo Park, 226–234, 1996.
21. D. R. Smith and K. Havelund: Automatic enforcement of error-handling policies. Technical report, Kestrel Technology, 2004. `http://www.kestreltechnology.com/`.
22. É. Tanter and J. Noyé: A versatile kernel for multi-language AOP. In: *Proceedings of the 4th ACM SIGPLAN/SIGSOFT Conference on Generative Programming and Component Engineering* (GPCE 2005). LNCS 3676, Tallin, Estonia, Springer, 173–188, 2005.
23. W. Vanderperren, D. Suvee, M. Cibran, and B. de Fraine: Stateful aspects in JAsCo. In: *Proceedings of Software Composition 2005*, Springer LNCS 3628, 167–181, 2005.
24. R. Walker and K. Viggers: Implementing protocols via declarative event patterns. In: *SIGSOFT Foundations of Software Engineering* (FSE04), ACM Press, 159–169, 2004.
25. M. Wand, G. Kiczales, and C. Dutchyn: A semantics for advice and dynamic join points in aspect-oriented programming. *ACM Transactions on Programming Languages and Systems* 26(5), 890–910, 2003.
26. S. Westfold and D. Smith: Synthesis of efficient constraint satisfaction programs. *Knowledge Engineering Review* 16, 69–84, Special Issue on AI and OR, 2001.

Program Transformations: Some Lessons from the 1980s

David S. Wile

Teknowledge Corp.,
4640 Admiralty Way #1010, Marina del Rey, CA 90292 USA
dwile@teknowledge.com

Summary. The 1980s were a fertile time in the development of program transformation systems. Much of the experience we gained during those years, what we learned about designing, implementing, and using transformation systems, is written down in obscure places that today's researchers probably do not have access to, or probably will not find if they do. Moreover, many practical tidbits of information were understood, but never written down. I do not really think the community is aware of some of the problems they are revisiting today and solutions they are reinventing (some that do not work!). Bob Paige was a major contributor to the body of transformation research; he was also very concerned that his many results not be lost to successive generations of researchers. Late in his life he produced a compendium of his results and he encouraged his colleagues to do likewise. This is my effort towards that goal.

Keywords: metaprogramming, automatic program development, program transformation.

1 Introduction

In 1987 at a transformation workshop in Bad Tölz, (then West) Germany [29], Martin Feather presented a survey of transformation system issues to that time [17]. He claimed that [17, page 165]:

> The common thread of all transformation research is the *formal development* from specification to implementation. Because it is *formal* it offers the potential for introducing extensive mechanized support into much of the programming process. Because it is a *development* it records the complete path from the descriptive nature of the specification (which outlines what is to be achieved), to the prescriptive nature of the program (which details how things are actually done).

He broke the approaches down into: (i) extended compilation — wholly automatic methods in which the compiler is given advice, (ii) metaprogramming — automated methods with varying degrees of interaction, and (iii) synthesis — automated methods that understand the algorithmic knowledge they are manipulating. (I would have added a 4th category: foundational — theories underlying transformation methods and processes.) Although I cannot claim to cover all three of these in any depth, I would characterize our research group's approach as an evolution from synthesis concerns, to metaprogramming concerns, to extended compilation, over the years. Below I indicate some of our successes, along with the mistakes we made, to draw lessons from the experiences that others may use today.

2 Historical Perspective

To set the stage allow me to indulge in a brief, rather egocentric, historical perspective. First, our entry into the field was from a fairly unusual direction: we — Bob Balzer, Neil Goldman

O. Danvy et al. (eds.), Automatic Program Development – A Tribute to Robert Paige, 264–272.

and myself, initially, led by Bob Balzer at the University of Southern California's Information Sciences Institute — were interested in *automatic programming*, by its very nature a synthesis activity, of programs from natural language descriptions! We discovered that an important "key" to its success was to translate natural language into a formal specification language that was as close to the models natural language support as possible [3].

Gist

These natural language metaphors included "free" reference to information, any implementation of which may require checkpointing — e.g. "the protection of the file before it was last edited" — or derivation — e.g. "the total number of bytes occupied by any version of this file". In addition, references to nouns correspond to types, attributes correspond to relationships accessed in a variety of ways, and functions correspond to processes. In particular, quantification over predicates is easily expressed — e.g. "every message that has been marked for action and not distributed". Moreover, one could specify triggered activities that were to be evaluated whenever a condition arose — e.g. "whenever two differently named files contain the same information (talk about free reference to information!), warn the user". The most distinguishing aspect of our language, called Gist, was that one could state constraints globally and specify behavior nondeterministically, *guaranteed that the behaviors that violate constraints were not part of the intended specified behavior.* This was very high level indeed, yet formalizable.

But the very high-level nature of the language guaranteed that some sort of compilation activity was going to be required to translate into efficient implementations. We had heard of John Darlington and Rod Burstall's pioneering efforts in program transformations [10] — originally only fold/unfold were identified, while what our group considered to be transformations were to them just equational reasoning (or something like that). This seemed to be the key to developing the implementation from our high-level specifications, so we worked out an example related to our automatic programming research and submitted a paper to the Software Engineering Conference in San Francisco in 1976 [4]. At that conference I recall Bob Balzer excitedly greeting me as I awoke — we were sharing a room and he had gotten up at 5:00 and had read all the conference papers from our session! Professor Bauer of the Technische Universitaet in Munich was also talking on transformations about his CIP system [6, 12]. He and his students became a major influence on formal underpinnings of the field, the foundation, in succeeding years [9].

At this same time, the mid 1970's, Cordell Green's group at Stanford — including Elaine Kant and David Barstow — was interested in very similar ideas for synthesis embodied in their PSI system [20]. In particular, their interests evolved into implementing control [25] and data structures [21]. Another related project was the Programmer's Apprentice at MIT by Rich, Waters, and Shrobe [38]. And the other very high-level language at the time (aside from APL and Prolog) was SETL; Bob Paige was just then inventing formal differentiation at Courant [32]. Tom Cheatham and Glen Halloway at Harvard were transforming EL-1, an extensible language [11]. But perhaps the greatest purveyor of useful transformations at the time was Jim Boyle, whose TAMPR system was still in use in 2000 [8]! Representatives from these groups of researchers met on the West Coast at Information Sciences Institute in 1979. (Naturally, there were some I am omitting simply from fading memory.) It was some time before the first meeting that we met and hired Martin Feather, who studied with Darlington; his thesis on the ZAP system concerned capturing fold/unfold strategies formally [15].

Many from this group then met at the Second NATO Workshop on Software Engineering in Munich in 1983 [33]. It was there that I met Bill Scherlis whose Stanford thesis [39] concerned a particularly fruitful tactic for program transformation that people should be aware of today, called *specialization*. The technique is simple: change the statement of the problem from transforming $f(type)$ to one of transforming $F(setoftype)$, such that $F(\{x\}) = f(x)$. This is a very elegant transformation that yields surprising results, because one can take advantage of the relationships between the pieces of what amount to be intermediate results in F's computation. His derivations of parsing algorithms, e.g. Earley's algorithm, by specialization showed how one can take advantage of sets of partial results effectively. This meeting was also where Bob Paige introduced his formal differentiation work to the community.

But back to the main thread of our research. In the beginning stages, we concentrated on synthesis [1] and foundational aspects of program transformation [44]. At Information Sciences Institute in a different group, Susan Gerhart studied the relationship of program transformations to program verification, establishing a formal method for determining when a transformation is valid [18]. However, fairly early on, we realized that the majority of transformation applications in real programs would be of rather limited algorithmic interest and focused on the transformation *process* instead. The human decision process of deciding what transformations to apply, where, and why, was exceedingly expensive in human time — the computer, to optimize even the simplest program, applied hundreds of transformations on behalf of the human. Having just hired Martin Feather, we capitalized on the idea he (and Darlington) germinated in the ZAP system, which was to capture the transformation process in a formal document that could be used to *replay* the transformations at a later time. I invented a language called Paddle [45], that worked within our syntax-based transformation system, Popart, that had been under development since 1977 [47].

Popart

Popart provides a tool suite with which transformation experts can prototype languages rapidly — both their syntax and associated semantics; it was a kindred spirit of Mentor [14]. Popart relies on the specification of concrete syntax in a BNF variant that supports directly repetition, optionality, and precedence. Annotations to the concrete syntax affect details of the abstract syntax used to represent "parse trees" compactly. Other declarations may be used to affect the lexical analyzer tables, which are produced automatically for each concrete syntax. The concrete syntax can be specified in such a way that formatting ("pretty-printing") information is derived from its specification. In addition, it can also be affected by tables much as the lexical analyzer. Analysis routines, simplifiers, transformers and translators can be expressed using so-called "syntax-directed experts" — rule sets whose patterns are written in an extension to a language's concrete syntax that introduces "pattern variables". (Today's LCF system exhibits many of these same characteristics [19].) These experts produce attributes of the parse trees to which they are applied often in a more expressive way (I believe) than using attribute grammars [37]. Often, such experts are easily written, and more importantly, easily read by novice programmers; generally, a deep background in the use of Common Lisp is unnecessary for generating a grammar and some of the associated semantics, quickly.

Paddle

In order to manipulate the abstract syntax trees back then we had two facilities: the sets of transformation rules used to implement analyzers, transformations, and translations mentioned above, along with a more primitive, syntax-directed editing mechanism. The Paddle metaprogramming language invoked these mechanisms as primitives. Paddle was goal-based, in the sense that activities could fail and mechanisms could expect and react to that failure. It was an imperative, sequential language with composition mechanisms to achieve compound, conditional, parallel (conceptually, implemented sequentially), or choices of goals. Because the partial effects of failing events were automatically undone, backtracking was inherent. At the time, the LCF tacticals were clearly related, although applied in the different context of theorem proving [19].

In the early 1980s, Balzer, Cheatham, and Green wrote their famous article on Knowledge-Based Software Assistants [2], and transformation systems were established as the key to the future of program development environments. Around 1982 I started attending the IFIP WG2.1 (the Algol committee) meetings, along with the CIP — Fritz Bauer, Peter Pepper, Manfred Broy, Helmut Partsch — and SETL groups — Jack Schwartz and Bob Paige — and

Jim Boyle, and met others interested in transformations that had not been brought to my attention before that; Alberto Pettorossi, Richard Bird, Robert Dewar, Lambert Meertens, and Michel Sinzoff, to name a few. Later members Doug Smith, Alan Goldberg, Doaitse Swierstra, Jeremy Gibbons and Ooge de Moor have done extensive work with transformations also.

Our group at Information Sciences Institute concentrated on specific kinds of transformations, such as those to remove nondeterminism, or those to remove historical references, and further specification techniques [16], However, even in the early 1980s we decided we were having little success convincing people of the power of transformation systems, in part because we were not using them to program our everyday programs ourselves. Hence, we decided to apply the technology to our own programming environment, in what became the Formalized System Development testbed at Information Sciences Institute. This system further stimulated systems such as Gail Kaiser's Marvel system and HP's Softbench, especially their reliance on triggered activities.

Unfortunately, the transformation process itself was lost! Instead, aspects of the Gist language were incorporated into the system infrastructure, including the use of relational abstraction and triggered activities [13]. The formal development process was captured, as a stylized "process program", but the grain-size of the transformations was large editing steps [30]. Part of the group began to emphasize requirements engineering and the use of "high-level editing commands" — transformations that preserved some properties of the specification, but not the functionality. Feather and Johnson designed the Aries system to incorporate these concepts [22, 23].

I on the other hand, diverged along a different track. My feeling was that the whole idea of a "wide-spectrum" specification language was flawed: every construct would need to be understood in concert with every other one. This might require reasoning with concepts as disparate as backtracking and register allocation at the same time, which is clearly impossible. I noticed that the way wide spectrum languages were being used was much more stylized in both our work and the CIP system: layers of concern were mapped away in a logical sequence. Essentially a process of simplify and then apply all the transformations required to remove, e.g. historical references, was used, followed by simplification and application of transformations to remove derived relationships, etc. This, together with the notions that Jim Neighbors was proposing in his Draco system [31], and Tim Standish with the Arcturus language [42], led me to develop the notion of Local Formalisms — formalisms tailored to just the specified problem area, to be implemented with a very specific metaprogram having the flavor just described [46]. Surprisingly, that characterization was exactly what Jim Boyle had been using in his TAMPR system for 15 years. We had "progressed" from synthesis and foundation work, to metaprogram development work, and finally ended up with problem domain-specific, extended compilers.

3 Lessons Learned or Observed

Although the previous section indicated broadly some of the lessons that caused us to shift our directions or emphasis, here I will try to cull out the essential lessons I learned, almost all from mistakes we made, but in some cases, observations of mistakes made by others. *These should be taken with a grain of salt: they are very idiosyncratic.* It must be admitted that our group might not have agreed among ourselves on any one of these lessons!

System Design Aspects

Automate Voraciously. The single biggest mistake we made consistently throughout our history was to assume that the user needs to micromanage the development process. There is a fine balance between complete automation, advice-driven automated activity, and purely manual actions. I have noticed that the tendency of almost all formally oriented people — from theorem prover types to rocket scientists programming deep space satellites (literally!) — is to put up with extensive repetition of extremely rote activities. For one theorem prover,

I was able to program a very small strategy in Paddle that proved the entire practice set of theorems in the system tutorial entirely automatically; yet no facility was ever introduced into the system to permit users to do this themselves. Instead, they had to repeatedly invoke the same manual procedures "as last time". An early version of the CIP system suffered similarly from this malady.

Despite our best efforts, we were guilty of being too cautious as well. The user was to make all the decisions regarding transformations. Fairly quickly we recognized a pattern of activity comprising the steps: (1) get the program into condition for the transformation pattern we really want to apply, by applying what we called "jittering" transformations ("conditioning" was a better term, but the game industry later came up with the ideal term: "morphing!"); (2) apply the transformation; and then, (3) clean up or simplify the result. Hence, Popart allowed one to supply the conditioning and simplifying transformations, which were then applied automatically when a user selected a transformation for the system to apply.

Although these facilities were helpful, in the end the user still had far too much autonomy. When the formal development process was replayed on slightly changed programs, the transformations failed to apply and the process halted. We actually needed to find ways to increase the automation of the development process, removing some of the ability of the user to make detailed decisions regarding sequences of transformation applications. So, the lesson to take away is to be diligent in seeking to support the normal case with automation or syntactic sugar; mathematically inclined folks like ourselves tend to be wary of losing control and, hence, are reluctant to look for such opportunities, since they are not intellectually challenging. Actually, it is probably more honest to admit that our expertise in such design is quite limited.

Take control of input and output formatting. This is somewhat related to the above lesson. Although the user should feel free to edit a textual representation of the specification, one should feel equally free to apply a transformation to the specification (equivalence preserving or not). When we switched from Interlisp to Common Lisp on the Symbolics machine, we lost the structural representation of the program and were set back seriously in the Lisp programming environment. We never regained the power of the Interlisp metaprogramming facilities (on Lisp programs themselves) for our everyday programming activity.

Another implication here is that the system should be doing the program/specification layout (pretty printing) and the association of comments with relevant parts of the program. Again, provide advice mechanisms to guide the automation of these activities rather than allow users to determine the exact layouts themselves.

Separate the API from the User Interface. Otherwise, one cannot possibly do replay, or rather an entirely different dialect must be developed that is equivalent to the gestures coming from the interface. This is a much better-recognized tenet of modern system design, so it might seem not to be worth mentioning, but an implication of this is that every time your program pops up a question with an answer box to be filled in by the user, be sure that the user could have provided the program previously — declaratively — with exactly the same information.

Describe, don't point! Capturing user gestures for replay is a bad idea for several reasons: the system never understands why the user pointed; it does not understand whether it should be done again if the program changes a bit; and it may not even know what to point to if it changes! The Aries system overcame the use of pointers by characterizing with predicates the situation in which a transformation would reapply.

Monolithic ("Stovepipe") systems are a bad idea. Back in the 1980s there were very few standards for systems that could be relied on. Lots of activity in programming environments meant different proposals for persistence, versioning, abstract syntaxes, tool invocation protocols, etc., so we tended to grow our own versions of all of these, even when they were not the focus of our interest. Popart itself was an instance of a solution to provide a foundation for transformation systems rather than an object of research in itself. So most systems failed due to the high cost of adopting the tools and the costs in time, if not money, to learn all the peripheral aspects around the tool that really constituted the important contribution. Today's software engineers are much more careful to make APIs very clean and use those standards that exist — e.g. Corba and DCOM — whenever possible. However, many systems introduced in recent years require an incredibly high buy-in price, in time invested to learn how to use them if not in literal cost. If we want our tools to be tried out by a variety of people on a variety of platforms, they must be simplified and their functionality decomposed into "bite-sized", easily adopted or adapted tools.

Technical Aspects

Yesterday's hard problems remain difficult. Many problems we had were due in part to resource limitations. "Jittering" (conditioning) is one such problem that may very well be more tractable today, so it should be revisited. Moreover, there are cleaner theories of metaprogramming, associative-commutative Knuth-Bendix and fixed-point fusion to use as starting points, for example.

Another was the need to delay transformation applications to allow further refinements to constrain the solution space — I needed this in my "type transformations" paper to handle the tricky merge that occurs in heapsort of the output with the input data structures [39]. This was generally not done, because it was too hard to keep track of. I think that both of these problems will benefit from Kibler's approach [26], of applying metaprograms to the transformations themselves to preanalyze what situations they should best be applied in. This idea was incorporated in the Aries system, where the transformations were indexed by their effects [22].

Metaprograms should be expected to fail. I do not think there is much controversy here. All the interesting strategic programs are failure-based. Yet there is precious little support for failure, especially in modern programming languages.

Capturing intent is very difficult. A necessary tenet of the use of replay is that the designer somehow expresses intent when doing the design. Unfortunately, all mechanisms heretofore have been *intrusive* and very brittle, due to the *unpredictable* nature of the changes that may be made to the specification. This is exactly why the preference to deal with layers of concerns, if possible, arose (Boyle; CIP layers; modern approach of Batory). Layers may also have structural assumptions built in, such as layer-specific transformations for conditioning and subsequent simplification.

4 Conclusions

Many modern results in transformation systems as used in problem-specific areas have put to rest the fear that we had in the early 1980s that the use of transformations would never become practical. For example, everyone should know about the amazing results in synthesizing solutions to real problems embodied in Doug Smith's KIDS system [40]. Moreover, research in metaprogramming is one of today's most active transformational areas; see Kestrel's Specware system [41], my metaprogramming calculus [48], OGI's use of Haskell [27], Batory's technology [5], and Visser's Stratego system [43]. The emphasis on metaprogramming is in part due to the increased interest in domain-specific language design and support [49,50]. Essentially, research in extended compilation has become focused on particular problem domains, where reusing the still-significant amount of work that goes into designing a metaprogram is effectively leveraged [36].

Spinoffs from early transformation research threads continue. For example, specialization led to results in partial evaluation [24] and early recursion removal strategies led to [34,35]. The transformation field is beginning to flourish again as evidenced by interest in recent workshops at OOPSLA 2003 and Dagstuhl in 2005 and by its reinvention in "refactoring" and re-engineering. Its traditional success as an explanatory mechanism for algorithm understanding and derivation has always formed a firm pedagogical basis for its use. Although Bob Paige was proud of his Raps system, one of his greatest abilities was to explain how intricate algorithms were really the result of the stepwise application of incremental transformations to a perspicuous problem solution. He applied these in a wide variety of problem domains, the last of which he reported on in a workshop on algorithmic languages and calculi [7]. His work also continues in the highly competent hands of Y. Annie Liu [28].

In summary, I have tried to present an egocentric version of the history of transformation systems through the 1980s as they affected me and our group, along with random lessons from the 1980s that can still be applied today, especially encouraging all not to repeat our mistakes! It would be interesting to measure how today's systems have progressed along the dimensions of Foundations, Synthesis, Metaprogramming, and Extended Compilation. I'm sure Bob Paige would have found that to be a fascinating and challenging exercise.

Acknowledgements

I want to thank Bob Balzer, Neil Goldman, Martin Feather, Bill Swartout, Lewis Johnson, Don Cohen, K. Narayanaswamy, Dennis Allard, and many others in Information Sciences Institute's Software Sciences Division who contributed over the years to these efforts!

References

1. R. M. Balzer: Transformational implementation: An example. *IEEE Transactions on Software Engineering SE*, 7(1):3–14, January 1981.
2. R. M. Balzer, T. E. Cheatham, and C. Green: Software technology in the 1990s: Using a new paradigm. *Computer Magazine*, 1983.
3. R. M. Balzer and N. M. Goldman: Principles of good software specification and their implications for specification languages. In *Specification of Reliable Software*, 58–67. IEEE Computer Society, 1979.
4. R. M. Balzer, N. M. Goldman, and D. S. Wile: On the transformational implementation approach to programming. In *Proceedings of the Second International Conference on Software Engineering*, 337–344, October 1976.
5. Don Batory and B. J. Geraci: Composition validation and subjectivity in GenVoca generators. *IEEE Transactions on Software Engineering SE*, February 1997.
6. F. L. Bauer. Programming as an evolutionary process: In *Proceedings of the Second International Conference on Software Engineering*, 223–234. IEEE Computer Society, October 1976.
7. R. S. Bird and L. G. L. T. Meertens, editors: *Algorithmic Languages and Calculi.* IFIP TC2/WG 2.1 International Workshop on Algorithmic Languages and Calculi, 17–22, Alsace, France. Chapman & Hall, February 1997.
8. J. M. Boyle: Program adaption and transformation. In *Practice in Software Adaption and Maintenance: Proceedings of Workshop on Software Adaption and Maintenance, Berlin*, 3–20. North-Holland, 1979.
9. M. Broy, H. Partsch, P. Pepper, and M. Wirsing: Semantic relations in programming languages. In *Proceedings of IFIP Congress, Amsterdam*, 101–106. North-Holland, 1983.
10. R. M. Burstall and J. Darlington: A transformation system for developing recursive programs. *Journal of the ACM*, 24(1):44–67, January 1977.
11. T. E. (Jr) Cheatham, G. H. Holloway, and J. A. Townley: Program refinement by transformation. In *of the Fifth International Conference on Software Engineering, San Diego, CA*, 430–437, March 1981.
12. CIP Language Group: *The Munich Project CIP. Volume I: The Wide Spectrum Language CIP-L.* Lecture Notes in Computer Science 183. Springer, 1985.
13. D. Cohen: Automatic compilation of logical specifications into efficient programs. In *Proceedings of the 5th National Conference on Artificial Intelligence AAAI-86, Philadelphia, PA, USA*, 20–25, April 1986.
14. V. Donzeau-Gouge, G. Kahn, B. Lang, and B. Mélèse: Documents structure and modularity in Mentor. In *Software Development Environments (SDE)*, 141–148, 1984.

15. M. S. Feather: A system for assisting program transformation. *ACM Toplas*, 4(1):1–20, 1982.
16. M. S. Feather: Language support for the specification and development of composite systems. *ACM Trans. Program. Lang. Syst.*, 9(2):198–234, 1987.
17. M. S. Feather: A survey and classification of some program transformation techniques. In L. G. L. T. Meertens, editor, *Proceedings IFIP TC2 Working Conference on Program Specification and Transformation, Bad Tölz, Germany*, 165–195. North-Holland, 1987.
18. Susan L. Gerhart: Correctness-preserving program transformations. In *2nd ACM POPL Symposium, Palo Alto, CA, USA*, 54–66, 1975.
19. M. J. Gordon, A. J. Milner, and C. P. Wadsworth: *Edinburgh LCF.* Lecture Notes in Computer Science 78. Springer, 1979.
20. C. Green: A summary of the PSI program synthesis system. In *Proceedings of the Fifth International Conference on Artificial Intelligence, Cambridge, MA, USA*, 380–381, August 1977.
21. C. Green and D. R. Barstow: On program synthesis knowledge. *Artificial Intelligence*, 10(3):241–279, 1978.
22. W. L. Johnson and M. S. Feather: Using evolution transformations to construct specifications. In M. Lowry and R. McCartney, editors, *Automating Software Design*, 65–91. AAAI Press, 1991.
23. W. L. Johnson, M. S. Feather, and D. R. Harris: Representation and presentation of requirements knowledge. *IEEE Trans. Software Eng.*, 18(10):853–869, 1992.
24. N. D. Jones, C. K. Gomard, and P. Sestoft: *Partial Evaluation and Automatic Program Generation.* Prentice-Hall, 1993.
25. E. Kant: On the efficient synthesis of efficient programs. *Artif. Intell.*, 20(3):253–305, 1983.
26. D. F. Kibler: *Power, Efficiency, and Correctness of Transformation Systems.* PhD thesis, Computer Science Department, University of California at Irvine, CA, USA, 1978.
27. R. Kieburtz, F. Bellegarde, J. Bell, J. Hook, J. Lewis, D. Oliva, T. Sheard, T. Walton, and T. Zhou: Calculating software generators from solution specifications. Technical Report CS/E-94-032B, Oregon Graduate Center, USA, 1994.
28. Y. A. Liu: *Incremental Computation: A Semantics-Based Systematic Transformational Approach.* Ph.D. thesis, Department of Computer Science, Cornell University, Ithaca, New York, 1996.
29. L. G. L. T. Meertens, editor: *Program Specification and Transformation. Proc. IFIP TC2/W.G. 2.1 Working Conference on Program Specification and Transformation, Bad Tölz, Germany.* North-Holland, 1987.
30. K. Narayanaswamy: Static analysis-based program evolution in the common lisp framework. In *Proceedings of the 10th International Conference on Software Engineering, Singapore*, 222–230, April 1988.
31. J. Neighbors: *Software Construction Using Components.* Ph.D. thesis, Computer Science Department, University of California at Irvine, CA, USA, 1981.
32. R. Paige: Transformational programming – applications to algorithms and systems. In *10th POPL Symposium, Austin, TX, USA*, 73–87, 1983.
33. P. Pepper, editor: *Program Transformation and Programming Environments, Workshop Report.* Nato ASI Series F 8. Springer, 1984.
34. A. Pettorossi and R. M. Burstall: Deriving very efficient algorithms for evaluating linear recurrence relations using the program transformation technique. *Acta Informatica*, 18:181–206, 1982.
35. A. Pettorossi and M. Proietti: Rules and strategies for transforming functional and logic programs. *ACM Computing Surveys*, 28(2):360–414, 1996.
36. J. C. Ramming and D. S. Wile: Guest editorial: Introduction to the special section "Domain-Specfic Languages (DSL)". *IEEE Trans. Software Eng.*, 25(3):289–290, 1999.

37. T. W. Reps and T. Teitelbaum: *The Synthesizer Generator*. Springer, New York, 1988.
38. G. Rich: A formal representation for plans in the programmer's apprentice. In *Proceedings 7th International Joint Conference on Artificial Intelligence*, 1981.
39. W. L. Scherlis: Program improvement by internal specialization. In *Proc. 8th ACM Symposium on Principles of Programming Languages, Williamsburgh, Va*, 41–49. ACM Press, 1981.
40. D. R. Smith: KIDS: A semi-automatic program development system. *IEEE Transactions on Software Engineering — Special Issue on Formal Methods*, 16(9):1024–1043, September 1990.
41. Y. V. Srinivas: Refinement of parameterized algebraic specifications. In R. S. Bird and L. G. L. T. Meertens, editors, *Algorithmic Languages and Calculi*, volume 95 of *IFIP Conference Proceedings*, 164–186. Chapman & Hall, 1997.
42. T. Standish: Arcturus. Seminar USC/Information Sciences Institute (Marina del Rey, CA, USA), Fall 1980.
43. E. Visser: Program transformation with Stratego/XT: Rules, strategies, tools, and systems. In C. Lengauer et al., editor, *Domain-Specific Program Generation*, Lecture Notes in Computer Science 3016, 216–238. Springer, 2004.
44. D. S. Wile: Type transformations. *IEEE Trans. Software Eng.*, 7(1):32–39, 1981.
45. D. S. Wile: Program developments: Formal explanations of implementations. *Commun. ACM*, 26(11):902–911, 1983.
46. D. S. Wile: Local formalisms: Widening the spectrum of wide-spectrum languages. In L. G. L. T. Meertens, editor, *Proceedings IFIP TC2 Working Conference on Program Specification and Transformation, Bad Tölz, Germany*, 459–481. North-Holland, 1987.
47. D. S. Wile: Popart: Producers of parsers and related tools. Reference manual, USC/Information Sciences Institute, Marina del Rey, CA, USA, 1993.
48. D. S. Wile: Towards a calculus for abstract syntax trees. In R. Bird and L. Meertens, editors, *Algorithmic Languages and Calculi*, 324–335, February 1997.
49. D. S. Wile: Supporting the DSL spectrum. *Journal of Computing and Information Technology. CIT*, 9(4):263–287, 2001.
50. D. S. Wile: Lessons learned from real DSL experiments. *Sci. Comput. Program.*, 51(3):265–290, 2004.

GPSR Compliance

The European Union's (EU) General Product Safety Regulation (GPSR) is a set of rules that requires consumer products to be safe and our obligations to ensure this.

If you have any concerns about our products, you can contact us on ProductSafety@springernature.com

In case Publisher is established outside the EU, the EU authorized representative is:

Springer Nature Customer Service Center GmbH
Europaplatz 3
69115 Heidelberg, Germany

Batch number: 10583998

Printed by Printforce, the Netherlands